KT-466-012

WITHDRAWN FROM
THE LIBRARY

UNIVERSITY OF
WINCHESTER

KA 0394891 9

Toward a Visually-Oriented School Mathematics Curriculum

Mathematics Education Library

VOLUME 49

Managing Editor

A.J. Bishop, *Monash University, Melbourne, Australia*

Editorial Board

M.G. Bartolini Bussi, *Modena, Italy*
J.P. Becker, *Illinois, U.S.A.*
M. Borba, *Rio Claro, Brazil*
B. Kaur, *Singapore*
C. Keitel, *Berlin, Germany*
G. Leder, *Melbourne, Australia*
F. Leung, *Hong Kong, China*
D. Pimm, *Edmonton, Canada*
K. Ruthven, *Cambridge, United Kingdom*
A. Sfard, *Haifa, Israel*
Y. Shimizu, *Tennodai, Japan*
O. Skovsmose, *Aalborg, Denmark*

For further volumes:
http://www.springer.com/series/6276

Ferdinand D. Rivera

Toward a Visually-Oriented School Mathematics Curriculum

Research, Theory, Practice, and Issues

Dr. Ferdinand D. Rivera
San Jose State University
Department of Mathematics
1 Washington Square
San Jose California 95192
USA
rivera@math.sjsu.edu

Series Editor:
Alan J. Bishop
Monash University
Melbourne 3800
Australia
Alan.Bishop@Education.monash.edu.au

UNIVERSITY OF WINCHESTER

ISBN 978-94-007-0013-0 e-ISBN 978-94-007-0014-7
DOI 10.1007/978-94-007-0014-7
Springer Dordrecht Heidelberg London New York

Library of Congress Control Number: 2010937858

© Springer Science+Business Media B.V. 2011
No part of this work may be reproduced, stored in a retrieval system, or transmitted in any form or by any means, electronic, mechanical, photocopying, microfilming, recording or otherwise, without written permission from the Publisher, with the exception of any material supplied specifically for the purpose of being entered and executed on a computer system, for exclusive use by the purchaser of the work.

Printed on acid-free paper

Springer is part of Springer Science+Business Media (www.springer.com)

Seeking the earliest use of the concept of logos, *Detienne . . . proposes that "on the one hand,* logos *was seen as an instrument of social relations. . . . Rhetoric and sophistry began to develop the grammatical and stylistic analysis of techniques of persuasion. Meanwhile, the other path, explored by philosophy led to reflections on* logos *as a means of knowing reality: is speech all of reality? If so, what about the reality expressed by numbers?" (Detienne, 1996, p. 17). If the question were posed today, "How can knowledge be verbalized?" no one would disagree that it is through* logos, *the word, whether spoken or written. But what if we pose the question differently and ask, "How can knowledge be visualized?"*

(Bier, 2006, p. 270)

The dialectic of word and image seems to be a constant in the fabric of signs that a culture weaves around itself. What varies is the precise nature of the weave, the relation of warp to woof. The history of culture is in part the story of a protracted struggle for dominance between pictorial and linguistic signs, each claiming for itself proprietary rights on a "nature" to which only it has access.

(Mitchell, 1984, p. 529)

The unity of perception and conception . . . suggests that intelligent understanding takes place within the realm of the image itself, but only if it is shaped in such a way as to interpret the relevant features visually. . . .. Visual education must be based on the premise that every picture is a statement. The picture does not present the object itself but a set of propositions about the object.

(Arnheim, 1971, p. 308)

Contents

Chapter 1
Introduction

A Reflection on Visual Studies in Mathematics Education: From Purposeful Tourism to a Traveling Theory

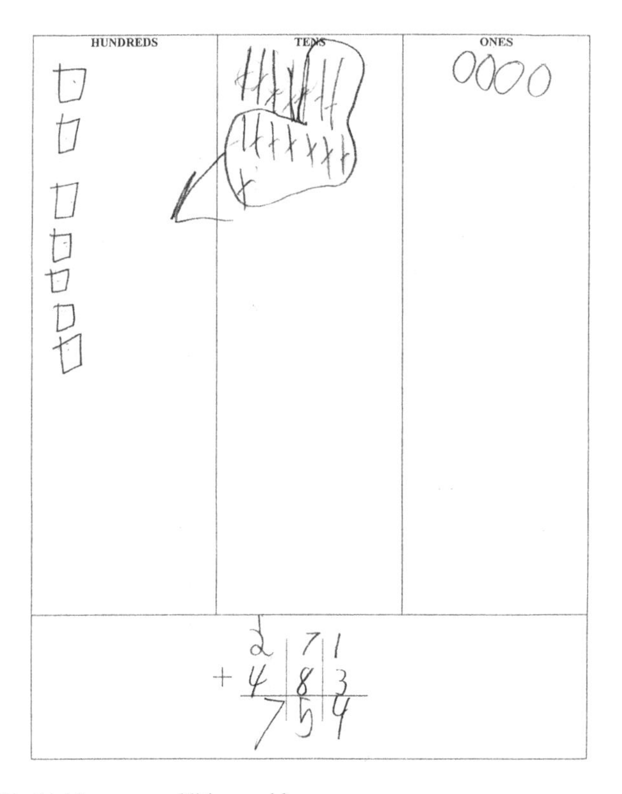

Fig. 1.1 Nikki's thinking on an addition problem

This is awesome.

(Nikki, Grade 2, 7 years old)

As classroom observations and teacher interviews continued, it seemed that [the ten middle grades] teachers described their use of manipulatives as fun and distinct from their regular teaching of mathematics. Although these distinctions emerged subtly, a very clear indication of this occurred halfway through the year. Describing a lesson with manipulatives, Joan said,

F.D. Rivera, *Toward a Visually-Oriented School Mathematics Curriculum*, Mathematics Education Library 49, DOI 10.1007/978-94-007-0014-7_1,
© Springer Science+Business Media B.V. 2011

"Sometimes I think that they are just having fun, but I don't mind because eventually we'll get to the real math part" (interview 2). Later in the same interview Joan stated, "When we're doing hands-on stuff they're having more fun, so they really don't think about it as being math" (interview 2). . . .. Not only did teachers appear to distinguish between "fun math" lessons where manipulatives were used and "real math" lessons where traditional paper-and-pencil methods were used, but they also made distinctions between parts of individual lessons. For example, the manipulatives may be used for exploration at the beginning or "fun math" part of a lesson, or they may be used in an activity or a game after the mathematics content was taught; but during the teaching of specific skills or content, paper-and-pencil methods were used to teach and practice the "real math."

(Moyer, 2001, p. 187)

The lesson of the dichotomies should now be clear: they demand the "and" of intersection. Geometrical intuition never gets far without analytic abstraction, and vice versa.

(Wise, 2006, p. 80)

In writing this book, I have definitely stood on the shoulders of giants whose works on various aspects of mathematical visualization have enriched our understanding of how students actually learn mathematics. The impressive critical syntheses of research studies on visualization in mathematics by Presmeg (2006) and Owens and Outhred (2006), which have been drawn from the annual peer-reviewed proceedings of the International Group for the Psychology of Mathematics Education (IGPME) over a period that spans three decades (1976–2006), provide a comprehensive list of important contributors whose thoughts are reflected in various places in this book. Twenty years ago, the Mathematical Association of America published a visual-driven monograph edited by Zimmermann and Cunningham (1991) that consists of reflective essays by, including references to other, researchers who then began the exciting task of exploring ways to visualize abstract mathematical objects via the power of computer software tools that could support and mediate in the development of advanced mathematical concepts and processes.[1] Many of their ideas have also been incorporated in this book.

---

[1] Situating the nature and context of thinking that was taking place in the 1990s outside the confines of mathematics education, Mitchell's (1994) influential book, *Picture Theory,* cushioned the decade point within the "*pictorial [or iconic or visual] turn* in contemporary culture, the widely shared notion that visual images have replaced words as the dominant mode of expression in our time," which should then necessitate "a new initiative called visual culture (the study of human visual experience and expression)" (italics added; Mitchell, 2005, p. 5). Mitchell, then, of course, was not recommending but problematizing the proposed wholesale shift from word to image by surfacing the reality-creating power of the latter via imaging technologies and the theories and attitudes people have about pictures/visuals/images and their relationships with the corresponding verbal representations. Brown's (2002) thoughts below, which basically ground mathematics as being prior to imagination, offer a complementary perspective to Mitchell's thoughts about the ontological power of images:

In this prolegomenon, I revisit many of the same references, however, with an eye on clarifying what and how I mean by my basic claim of *Toward a Visually Oriented School Mathematics Curriculum*, which is also the title of this book. My recent elementary and middle school classroom experiences (2005–2010) have also significantly influenced the manner in which I developed many of the ideas that contributed to this basic claim. In these classrooms, I had the immense pleasure of working with students like second-grade Nikki in the opening epigraph who constructed and appropriated mathematical knowledge and reasoned mathematically through an effective process that Barwise and Etchemendy (1991) have aptly called *heterogeneous inference*, that is, the use of a "wide range of types of reasoning" (p. 10) – from linguistic to nonlinguistic or visual – in sense making, representing, and problem solving. Seven-year-old Nikki's enthusiasm toward her visual process in obtaining the sum of two whole numbers reminds me of the celebrated English art critic and painter Berger who pointed out that

> it is a platitude that what's important in drawing is the process of looking. A line drawn is important *not for what it records so much as what it leads you on to see*. Each confirmation or denial brings you closer to the object, until finally you are, as it were, inside it: the contours you have drawn no longer marking the edge of what you have seen, but the edge of what you have become . . . a drawing is an autobiographical record of one's discovery of an event, seen, remembered, or imagined.
>
> (italics added for emphasis; Berger quoted in Taussig, 2009, p. 269)

When Nikki drew her figures, regrouped a set of 10 sticks to form a new square, and then recorded her actions symbolically, her visually derived mathematical understanding exemplifies in quite poetical terms how she has "become" and felt closer and "inside" both the visual and symbolic forms she constructed for herself. Not only with Nikki, I should say, but also with many of my young and older participants, I marveled at their "incorrigible tendency to lapse into vitalistic and animistic ways of speaking" (Mitchell, 2005, p. 2) about the objects and images they use to learn mathematics.

The 2-year professional development work that I conducted with in-service teachers of Grades 5 through 8 in 2 school districts in Northern California, which occurred during the time I was in my middle school classroom implementing my longitudinal design research investigations, further strengthened many of the arguments I present in favor of the basic claim. Like the teachers in Moyer's (2001) study in the opening epigraph, many of my teachers echoed the same perennial view about the "fun" part of manipulatives and other visual strategies and tools without connecting and embedding them to serious aspects of learning and doing mathematics. During the many hours I spent with them, I found myself repeatedly asking why they held such a different view, which appears to be devaluing the epistemological

---

> It is mathematics which inspires the imagination, as totally new forms and almost unbelievable patterns and structures are revealed. These fruits of the imagination, verified through the tough testing of logic and calculation, have what has been described as an "unreasonable effectiveness in the physical sciences."
>
> (Brown, 2002, pp. 157–158)

significance of visualizing in mathematical learning, including how I could assist them to (help their students) see the power of structured visualizing in contributing to a better and more meaningful understanding of school mathematical knowledge (cf. Rivera, 2010b). At the outset, let me say that in the history of the present, such views, which include mine, are emerging within the context of a struggle between *rational* and *sensuous modes of mathematical knowing*. In the USA, in particular, the tight alignment between and among mathematics instruction, classroom assessment, and state mathematics standards and testing seems to imply the need for definite and more predictable forms of mathematical knowledge that gestate and ossify in the alphanumeric (i.e., rational). But recent classroom research knowledge in mathematics education drawn from purposeful design experiments provides validatory empirical and complementary proof of the central role of gestural, tactile, and visual activity (i.e., sensuous) in rational mathematical activity.

In organizing such a re-visitation of referenced work, I have found it useful to think in terms of the distinction that Damarin (1995) has pointed out concerning the words *traveler* and *tourist*. Damarin writes:

> A traveler and a tourist can visit the same city, but experience it very differently. A tourist's goals are typically to see all the sights, learn their names, make and collect stunning pictures, eat the foods, and observe the rituals of the city. A traveler, on the other hand, seeks to understand the city, to know and live briefly among the people, to understand the languages, both verbal and non-verbal, and to participate in the rituals of the city. At the end of equally long visits, the tourist is likely to have seen more monuments but the traveler is more likely to know how to use the public transportation.
>
> (Damarin, 1995, p. 29)

Hence, in this introduction, I engage in at least two complementary tasks. Section 1 rereads many of the findings drawn from various research investigations in visualization and mathematics from the point of view of *purposeful tourism*, that is, as comprising important, oftentimes disparate, pieces of work in an emerging landscape. Presmeg (2006) articulates this viewpoint in the following thought-provoking passage that she expressed near the end of her comprehensive synthesis:

> Where have we been and where are we going? At this point the diffused nature of the continuing research on visualization would seem to be a disadvantage – but it is probably necessary, as puzzle pieces are necessary in the completion of a whole picture.
>
> (Presmeg, 2006, p. 225)

In Section 2, I begin to map out a *traveling theory*, one that attempts to resolve Presmeg's (2006) 13th Big Research Question in visualization in mathematics education, that is,

> What is the structure and what are the components of an overarching theory of visualization for mathematics education?
>
> (Presmeg, 2006, p. 27)

A full response, of course, is reflected in the succeeding chapters that focus on the nature, role, and significance of visualization in various aspects of mathematical activity. Section 3 provides an overview of the ideas that are pursued in Chapter 2.

I hope that this introduction and the next one encourage readers to plough through several hundreds of pages of thinking devoted primarily to issues in visualization in mathematics learning and armed with the same excitement that McCormick, DeFanti, and Brown (1987) felt in the late 1980s when they perceived (scientific) visualization as having the capacity to "transform the symbolic into the geometric," "offer a method of seeing the unseen," "enrich the process of scientific discovery," and "foster profound and unexpected insights" (p. 1). Section 4 provides a description of the students in my studies who benefited from my own excitement in this research field and whose thinking and visual experiences provided much insight and motivation that allowed me to test, refine, and test again many of the ideas I pursue in this book.

1 Purposeful Tourism

As I was rereading the edited collection of essays by Zimmermann and Cunningham (1991), I found myself generating the binary terms in Fig. 1.2a with the words on the left of the slash seen together as characterizing the visual as it was perceived back in the 1990s in mathematical discourse.[2] Using what were then considered current state-of-the-art educational software tools, most of the authors in this collection showcased the power of such tools in supporting growth in rigorous academic or formal knowledge that covers concepts and processes in calculus, spatial geometry, differential equations, differential geometry, linear algebra, fractals, complex analysis, and stochastics and random phenomena. As editors of the monograph, Zimmermann and Cunningham (1991) interpreted the role of visualization as a precursor for something more. In "process[es] of producing or using geometrical or graphical representations of mathematical concepts, principles or problems, whether hand drawn or computer generated" (p. 1), visualization could be seen

> not [as] an end in itself but [as] a means toward an end, which is understanding. .... Mathematical visualization is the process of forming images (mentally, or with pencil and paper, or with the aid of technology) and using such images effectively for mathematical discovery and understanding.
>
> (p. 3)

There is, thus, a sense in which visual processes were viewed as being conceptually prior to understanding and discovery. In the same volume, the widely cited essay by

[2]Certainly the visual/symbolic divide, O'Halloran (2005) notes, "has a long history" (p. 129). O'Halloran's trace begins with traditionalists during the time of Descartes who perceived the visual as a heuristic tool and the symbolic as the rigid road toward a valid proof. She then points to Davis (1974) as representative of mathematicians who hold the opposite view that visuals could be considered a valid form of proof leading to theorem-of-the-perceived kind of arguments. O'Halloran also draws on the work of Galison (2002) who sees the divide as an oscillating phenomenon and Shin (1994) who interprets mathematicians' incredulity toward the visual medium as being all about constructing diagrams that are loaded with errors, imprecise, narrowly generalizable, and incomplete.

a

informal and experimental/formal and proof-driven
intuitive/logical and analytic
intuitive/counterintuitive
experiential/algorithmic
means/end
forming and using images/making discoveries and developing understanding
geometric/symbolic
unseen/seen
spatial and kinesthetic/verbal, arithmetic, and algebraic
diagrammatic/symbolic and sentential
pictorial/linguistic
higher/lower cognitive demand
illustrative/generalizable
generative and mnemonic/proof
multiple paths/sequentially linear
generative/proof

b

Visual	Verbal-algebraic
Abstracts spatial properties such as shape, position	Abstracts properties which are independent of spatial configuration such as number
Harder to communicate	Easier to communicate
May represent more individual thinking	May represent more socialized thinking
Integrative, showing structure	Analytic, showing detail
Simultaneous	Sequential
Intuitive	Logical

Fig. 1.2 **a** Binaries associated with the visual/symbolic in mathematics. **b** Skemp's binaries associated with the visual/verbal in mathematics (Skemp, 1987/1971, p. 79)

Eisenberg and Dreyfus (1991) provided sufficient empirical evidence, which they have drawn from their own and several other research investigations, about undergraduate students' "reluctance to visualize" in calculus in favor of more readily manipulable and algorithmic-driven symbolic forms that appeared to characterize the dominant content of mathematics in the 1990s.[3]

Skemp (1987/1971) has also noted the visual/verbal opposition in mathematics in the 1970s. In one section (pp. 66–82) of his classic book, *The Psychology of Learning Mathematics*, Skemp spoke about the opposition in terms of the advantages and disadvantages of the two imagery systems and the various contexts in

[3]In the US history of school mathematics, in particular, it is interesting to note how nineteenth century textbooks have found the use of manipulatives to be a valuable complementary pedagogical tool to the inductive method espoused in arithmetic and problem solving (Michalowicz & Howard, 2003). But then its status, including visualization, more generally, seemed to have been afloat in twentieth century algebra and geometry textbooks (Donaghue, 2003). Rousseau and Pestalozzi, as well as the emphasis on real-life problem solving, influenced mathematics pedagogy in the nineteenth century, which could not be upheld at least in the twentieth century context as a result of the mathematics community's preoccupation toward developing an axiomatic structural approach to teaching mathematics.

which we use them. Figure 1.2b is his list of such differences that he nevertheless saw as being complementary.

Drawing on Presmeg's (2006) synthesis of more than 150 relevant studies in visualization in mathematics education, research reported at the IGPME appears thus far to have been pursued along the following four (overlapping) dimensions, namely: psychological; curriculum and instruction; technological; and semiotic. Presmeg's characterization of visualization involves simultaneous acts of creating a spatial arrangement and constructing its image, that is,

> when a person creates a spatial arrangement (including a mathematical inscription) there is a visual image in the person's mind, guiding this creation. Thus visualization is taken to include processes of constructing and transforming both visual mental imagery and all of the inscriptions of a spatial nature that may be implicated in doing mathematics.
>
> (Presmeg, 2006, p. 206)

The term *mathematical inscription* refers to graphical representations (e.g., *symbols* and *diagrams* in this book) that are necessary in mathematical practice and communication. *Visual images* are mental constructs that represent visual or spatial content. Thus, a visualizer in Presmeg's sense above engages in a mutually determining process of image construction and externalization. That changes in the visualizer's image of some target content meant changes in the corresponding inscriptions as well.

1.1 Psychological Dimension

This dimension emerged from Bishop's work in the 1970s on visualization and spatial thinking and prevailed through the late 1990s when much of theory and research in mathematics education were immersed in the constructivist phenomenon. Figure 1.3 provides a sample of topics that were reported under this dimension. Three findings are worth noting.

First, developing mental images of mathematical objects, concepts, and processes actually took more time and required greater cognitive load among learners than using analytic methods and processes.

Second, encouraging the use of imagery in conceptualizing mathematical concepts and processes has opened up a farrago of different types of images and imagery systems or schemes, some more effective than others, including the view that more sophisticated ones could be acquired or revised and improved through more learning. Also, it appears that while prototypical images were documented to exist and could either support or hinder mathematics learning among different grade-level groups, longitudinal studies could not establish empirical evidence for possible developmental models or trends.

Third, the perceived phenomenon of reluctance toward visualizing in mathematics among students in various grade levels could not be analyzed in simple terms since, Presmeg writes,

1. Levels of complexity of images in relation to geometric solids and nets
2. Effects of parallel axes representations involving linear functions in visualizing the concept of slope
3. College students' reluctance toward visual thinking and processing
4. Role of imagery in mathematical reasoning of elementary students
5. Different kinds of imagery schemes (figurative, operative, relational, and symbolic)
6. Different types of imagery (concrete, kinesthetic, dynamic, memory image of formulas, and pattern imagery)
7. Individual children's visualization of various aspects of content and processes in mathematics
8. Imagistic systems as a category of cognition and affect in mathematical learning
9. Interpreting figural information and visual processing
10. Interaction of students' beliefs about mathematics, problem solving, and visualization
11. Role of personal imagery and affect in learning mathematics and their relationships to visualization
12. Types of representations, representational systems, and visualization
13. Gender differences in visualization
14. Visualization in problem solving
15. Expert/novice views and uses of imagery in doing mathematics
16. Relationships between visualization, reification, generalization, and abstraction in mathematical learning

Fig. 1.3 Psychology-driven topics in visualization

> (f)or most people, the task itself, instructions to do the task a certain way ... and sociocultural factors including teaching situations ... influence the use of visual thinking in mathematics. However, there are a few people for whom visualization is not an option ... whereas some others do not feel this need at all.
>
> (Presmeg, 2006, p. 216)

1.2 Curriculum/Instruction Dimension

There were very few studies that tackled this dimension. Presmeg used her 1991 study with 13 high school mathematics teachers and the role of visualization in their daily instruction in noting the following three important findings: (1) teachers with strong mathematical visual skills were not solely teaching visual skills to their students; (2) teachers along the middle point of the mathematical visual skills continuum taught their students to value visual skills and strategies but only as a means toward the end, which is generalization; and (3) teachers with weak to no visual skills taught their students to silence their visual capacities in favor of rote memorization and other procedural, symbolic skills. Those teachers who were strong visualizers tried to connect mathematics with other areas of thinking and aspects relevant to the real world and valued "creativity," "openness to external and internal experience, self-awareness, humor, and playfulness" (Presmeg, 1991, p. 211). Attention to curriculum was focused on how students effectively interpreted, drew, and understood graphs appropriate at their grade levels by relying on their intuitions and images, including the important role of individually generated metaphors in both the construction of generalizations and the development and retention of meanings.

1.3 Technological and Semiotic Dimensions

Advances in technological tools basically provided the impetus for conducting more research in visualization in mathematics. While a few studies focused on students' interactions with educational software, the *dynamic* aspect of emerging tools was seen primarily as a means for "facilitating visualization processes" (Presmeg, 2006, p. 220). Interest in the semiotic dimension began when researchers saw the significant role of (representational) gestures in externalizing (mathematical) mental images. In light of Presmeg's discussion of research in these two dimensions, it seems there is still much work that needs to be accomplished especially about ways in which emergent findings could facilitate better mathematics learning for all students. For example, under the technological dimension, we still need to understand the important role of visualization in, say, instrumental genesis. Under the semiotic dimension, we need to know how visual images and/or processes change as learners transition from one type and/or phase in semiotic representation to another.

In this book, I address many of the same fundamental issues raised in each dimension. Under the cognitive dimension, I am interested in ways in which visual images transform from personal images (i.e., *imaginals*) and images-in-the-wild in their initial states to their final state that I classify as visual representation. While imaginals are subjectively constructed, *images-in-the-wild* have an Hutchinean[4] content in them, which means to say that they emerge naturally among most students during classroom activity. They also interact with, and undergo changes in, form as a consequence of more structured learning relevant to some intended knowledge. Goldin's (2002) notion of *representational systems* or, simply, *representation*, involves the presence of signs (or characters or inscriptions) with well-defined configurations (or rules) for combining them and a structure that basically provides meaning and fundamentally sets up the relationship between the relevant signs and their rules. Clearly, computer-generated or manipulatives-based images exemplify visual representations in Goldin's sense, that is, they are structured visual images that operate within well-defined schemes. However, everyday visual images (imaginals) or visual-images-in-the-wild do not have this formal or structured character, at least not yet. Hence, it is interesting to understand how such images support, or transition into, representational systems. Under the curriculum/instruction dimension, what instructional scaffolds enable the formation of visual representations in mathematics? Under the technological and semiotic dimensions, how do we characterize the different visuals that define various semiotic representations? Duval (2006) gives an impressive account of the embedded nature of various semiotic representations in any intended geometric knowledge, including transformations that take place in the corresponding visual forms that accompany the representations. It would be interesting to see how we can use such a model in other areas of school

[4]Hutchins (1995) spoke about "cognition in the wild" in ecological terms as a type of adaptive thinking that occurs in its natural state within the context of culturally constituted activity (or "lifeworld-dependent" as we refer to it in Chapter 2 and in succeeding chapters).

mathematics such as number theory and algebra that are widely taught across the globe from a symbolic and nonvisual context.

2 A Traveling Theory

The following seven chapters in this book address various aspects of a proposed traveling theory that has guided my recent funded research in the classrooms. In this section, I talk about background issues that led me to the notion of *progressive modeling*.

2.1 Exploring Implications of Purposeful Tourism

Figure 1.4 illustrates how I perceive the relationships between and among the four interpretive dimensions of research in visualization in mathematics learning and teaching. At the outset, I see issues in the technology and semiotic dimensions as fundamentally determining concerns that are then pursued in the cognitive and instruction/curriculum dimensions insofar as learning, delivery, and packaging of content matter. Various discussions and research investigations concerning visual processes in mathematics are now brought to the table because (more) mathematics learning is taking place in technological settings. Further, changes in technological tools drive changes in semiotic resources and representations, and vice versa. Consequently, the manner in which instruments and relevant knowledge (mediated in signs) are learned is a concern in the cognitive dimension.

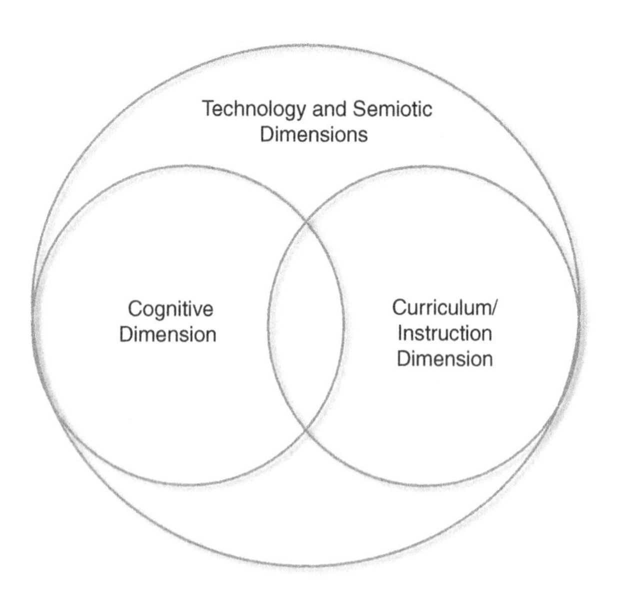

Fig. 1.4 Relationships of the four interpretive dimensions of research in visualization in mathematics learning

2.2 Exploring Implications of Instrumental Genesis

Several years ago, I collaborated with a high school mathematics teacher and together we taught a precalculus class of 30 students concepts and processes in algebra with the aid of a TI 89, a powerful graphic calculator that could model graphs, simplify algebraic expressions, and solve equations. I refer readers to Rivera (2007a) for an extensive discussion of findings since my primary aim at this stage is to establish a connection between this prior work and the proposed traveling theory that I pursue in this book. The underlying theoretical framework of the Rivera (2007a) study is *instrumental genesis* (IG), a Vygotskian-driven view in which students' mathematical experiences are, in fact, mediated by, and structured through, physical, semiotic, and other technological systems. Bruner (1978) also notes how "instrumental action is at the core of Vygotsky's thinking action that uses both physical and symbolic tools to achieve its ends" (p. 2). In the study and, in fact, throughout the rest of the school year, I held the strong view that each time a student acquired facility and competence with any visualizing tool, he or she would come to develop appropriate schemes reflective of growth in knowledge. In relation to the TI 89, I structured our classroom activities in such a way that as the students were competently learning to use the TI 89 and its various command menus and mathematical functions, they were also developing a deep understanding of the intended mathematical structures. My initial interest in IG was sparked by the theorizing offered by researchers like Artigue (2002), Bartolini-Bussi and Boni (2003), and Mariotti (2000, 2002) who saw the mediating capability of technological tools in enabling the *progressive evolution of mathematical thought*, from the concrete to the abstract or from material to theoretical knowledge. Knowledge is, thus, conceptualized as a tool-mediated activity. Schemes are constructed as a result of learners' expertise in using a tool (i.e., instrumental schemes) and, thus, reflect structures that interweave instrumental and intended mathematical knowledge.

In the Rivera (2007a) study, I gave an IG account of how my students' schemes in solving polynomial inequalities evolved over three phases that enabled them to successfully transition from a visual, calculator-dependent process to a visual-algebraic method that was as rigorous and valid as standard alphanumeric methods for solving such inequalities. The three phases relative to the students' relationships with the TI 89 were as follows: (1) full attachment; (2) partial detachment; and (3) full detachment.

In the full attachment phase, their graphical method for solving polynomial inequalities faithfully mirrored how they would actually solve them on their TI 89. An illustration of a student's work is shown in Fig. 1.5a, which shows details concerning the graph of the relevant polynomial function but were really unnecessary in solving the corresponding inequality.

In the partial detachment phase, their graphical solution method contained only a rough sketch of the relevant function and its zeros on a drawn Cartesian plane (see Fig. 1.5b for an illustration). This phase was facilitated by their deep understanding of the general structure of graphs of odd- and even-powered polynomial functions.

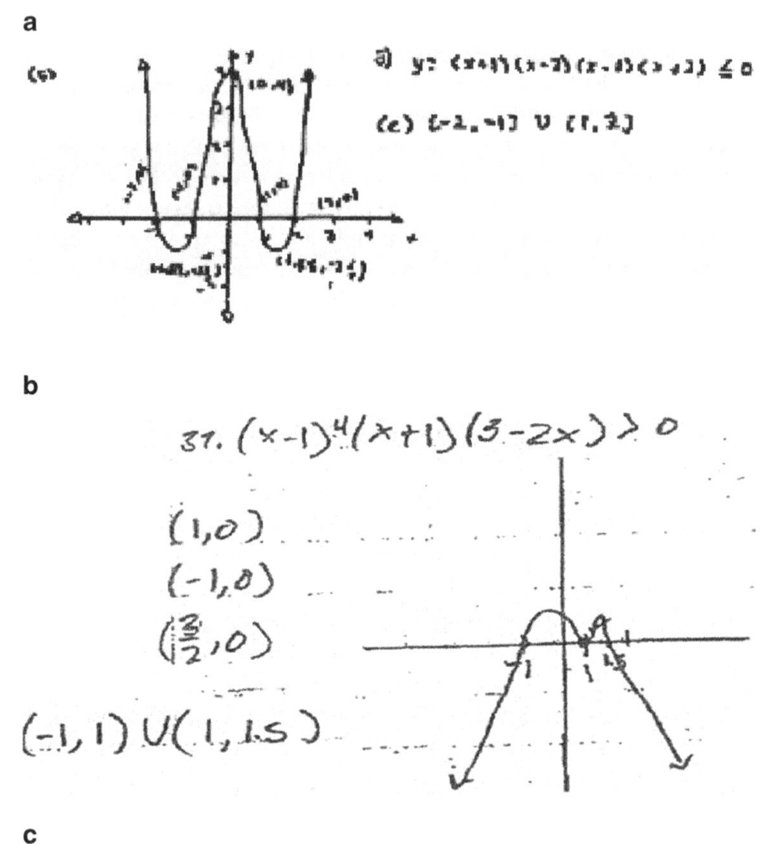

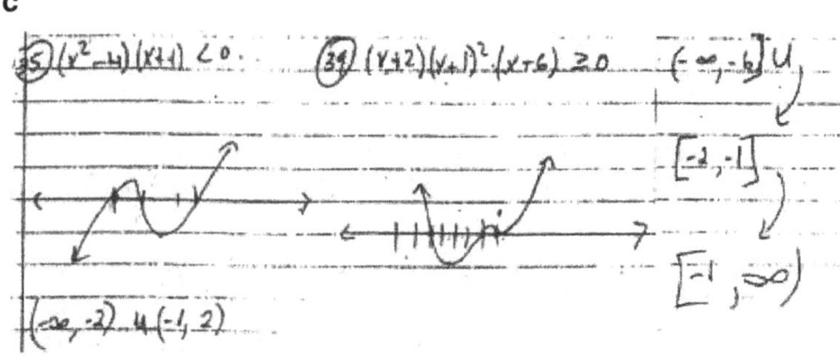

Fig. 1.5 **a** Danica's solution of $(x+1)(x-2)(x-1)(x+2) \leq 0$ (Rivera, 2007a, p. 295). **b** Kien's solution of $(x-1)^4(x+1)(3-2x) > 0$ (Rivera, 2007a, p. 299). **c** Reuben's solutions to two polynomial inequalities (Rivera, 2007a, p. 299)

In the full detachment phase, their graphical method exemplifies a full transition to an abstract process that contained only the essential elements necessary in solving polynomial inequalities. As demonstrated in Fig. 1.5c, Reuben used an empty number line to produce a rough sketch of some relevant graph with its zeros in order to generate his solution to the inequality. Also, the dot above the x-intercept -1 in the second graph in Fig. 1.5c symbolized for him that the graph of the function needed to be tangent to the x-axis at the given point.

At the end of the teaching experiment, which took 21 sessions of 55-min each to accomplish, both the teacher and myself felt awestricken by the event that took place among our students in mathematical activity. Instead of witnessing students who would simply establish a connection between visual and graphical methods for solving a polynomial inequality, which exemplifies the view that the visual phase basically provides the means toward the end, the symbolic phase, they demonstrated a capacity for *progressive visualization* and *progressive mathematization*. The conceptual transition that occurred in their mathematical thinking from the visual to a classification I label *visuoalphanumeric* illustrates what I interpret from Goldin (2002) to mean the transformation of visual images into more structured visual representations.

The idea of progression in mathematical thought is not a novel one.[5] Freudenthal explicitly used the terms *progressive schematizing* and *progressive formalizing* in 1981. In 2002, the book edited by Gravemeijer, Lehrer, van Oers, and Verschaffel on symbolizing, modeling, and tool use in mathematics education contained empirical reports that exemplify the notion of *progressive formalization* or *progressive understanding* in mathematical activity. The *piece de resistance* in the theory of, and empirical research surrounding, emergent constructivism by Cobb and colleagues (Cobb, 2007) involves detailed evolutionary accounts of classroom mathematical practices that fit squarely within the notion of progression of mathematical thought. Fundamental to these researchers' notion of progression is the view that mathematical signs are neither pre-determined nor arbitrary but negotiated and conventional. In geometry, Pedemonte's (2007) empirical work with 12th and 13th grade students in Italy and France exemplifies the possibility of *progressive argumentation* that sees a continuous relationship between conjecture and proof via abduction, induction, and deduction. Thirty years ago, Mason's (1980) notion of the spiraling of enactive, iconic, and symbolic expressions already hinted at the significance of *progressive symbolization*, where symbols are symbolic only when they have become enacted and effectively iconicized. He writes:

> Turning to enactive elements to explore the meaning of symbols or concepts; Using enactive elements to try to get a sense of pattern; Asking for images, metaphors, diagrams to illustrate what is going on; Crystallizing understanding in symbolic form; Practicing with examples to move the symbolic form into enactive elements.
>
> (Mason, 1980, p. 11)

---

[5] A similar progressive view is emerging in accounts of growth in scientific knowledge. For example, I refer readers to Maienschein's (1991) interesting analysis of drawings involving cell development, which depicts visualizations that progressively shift from photographs (as presented data) to diagrams (as represented data).

Even Skemp in his 1971 book wrote about the process of *progressive abstraction* in relation to the formation of mathematical concepts. He notes that our previous experiences consisting of examples and nonexamples enable us to abstract similar and invariant properties, which then "progress rapidly to further abstractions" (p. 10) that are themselves "becoming more functional and less perceptual – that is, less attached to the physical properties" (p. 10).[6] In 1974, Leushina (1991), Soviet psychologist, discussed the role of the visual principle in teaching (preschool) mathematics and, especially, the critical "role of language when using visuality in mathematics lessons" (p. 166). Leushina also hinted at progression in mathematical thought when teaching is able to achieve "unity" in "words and visual aids" which, consequently, "brings about a close connection and proper correlation of the visual and the abstract, [and] the specific and the generalized" (p. 166). In the passage below, Leushina anticipates changes in visual representations as a consequence of changes in young children's development and the mathematics curriculum:

> Visual aids should be used ... by gradually increasing their complexity; proceeding from specific, narrative objects to non-narrative objects; from material visual aids to materialized aids, e.g., to conventional tables (such as a "number ladder"), models (e.g., a string of ten beads), diagrams, etc. Visual aids should change not only from one age group to another, but also based on the correlation between the concrete and the abstract at different stages in the children's mastery of the curriculum material. For example, at a certain stage specific sets can be replaced by number tables or by numerals.
>
> (Leushina, 1991, pp. 167–168)

But what is apparently a lacuna in both the research and professional literature, which is addressed in this book, are sufficient longitudinal accounts of this progressive modeling view, especially the changing roles and structures of visualization in such a model. Also, my recent encounter with Millar's (1994) longitudinal studies with blind subjects enabled me to articulate, at least in metaphorical terms in this book, what I thought was missing in progressive modeling accounts of mathematical learning – that is, progression not only in form, content, and mathematical practices but progression that results from, in Millar's (1994) words, "the convergence and overlap of complementary inputs" (p. 15) as well. Millar expressed this view based on how her participants constructed some target knowledge by drawing on their senses that neither "operate as completely separate systems" nor "convey precisely the same information" (p. 15). That central to this effective progressive process is the "integration of convergent multimodal information" (1994, p. 16) leading to the acquisition of the relevant knowledge.

This book also addresses other fundamental concerns that are still in line with our progressive modeling theory. Certainly, it is easy to persuade learners to visualize, but there is a concern about how to instrumentally orchestrate the transition toward more structured visualizing through sustained mathematical activity at both

---

[6]Skemp's progressive model reflects inductive approaches to word learning whereas more recent progressive accounts of Cobb and Gravemeijer ground their model in authentic (real and experientially real) mathematical activity that sees changes in notating, symbolizing, and abstracting as effects of mathematical reality construction.

individual and classroom level. Here I am reminded of the important findings drawn from the empirical work of Ng and Lee (2009) with 10- to 11-year-old Singapore students, totaling 151, concerning the use of structured visual representations (i.e., rectangular models; see, for example, Fig. 2.11a–c) in solving simple and nonroutine arithmetical problems. At the very least, Ng and Lee found that especially among average-ability students, their ability to use such models should not be viewed as "an all-or-nothing process" (p. 311). They have to be taught to "learn to exercise care" in constructing such structured visuals and at the same time "deepen their conceptual knowledge" of the relevant ideas and skills. Further, they should not develop the practice that using such visual analogues is tantamount to replacing a symbolic procedure and reproducing a pictorial-driven algorithm that works by rote but instead should use them as "a problem-solving heuristic" (p. 311). Adding my own interest in this issue, I refer readers back to the binaries listed in Fig. 1.2a, b and Wise's (2006) comment in the opening epigraph. It certainly would be interesting to find ways that would allow us to go about blurring and moving past the conceptual paralysis brought about by the perceived boundaries. Following Wise (2006), in this book I demonstrate "the import of visualization" in school mathematics, *not* in the context of "illustration but argument" (p. 81), that constructing and using visual representations plays a central role in the constitutive practice of school mathematics. I have already illustrated above one progressive account that did not jettison visual images and processes in the symbolic domain of mathematical knowledge. At least among those of us who participated in that particular study, it was seeing the visual in the symbolic and the symbolic in the visual that characterized the progression leading to the graphical solution process. A stronger view of this convergent mode of seeing is exemplified in a recent study by Haciomeroglu, Aspinwall, and Presmeg (2010) in which they demonstrate in the context of their clinical interviews with three high-achieving university calculus students how a synthesized visual-symbolic approach appears to be more powerful than either a solely visual or symbolic process to learning and understanding mathematics.

I should also point out that my appropriation of the progressive modeling view as it is used in this book has less the characteristics of typical emergent constructivist projects as it is more sociocultural in context. Being a classroom teacher, I note my openly partial disposition toward instructional approaches and orchestrations that enable socioculturally embodied visualizing in producing meaningful interpretive conditions and, consequently, structured visual representations, which lead to growth and development in mathematical knowledge among my students. An example of this "structured visual representation" is shown in Fig. 1.6, which shows two institutionally constructed "illustrations that argue" why slopes of perpendicular lines are negative reciprocals of each other. While graphing tools enable beginning algebra students to individually (and socially) "detect a pattern in their slopes" (Tucker, 2010, p. 604) and verify an invariant property at least in the initial stage of their learning, the purposefully drawn representations of Fig. 1.6 allow them to understand the intentional relationships more rigorously based on institutional expectations. Beyond recent exciting work on instrumental genesis in mathematical contexts, it is worth noting that the decision-making strategies of expert players

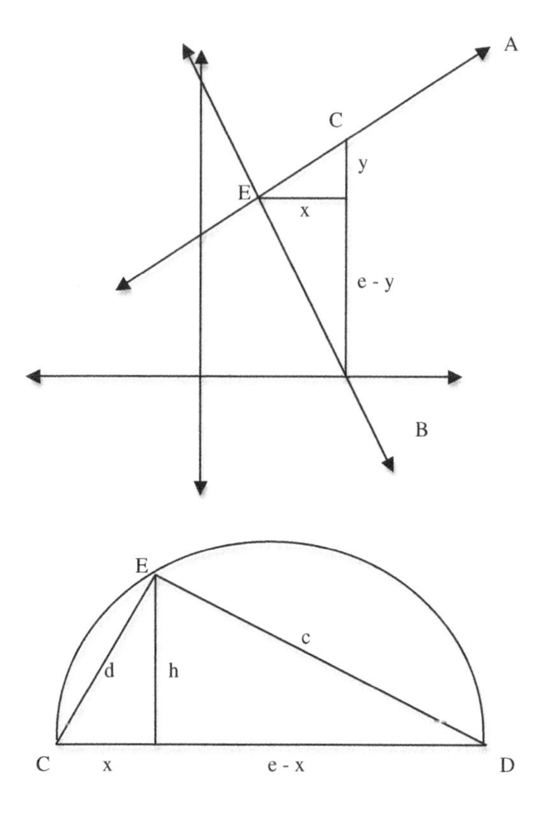

Fig. 1.6 Visual-symbolic proof concerning the slopes of two perpendicular lines (Parker & Baldridge, 2008, p. 164; Tucker, 2010, pp. 605–605)

in areas such as chess and other sports such as soccer, which draw heavily on sociocultural practices, have also been documented to be working within a similar progressive modeling process (cf. Poplu, Ripoll, Mavromatis, & Baratgin, 2008). Thus, the proposed traveling theory that I explore in this book represents a bricolage of various theoretical perspectives drawn from seeing mathematics and other relevant professional disciplines in terms of culturally constituted activity.

2.3 Exploring Implications of O'Halloran's Grammar of Mathematical Visual Images

For O'Halloran (2005), mathematical discourse consists of three semiotic resources, namely: language, mathematical symbolism, and visual images. Mathematics could then be interpreted as a (social) multi-semiotic construction in which each resource makes a significant contribution in the development of meaning. O'Halloran also distinguishes between intra- and inter-semiotics. Intra-semiotics pertains to "the grammars and functions" that are specific to individual resources, while inter-semiotics refers to meanings that are drawn "from the relations and shifts between the three semiotic resources" (p. 11).

In relation to progressive models, at least based on how we describe them above, it is interesting to document changes that take place at both intra- and inter-semiotic levels. For example, an intra-semiotic account of progressive visualization could focus on the convergence and transformations in relationships between and among different types and levels of visual images, while an inter-semiotic account could dwell on "code-switching" conditions between, say, the symbolic and visual components of the relevant mathematical knowledge. In the above study on the emergence of a solution process for solving a polynomial inequality, Fig. 1.5a–c from an intra-semiotic perspective could be interpreted in terms of conceptual convergences and shifts in visual image construction based on the students' developing relationships with the TI 89. However, there was not any noticeable change insofar as transitions at the inter-semiotic level were concerned.

Bakker's (2007) classroom-based research with a group of 30 eighth grade students in the Netherlands provides an exemplar of multi-semiotic progressions in both inter- and intra-levels. Based on data collected in a teaching experiment, Bakker documents how the students developed their formal understanding of statistical concepts relevant to sampling and distribution by initially constructing graphs of their own choice (bar graphs) and then producing a standard distribution model (dot plots resembling a pyramid and then a bell shape) over classroom time with meaningful intervention from their teacher. Consequently, the linguistic expressions they employed also underwent convergences and transitions. Initially they used indices, metaphors, and other concrete predicates in describing graphs of sample outcomes such as "more together, clumped, spread out, further apart," which then underwent some type of abstraction. In the final phase, they used the appropriate statistical terms such as "spread, average, and bell shape." Here, it is worth pointing out that Bakker's exemplary work pushes O'Halloran's multi-semiotic model of mathematical discourse further in an interesting way. Following Peirce (1976), Bakker roots the transitions in terms of a series of *hypostasized abstraction* (or simply *hypostasis*), a process that involves "making a subject out of a predicate" (Peirce, 1976) that consequently "simplifies mathematical thinking" (Levy, 1997, p. 102). Like Skemp's (1987/1971) account of progressive abstraction, a hypostasis substitutes "an abstract noun in place of a concrete predicate" (Peirce, 1976, p. 160) or "a predicate of a collection" of instances (Peirce, 1934, p. 373), which explains the occurrence of transitions in various phases of mathematical thinking. For Bakker, progressive diagrammatization, which is an instance of effective use of visualizing in concept development, signals growth in students' development of mathematical reasoning and discourse. The changing visuals are accompanied by a series of (hypostasized) abstractions.[7]

[7]Peirce's hypostasized abstraction, this "essential part of almost every really helpful step in mathematics" (1976, p. 160), is similar to Skemp's (1987/1971) account of concept or word acquisition, which involves the formal "naming" of a category (i.e., the subject) that represents a set of instances having a shared common property or attribute (i.e., the predicate). For example, "twoness" is a

3 Overview of Chapter 2

In Chapter 2, we begin to explore relationships between visualization and progressive schematization. We develop initial characterizations of the general nature of cognitive activity and link ongoing modeling action with progressive mathematization. We discuss the distributed epistemic relationship between internal and external representations, which is our initial bold attempt at conceptual molting past the binaries (e.g., Fig. 1.2a, b) that have dominated earlier conversations regarding the role of visualization in symbol-driven mathematical activity. We articulate tensions in traditional conceptualizations of numerical- and visual-driven cognitive processing of mathematical ideas and of everyday and mathematical ways of seeing as we attempt to carve a more productive middle ground via the notion of interanimation.

Three general fundamental principles of visualization are stated, discussed, and connected to the literature base on visualization in fields outside of mathematics education. The connection to external disciplines (cognitive science, philosophy of mathematics, science education, etc.) is an exciting and productive course of action since the knowledge we acquire from various reported findings in these fields will further enrich our understanding of the topic. The remaining sections address contexts, forms, and levels of visual representations in the general literature, including instructional implications. Interspersed throughout the chapter are overviews of the succeeding chapters that hopefully serve to clarify issues relevant to, and further deepen readers' understanding of, the proposed traveling theory.

4 A Note on Participants in My 4-Year Study

Throughout the book, readers will come across many instances in which I talk about three different cohorts of students that participated in my longitudinal studies. Some background context is needed for readers to understand why I was able to accomplish so much at a time when US school districts and educational policy

---

term – a "new abstract object" (Otte, 2003, p. 219) – that hypostasizes concrete instances involving any two objects (indeed for Peirce cardinal numbers represent hypostasized abstractions drawn from a predicate of a collection. Peirce writes:

> A term denoting a collection is singular, and such a term is an "abstraction" or product of the operation of hypostatic abstraction as truly as is the name of the essence. ... Indeed, every object of a conception is either a signate individual or some kind of indeterminate individual. Nouns in the plural are usually distributive and general; common nouns in the singular are usually indefinite.
>
> (Peirce, 1934, p. 299)

Otte (2003) provides additional examples such as the notion of a set "whose mode of existence depends on the existence of other fundamental things ... which is based on the existence of its own elements" (p. 218). Imaginary numbers as an object were used to develop the theory of complex functions. "Again and again," Otte (2003) notes about hypostasized abstractions, "a construction or an algorithmic procedure is taken as an object to be incorporated into another construction or procedure" (p. 220).

are too focused on rational-driven high-stakes testing that emphasizes "product over process" approaches to mathematics learning at the expense of quality and more meaningful experiences for K-12 students.

In Fall 2005, with funding from the National Science Foundation, I began a study on pattern generalization involving increasing linear patterns among a group of eager 29 sixth grade students (12 males and 17 females; mean age of 11 years) whose mathematical competence by the California assessment standards was at basic and below proficiency levels (i.e., not meeting previous grade-level state standards). Most of the student-participants in this class were of Southeast Asian origin. I spent one full semester (close to 5 months) with them and, in collaboration with the assigned sixth grade teacher, we explored activities that focused on integers and increasing linear patterns.

In Fall 2006, 3 students from the above class moved to a different school and were replaced with 6 seventh graders (14 males, 18 females). In collaboration with the assigned seventh grade teacher, we explored once again concepts and processes relevant to integers. Also, the study on linear patterns was extended to increasing and decreasing sequences.

From Fall 2007 to Spring 2008, 15 students from the earlier 2-year study were allowed to participate in the third (and final) year of the study. In this book, I refer to them as my Cohort 1. Cohort 1 was then mixed with a mathematically strong group of 19 seventh and eighth graders (12 males and 22 females of Southeast Asian origin; "Cohort 2"). Together, they comprised an Algebra 1 class that I taught the entire school year. Each student in this class had access to a graphing calculator and all the relevant manipulatives and visual tools (e.g., paper folding) that were needed to accomplish various design-driven tasks.

From Fall 2009 to Spring 2010, 21 Grade 2 students (7 girls, 14 boys; 20 Hispanic-Americans, 1 African-American; mean age of 7.5 years) participated in a yearlong study on visualization and prealgebraic thinking. Their first-grade mathematical competence on the basis of four periodic district assessments that were aligned with the California state standards indicate that among 18 of the 21 students who came from the same school, 2 were advanced (performing above first-grade state standards), 8 were proficient (meeting first-grade state standards), 4 basic (not meeting first-grade state standards), 2 below basic, and 2 far below basic. The remaining three students came from different school districts, so the only data available were the results of the first benchmark conducted in second grade, which indicate that 1 was advanced proficient, 1 basic, and 1 far below basic. In this class, I worked with two Grade 2 teachers and together we implemented activities that provided all the students with every opportunity to learn both symbolic and visual representations associated with the state-recommended second-grade mathematics curriculum. The activities involved concepts and processes in number sense, prealgebra, geometry, and statistics and data analysis.

Acknowledgment and Dedication I wish to express my gratitude to the National Science Foundation (NSF) that provided funding for me to engage in longitudinal classroom work from 2005 to 2010 (under NSF Career Grant #0448649). Results that are reported in this book are all mine and do not reflect the views of the foundation. This book is dedicated to the students and teachers in my 2005–2010 study.

Chapter 2
Visualization and Progressive Schematization: Framing the Issues

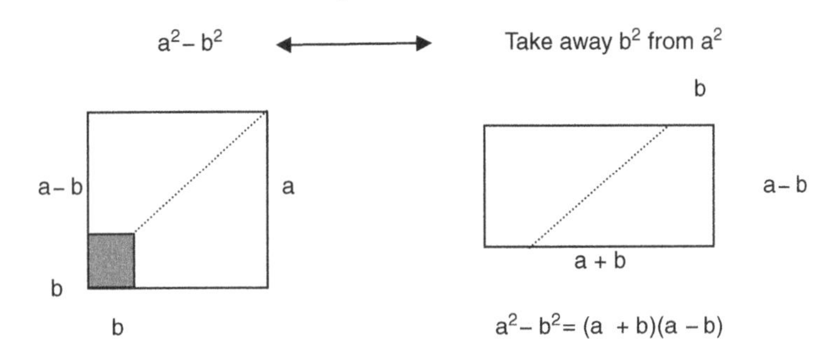

Fig. 2.1 A paper-and-scissors activity that models the difference of two squares

> *Something clicked in my head right this minute when we did that. I like factoring and it's nice to know why it makes sense like that.*
>
> (Jackie, Grade 7, 12 years old).
>
> *We can build visual images on the basis of visual memories but we can also use the recalled visual image to form a new image we have never actually seen. Certainly, imagery is used in everyday life, . . ., nevertheless imagery has to be considered as a major medium of thought, as a mechanism of thinking relevant to hypothesis generation. Some hypotheses naturally take a pictorial form.*
>
> (Magnani, 2001, p. 98)
>
> *(C)asting the virtual into physicality forces the illusion to withstand the light of day – to test its honesty. Experiencing a physical object . . . is a different sense of apprehension of the object . . . . Viewing the physical object, we have a more integrated idea of the whole object.*
>
> (Dickson, 2002, p. 221)

Jackie (Cohort 2) was in seventh grade when she joined my Algebra 1 class to participate in a yearlong teaching experiment involving various aspects of algebraic thinking at the middle school level. Her earlier mathematical experiences had solidified for her the impression that mathematics was something that she merely

F.D. Rivera, *Toward a Visually-Oriented School Mathematics Curriculum*, Mathematics Education Library 49, DOI 10.1007/978-94-007-0014-7_2,
© Springer Science+Business Media B.V. 2011

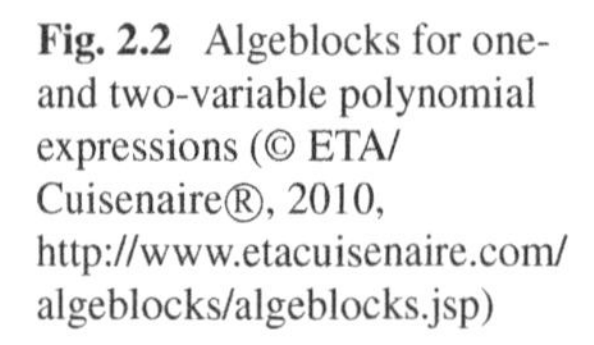
Fig. 2.2 Algeblocks for one- and two-variable polynomial expressions (© ETA/Cuisenaire®, 2010, http://www.etacuisenaire.com/algeblocks/algeblocks.jsp)

followed on the basis of rules that occasionally made sense to her. Her comment in the opening epigraph came early in the fall semester when our class was exploring factoring simple polynomials initially from a visual perspective. In a follow-up activity, when I asked the class whether it was possible to factor a sum of squares, all of them said no and explained that there was no way of reconfiguring the totality of two squares having two different dimensions into a single rectangle (i.e., with no overlaps).

Later, when the students learned to use algeblocks (Fig. 2.2), a three-dimensional version of algebra tiles, the manner in which they explained the process of factoring simple quadratic trinomial and cubic binomial expressions reflected a firm grasp of a visual strategy. For example, the responses of two Cohort 1 Grade students, Cheska and Jamal, in Fig. 2.3 had them associating the task of factoring a simple polynomial expression in terms of whether it was possible to reconfigure the relevant algeblocks into a single rectangle whose dimensions represent the factors of the given expression. Also, when I asked them to factor $x^3 + 2x^2$, their initial approach had them gathering algeblocks consisting of a cube and two squares and combining them to form a cuboid with a square base x^2 and a height of $(x + 2)$. The visual action allowed them to conclude the equivalent form $x^3 + 2x^2 = x^2(x + 2)$.

1 Nature of Cognitive Activity and Cognitive Action

I have found it interesting to recast the cognitive actions of Jackie, Cheska, and Jamal from a Peircean perspective – that is, interpreting their thinking and inferring as matters that involve signs and sign activity, respectively, in relation to the concept and process of factoring. The signs they used, as a matter of fact, convey

Cheska:

1. Draw algeblocks corresponding to $x^2 + 2x + 2$ below. How do you use the blocks to help a friend determine whether the polynomial is factorable or not?

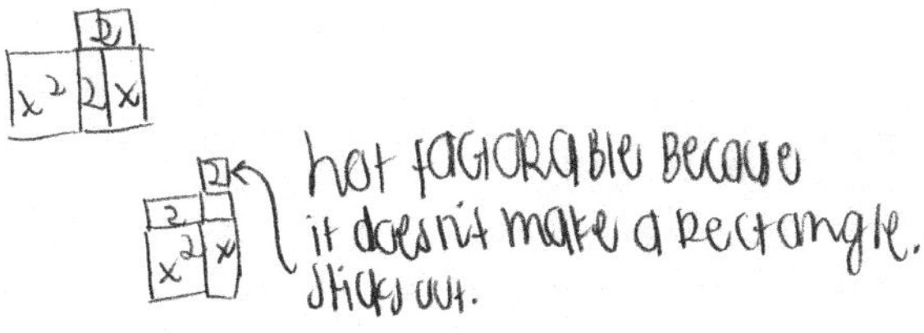

Jamal:

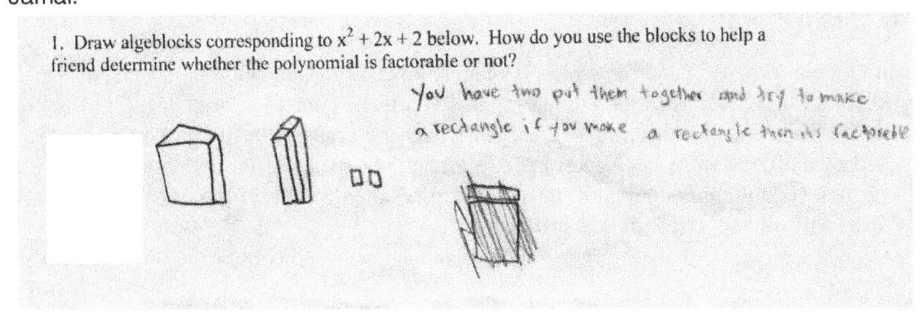

Fig. 2.3 Responses of Cheska and Jamal in relation to the primeness of $x^2 + 2x + 2$

the meaning and content of their experiences through an impressive appropriation of both internal and external semiotic resources. Jamal and Cheska used their externally drawn knowledge of algeblocks to explain the nonfactorability of a quadratic trinomial expression in terms of an internally derived diagram (Fig. 2.3). Jackie was successful in factoring all binomials of the difference-of-squares type because the paper-and-pencil activity in Fig. 2.1 helped her construct a canonical image of a rectangle with dimensions corresponding to the factors of a difference of squares.

The students' cognitive actions could also be interpreted as exemplifying what Magnani (2004) calls *model-based reasoning* that, at the very least, involves the "construction and manipulation of various kinds of representations" (Magnani, 2004, p. 516), which includes visual representations. Central to scientific work, in particular, conceptual change processes, Magnani (2004) notes how scientists frequently engage in both theoretical (internal) and practical (external) activities that allow them to gain a concrete experience of a phenomenon undergoing observation and analysis. However, the reasoning that accompanies the manipulations has oftentimes been analyzed solely as an internal phenomenon with very little value accorded to actions such as thinking through doing and/or thinking with the use of external representations.

Recent empirical work in school mathematics education has also begun exploring a similar modeling perspective. For example, central to instrumental genesis in technology-mediated learning of mathematics is the acquisition of intended mathematical knowledge as a consequence of learning more about ("instrumentation") and beyond ("instrumentalization") the tools and their built-in functions (e.g., Rivera,

2005, 2007a; Trouche, 2003, 2005). Researchers who work in an emergent modeling perspective ground initial mathematical activity in real or experientially real settings, which provide learners with an opportunity to develop a "model of" some (informal) intended mathematical knowledge that then becomes their basis in producing a "model for" more (formal, general) mathematical concepts, processes, and reasoning (e.g., Gravemeijer, Lehrer, van Oers, & Verschaffel, 2002; Stephan, Bowers, Cobb, & Gravemeijer, 2003).

The following two related points below are worth noting early at this stage in this book in light of how I situate the role of visualization in cognition involving mathematical objects, concepts, and processes. *First*, cognitive activity is a dynamic and distributed phenomenon, that is,

> [it] is in fact the result of a complex interplay and simultaneous coevolution, in time, of the states of mind, body, and external environment. Even if, of course, a large portion of the complex environment of a thinking agent is internal, and consists in the proper software composed of the knowledge base and of the inferential expertise of the individual, nevertheless a "real" cognitive system is composed by "distributed cognition" among people and some "external" objects and technical artifacts.
>
> (Magnani, 2004, p. 520)

One important implication of this distributed view on visualization in mathematical learning involves seeing "the cognitive system" as "instantiating the process rather than cognition simpliciter," that is, the product (Giere & Moffatt, 2003, p. 303).

Second, when a mathematician employs diagrams in his or her reasoning, the visual process is part of this distributed epistemic phenomenon – that is,

> a kind of epistemic negotiation between the sensory framework of the [mathematician] and the external reality of the diagram. This process involves an external representation consisting of written symbols and figures that for example are manipulated "by hand." The cognitive system is not merely the mind–brain of the person performing the [mathematical] task, but the system consisting of the whole body (cognition is embodied) of the person plus the external physical representation. In [mathematical] discovery the whole activity of cognition is located in the system consisting of a human together with diagrams.
>
> (Magnani, 2004, p. 520)

Thus, the term *visualization* as it is used in this book should not be viewed as a thaumaturgical concept or a tool that is isolated and analyzed for the sole purpose of developing and justifying an alternative mathematical practice distinct from current institutional practices that continue to favor alphanumeric competence. I see it as functioning within a distributed system that Magnani (2004) has clearly described in the above two passages. Further, the dynamic nature of the system is compatible with, and adaptive to, changes in representational competence or, in Freudenthal's (1981) words, *progressive schematization.*

In his August 1980 plenary address at the Fourth International Congress of Mathematics Education, Freudenthal explained his idea of progressive schematizing, which I consider to be an integral element in any psychological account of visual thinking in mathematics, in the following manner:

> The history of mathematics has been a learning process of *progressive schematizing*. Youngsters need not repeat the history of mankind but they should not be expected either to start at the very point where the preceding generation stopped. In a sense youngsters should repeat history though not the one that actually took place but the one that would have taken place if our ancestors had known what we are fortunate enough to know. Schematizing should be seen as a psychological rather than a historical progression. . . . The idea that mathematical language can and should be learned in such a way – by progressive formalizing – seems even entirely absent in the whole didactical literature.
>
> (Freudenthal, 1981, p. 140)

In present history, we are certainly fortunate enough to witness, and benefit from, theoretical and methodological advances in the cognitive science of visualization and learning technologies that we now find ourselves in a much better position to realize Freudenthal's vision of a mathematical language that could be learned from a visually oriented progressive formalization perspective. In this context, visualization is framed within, and pursued as, an epistemological process that lies at the kernel of a conceptual and developmental account of alphanumeric competence.

Following O'Halloran (2005) and Millar (1994), *progressive schematizing involves a complex conceptual evolution and coordination of, and convergence among, systems of semiotic resources.* Progression at the intra-semiotic level occurs within mathematical language, mathematical symbols, and visual images, while progression at the inter-semiotic level occurs between and among the three systems. Since each type of semiotic resource contains information and several resources that together convey redundancy of the same information in varying form and content, progression is a marked indication that some type of convergence is occurring. To take a case in point, Jackie, Cheska, and Jamal acquired their algebraic competence in factoring in a progressive manner by drawing on their visual experiences with the algeblocks that provided them with the necessary "insight, understanding, and thinking" before they finally transitioned to "rote, routines, drill, memorizing, and algorithms" (Freudenthal, 1981, p. 140). Factoring on a quadrant mat with the algeblocks contained sufficient visual structural information that enabled them to construct an analogous symbolic structure, which they conveyed through a diagram. Progressively, the diagram emerged as a symbolic (algebraic) process that could then handle all cases of factoring involving quadratic trinomial squares.

Following Bakker (2007), the use of symbolic forms associated with procedures basically signifies successful hypostasis and increased "structuring of thought" (O'Halloran, 2005, p. 2). Further, familiarity with the relevant concepts does not necessarily imply the absence of visualization. For example, Fig. 2.4a–c shows the written work of Jackie, Cheska, and Jamal on three assessment tasks that were given to them toward the end of a teaching experiment on factoring. Clearly, they fully transitioned to a routinized algebraic strategy they hypostasized as tic-tac-toe that reflected their structural experiences with the algeblocks. The tic-tac-toe strategy is a coded symbolic knowledge in algebraic form; it exemplifies a *diagram*, a type of visual spatial representation in which the symbols, expressions, and other related inscriptions are positioned and combined in a particular way that makes sense within some stipulated rule.

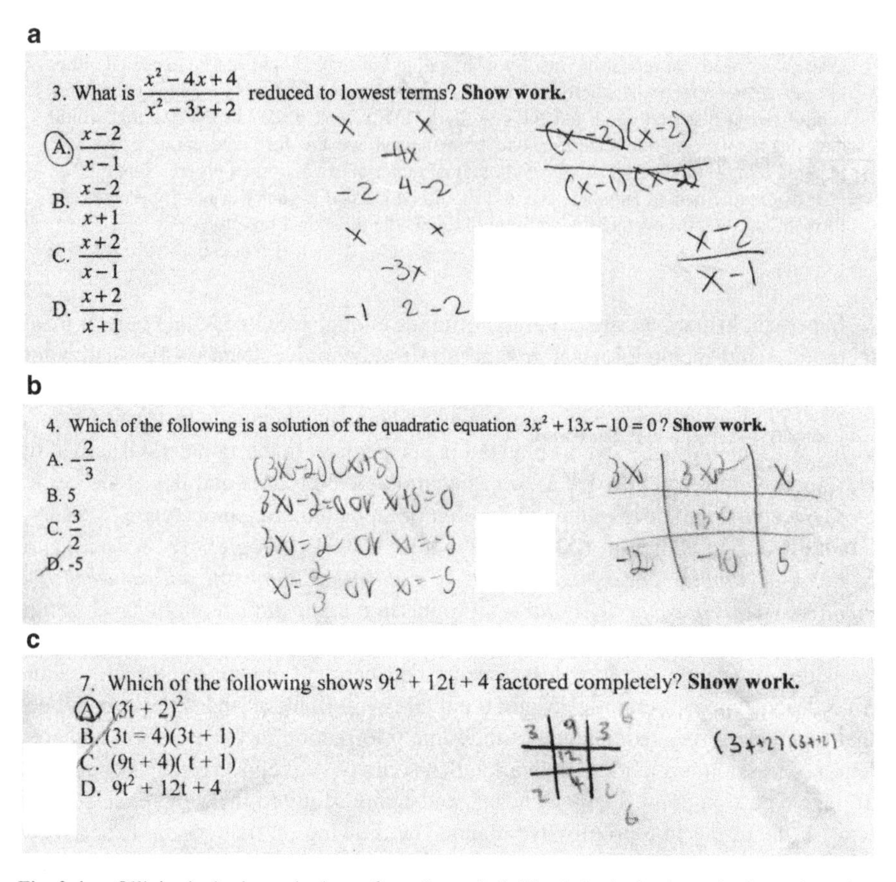

Fig. 2.4 **a** Ollie's algebraic method on a factoring task. **b** Cheska's algebraic method on a factoring task. **c** Jamal's algebraic method with no variables on a factoring task

In making visualization an integral element of progressive schematization in the epistemology of school mathematics, I may have given the impression that I am denying the possibility of different styles, abilities, preferences in learning, and cultural variations in visual attention, thinking styles, and other perceptual processes (cf. Mayer & Massa, 2003; Pylyshyn, 2006; Nisbett, Peng, Choi, & Norenzayan, 2001). Indeed, some learners are more verbal than visual, others more numerical than pictorial, and still others preferring more abstract than concrete approaches. In my Algebra 1 class, when I asked them to analyze the assessment task shown in Fig. 2.5a at the end of a visual-driven teaching experiment on linear systems of equations and inequalities, about 44% suggested a visual approach (Fig. 2.5b) and about 56% offered a nonvisual (algebraic, numerical; Fig. 2.5c) solution. Hence, it is hard to escape the ubiquity of the dual coding phenomenon – the verbal versus nonverbal processing reality among individuals – that Paivio and his colleagues (e.g., Clark & Campbell, 1991; Clark & Paivio, 1991; Paivio, 1986, 2006; Paivio & Desrochers,

Consider the following linear system:

$$\begin{cases} y = 2x - 3 \\ y = 2x + n \end{cases}$$

a

For what value or values of n will the linear system have no solution? Explain.

For what value or values of n will the linear system have an infinite number of solutions? Explain.

b

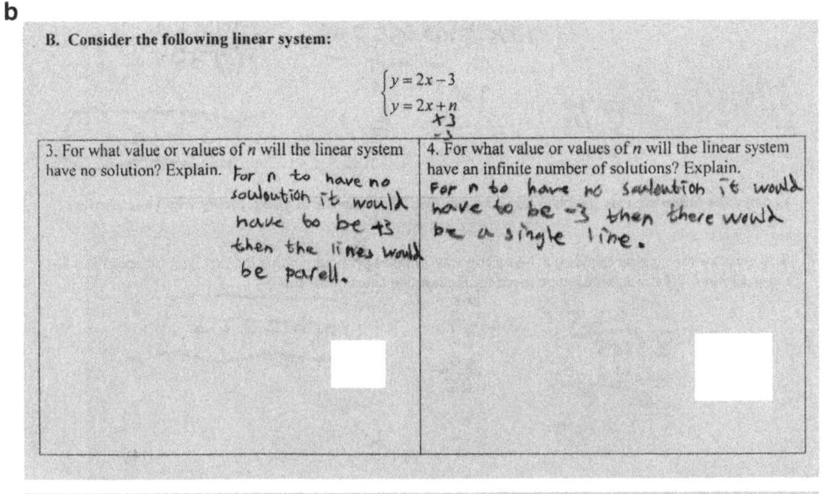

c

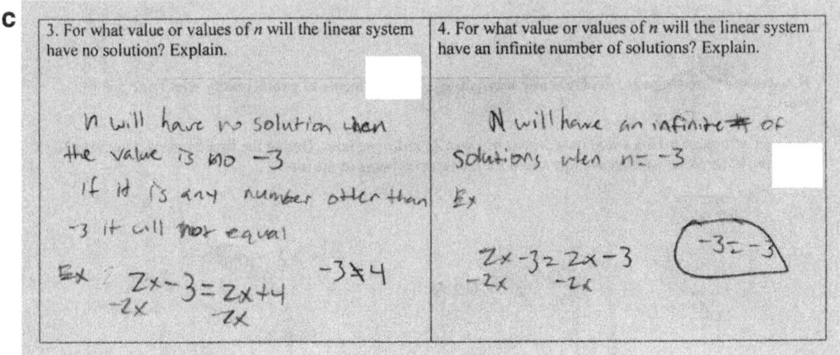

Fig. 2.5 **a** An end-of-a-unit Algebra 1 assessment task involving a linear system of equations in two variables. **b** Kirk's visual approach to the Fig. 1.5a task. **c** Earl's numerical solution to the Fig. 1.5a task

1980) have empirically demonstrated in various activities, linguistic, mathematical, and otherwise.

The claim concerning the existence of dual coding reminded me of three things. *First*, I had students like Dina (eighth grader, Cohort 1) who consistently used a numerical approach to generalizing patterns from sixth through eighth grade. In sixth grade, when she extended and obtained a generalization for the pattern in Fig. 2.6a, she initially noticed how each succeeding stage after the first kept increasing by two circles. In reconstructing the pattern on the table with actual chips and extending it to include two additional stages (Fig. 2.6b), she apparently saw nothing else (e.g., shape of the pattern) beyond the additive relationship. Her preference in dealing with patterns numerically was, in fact, carried through all the 3 years she was involved in our study. For example, Fig. 2.7 shows her written work on a

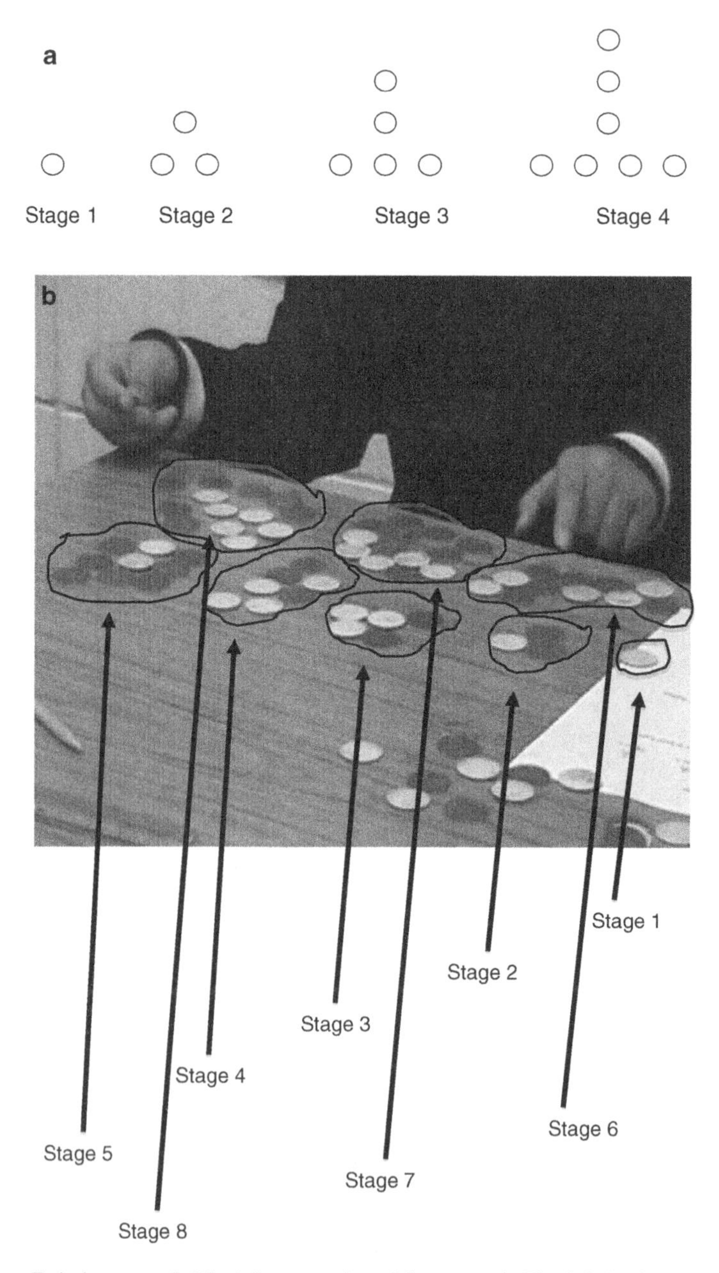

Fig. 2.6 **a** T circle pattern. **b** Dina's interpretation of the pattern in Fig. 2.6a in sixth grade

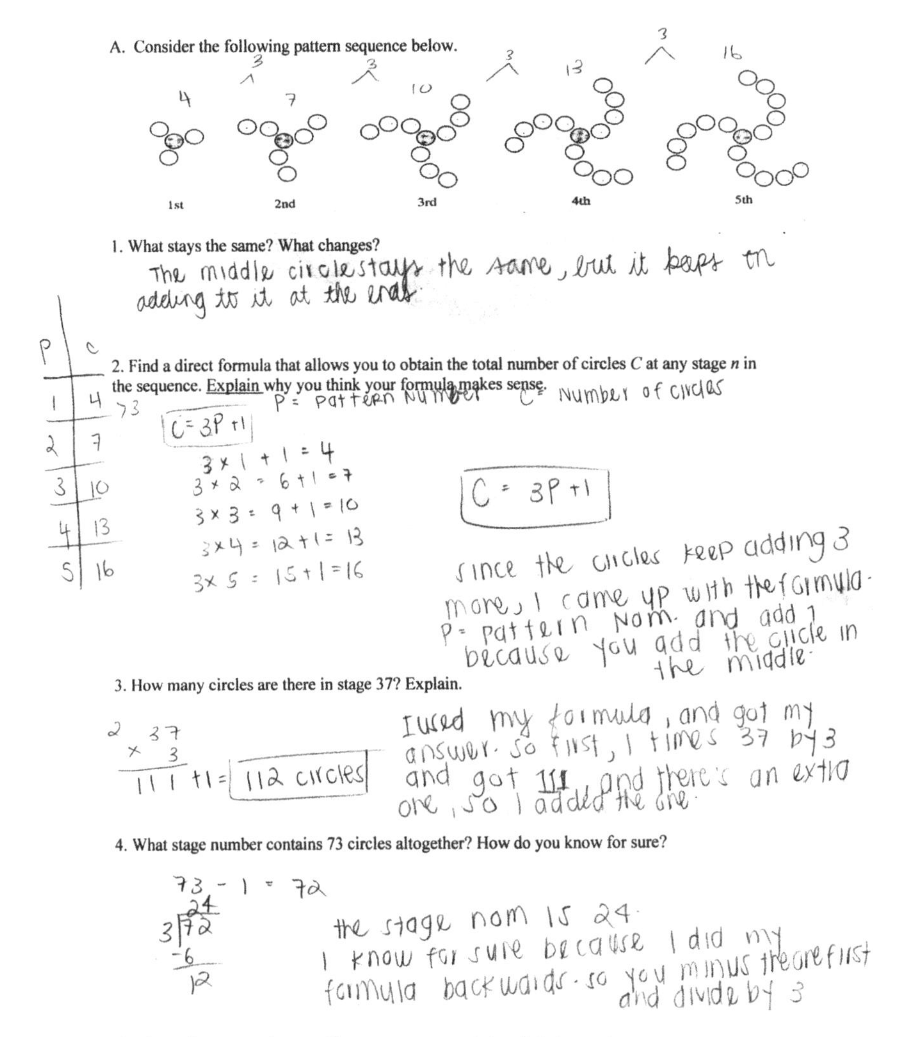

Fig. 2.7 Dina's written work on a linear pattern task in eighth grade

linear pattern in Grade 8. Her written response to item 1 in Fig. 2.7 conveys a surface knowledge of the basic features of the pattern. In explaining her direct formula, $C = 3P + 1$, which she established numerically, she associated the coefficient 3 with the constant addition of three circles and the constant, 1, to the "circle in the middle."

Second, a neural basis for both dual coding in mathematics and visuospatial processing relevant to various aspects of numerical calculations has now been established following rigorous methodological protocols. The results provide very interesting insights that enable us to understand the fundamental, genetic role of visual thinking in mathematics. For example, a recent brain imaging study by Tang et al. (2006) has shown that arithmetical processing in the brain appears to be shaped by cultures. When the thinking processes of 12 adult Western participants and 12 university Chinese students were compared in relation to a simple visually presented arithmetical task that asked them to determine whether a third digit was greater than the bigger one of the first two in a triplet of Arabic numbers, they actually found "differences in the brain representation of number processing between" the two groups (p. 10776). On this particular task, the Chinese participants processed visually, while the Western participants processed verbally. Tang et al. (2006) hypothesized that the visual dominance in number processing among the Chinese participants could be explained by their reading experiences in school, which involve repeatedly learning Chinese characters, and their early experiences in using an Abacus that activated the production of mental images that are all visual in form.

A recent review of research studies on the role of visuospatial processing in various aspects that matter to arithmetical computing by de Hevia, Vallar, and Girelli (2008) presents converging evidence that infers a "close relationship between numerical abilities and visuospatial processes" (p. 1361). For example, individuals have been documented to be relying on a spatial mental number line when comparing and obtaining differences of two numbers. The line, which research shows appears to be oriented from left to right, assists subjects in dealing with numerical quantities and their relationships and magnitudes at least in approximate (and not in exact) terms. Certainly, with more cultural learning, the relevant spatial properties are expected to further develop and undergo refinement. Beyond studies involving the mental number line, de Hevia et al. (2008) also underscore the significant role of visuospatial representations in memory and arithmetical problem solving. Findings include the following: (1) learners tend to keep numbers in an active state by drawing on some form of visuospatial support; (2) mathematical prodigies develop and use internal visual calculators that become evident when they solve numerical tasks that require complex computations; (3) individuals perform arithmetical procedures (e.g., obtaining the product of multi-digit numbers) by coordinating their stored arithmetical facts and their spatial image of the process relevant to the operation, a view that refines earlier interpretations which trace the process to be primarily working within a language-based semantic network (e.g., verbal rote learning of facts).

Third, in their recent extensive review of research concerning the imagery and spatial processes of blind and visually impaired individuals, Cattaneo et al. (2008) clearly articulate what I interpret from research among nonblind learners of mathematics to be a problematic aspect of the constitutive nature of the "visual experience." Cattaneo et al. write:

> The perceptual limitations of the congenitally blind are reflected at a higher cognitive level, probably because their cognitive mechanisms have developed by touch and hearing, which only allow a sequential processing of information. As a consequence, cognitive functioning of blind individuals seems to be essentially organized in a "sequential" fashion. On the contrary, vision – allowing the simultaneous perception of distinct images – facilitates simultaneous processing at a higher cognitive level.
>
> (Cattaneo et al., 2008, p. 1360)

One (pedagogical) concern about using visual thinking in mathematics as a cognitive strategy deals with the issue of how to cope with the "simultaneous perception of images," which could be an overwhelming experience for nonblind learners. From a practical standpoint, unlike alphanumeric representations that tend to bolster some form of sequential learning, a visual approach to understanding mathematical objects, concepts, and processes necessitates extensive assistance in the zone of proximal development. However, my own experiences with students like Jackie in the opening epigraph provide me with sufficient empirical proof to say that encouraging visual thinking in mathematics is worth pursuing due to its positive effects both in the short and the long haul.

Chapter 5 discusses the central role of visual thinking in the construction and justification of simple and complex algebraic generalizations relative to linear and nonlinear figural patterns. Chapter 7 addresses both physiological and sociocultural dimensions of visual representation in mathematical knowledge acquisition. At this stage, it simply suffices to say that while numerical-driven processes help students fulfill aspects of school mathematical activity such as factoring and constructing a direct formula, which may be institutionally sufficient (i.e., on the basis of the minimal competence required in state tests to be labeled mathematically proficient), the absence of any visual-based processing might prevent many of them from tapping onto their "natural" or "cultural" systems that would allow them to develop what Freudenthal above considers to be valued noble acts of insights, understanding, and thinking. Results of preinstructional clinical interviews with my Grade 2 students, for example, indicate that young children seem to have an early visual-based understanding of functions on the basis of their success in dealing with grade-level appropriate visual growth-patterning tasks (Rivera, 2010c; see Rivera (2006) for a brief synthesis of relevant work). If tapped effectively, the visual experience could assist them in hypostasizing the symbolic meaning of functions early in their school mathematics education. Here I am also reminded of Davydov (1990), who astutely pointed out that alphanumeric expressions remain senseless entities unless they are "placed under" a "real, object-oriented, sensorially given foundation" (p. 34). Paivio (1971), in fact, has noted the flexible capacity of visual imagery in dealing with several amounts of information at the same time, a view echoed by Cattaneo et al. (2008) in the case of vision in neuropsychological terms.

2 Basic Elements and Tensions in a Visually Oriented School Mathematics Curriculum

A thoroughgoing visually oriented school mathematics curriculum requires an explanation. Traditionally, visual thinking has been associated with concrete representations within a concrete-to-abstract continuum – that is, the visual or the concrete phase assists in meaningful construction and appropriation leading to an intended abstract knowledge in symbolic form. Figure 1.2a, b lists other binaries associated with this continuum phenomenon. It is a fact that some mathematical concepts (e.g., irrational and imaginary numbers) and processes (e.g., matrix multiplication) make sense only at the operative level in which they come to exist as a result of having been constructed extensions at the abstract level. Dörfler (2008) makes a useful distinction between *referential* and *operative* views of algebraic notation. In the referential view, there is a close mapping between rules and symbols and the arithmetical laws and experiences that govern them. In the operative view, there is no immediate appeal to any numerical arithmetical experiences. In the referential domain, letters are generalized numbers. In the operative domain, they are primarily signs without a referent but are capable of being manipulated according to agreed upon rules at the abstract level (Dörfler, 2008, pp. 144–145).

However, in this book, I take the position that the abstract phase in the continuum need not convey the absence of a visual representation. A more productive view involves the necessity of employing a different kind of seeing. In fact, many canonical forms and diagrams in almost all areas of mathematics are visual (hypostasized) abstractions of generalized concepts.

Figure 2.8 is a diagram of two-triangular relationships that illustrates the taken-as-shared view of a smooth transition from the visual to the abstract in the case of all referential content but a more perilous transition in the case of most operative content. In fact, much of school mathematics content operates within the solid, referential triangle. The conceptual tension begins when an operative content moves through the continuum in which some mathematical concepts and processes could not be sufficiently captured and described in visual terms unless, of course, some constraints are imposed on them.

For example, in my Algebra 1 class, I used the pebble arithmetic activity in Fig. 2.9a to explain the irrationality of $\sqrt{2}$. Unfortunately, my students never settled their doubt about the number being nonterminating and nonrepeating when alternatively represented by digits. While the activity in Fig. 2.9b convinced them that $\sqrt{2}$ was indeed a real number between 1 and 2 with an exact location on the number line and that it made sense to visually see, manipulate, and operate numbers that involve $\sqrt{2}$ and other irrational numbers for that matter, they were more interested in seeing how one would go about constructing the alternative representation below.

1.4142135623730950488016887242096980785696718753769480731766797379907324784621070388503875343276415727350138462309122970249248360558507372126441214970999358314132226659275055927557999505011527820605714701095599716 0...

A follow-up activity we used in class involves finding the sum of the harmonic series

$$\sum_{n=1}^{\infty} \frac{1}{n} = 1 + \frac{1}{2} + \frac{1}{3} + \frac{1}{4} + \cdots.$$

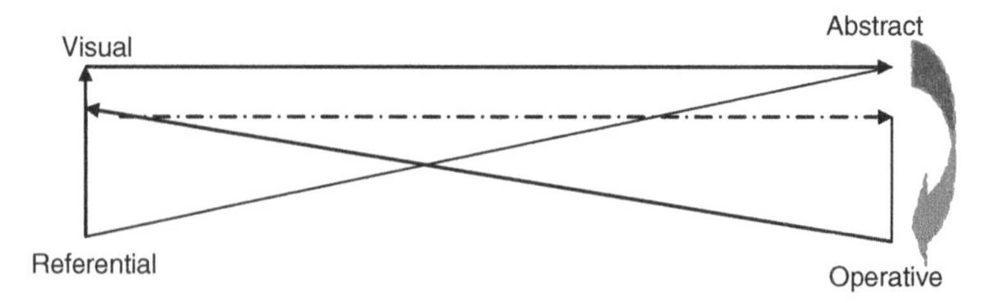

Fig. 2.8 A two-triangular model of content in school mathematics

a

Pebble Arithmetic Proof of the Irrationality of $\sqrt{2}$

An even number 2m means 2 horizontally aligned rows of m dots, say,

....

A square array of 2m x 2m means 4 square arrays, each consisting of m x m dots. Hence, every *even* square number is the sum of 4 square numbers.

....
....
....
....
....

....
....
....
....
....

An odd number 2m + 1 means 2 horizontally aligned rows of m dots plus an isolated dot, say,

....

A square array of (2m+1) x (2m+1) means 4 square arrays, each consisting of m x m dots, plus 4 groups of m dots, plus a central dot. Hence, every *odd* square number is odd (why?).

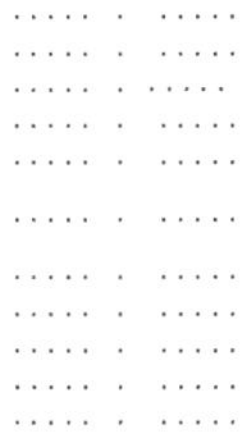

....
....
....
....
....

....

....
....
....
....
....

Hence, even squares are even and odd squares are odd. But every even square is a sum of four squares. We are now ready to prove that $\sqrt{2}$ is an irrational number.

Proof:

If $\sqrt{2} = \frac{a}{b}$, (a, b$\neq$0 integers), then $2 = \frac{a^2}{b^2}$ or $a^2 = 2b^2$. So, a^2 is twice another square number.

Then there would have to be a *smallest* square number that is twice another square number since there is no unending sequence of ever smaller square numbers. Let k^2 be the smallest. So

$$k^2 = 2(h^2).$$

Then k^2 is an even square. So, there is some number n such thet

$$k^2 = 4(n^2).$$

Also, $h^2 = 2(n^2)$.

Since h^2 is smaller that k^2, it means that a smaller square number than k^2 is twice another square number, but this contradicts the assumption we made about k^2 being the smallest.

Thus, there is no square number that is twise another.

b

Exploring $\sqrt{2}$

1. Use a construction paper to outline a unit square. Then draw a diagonal to the square and cut along the diagonal. What is the length of the diagonal?

2. Obtain a piece of adding-machine tape and draw a number line that is sufficient enough to construct and label integers from –10 to 10. Then construct and label –9.5, –8.5,–7.5, ..., 9.5.

3. Starting from 0, lay out the diagonal whose length is $\sqrt{2}$ on the number line. Estimate the numerical value of $\sqrt{2}$. Can the value be 1.5? Why or why not? Which number is it closest to?

4. Construct $2\sqrt{2}$, $3\sqrt{2}$, and $5\sqrt{2}$ on the same number line. Estimate their values.

5. Construct $2 + \sqrt{2}$. Is it the same as $2\sqrt{2}$? Explain.

6. Construct $\frac{2+\sqrt{2}}{2}$. Marian thinks that $\frac{2+\sqrt{2}}{2} = 1 + \sqrt{2}$. Is she correct? Why or why not?

Fig. 2.9 **a** Pebble arithmetic proof of the irrationality of $\sqrt{2}$ (Giaquinto, 2007, pp. 139–141). **b** Activity involving $\sqrt{2}$ (adapted from Coffey, 2001)

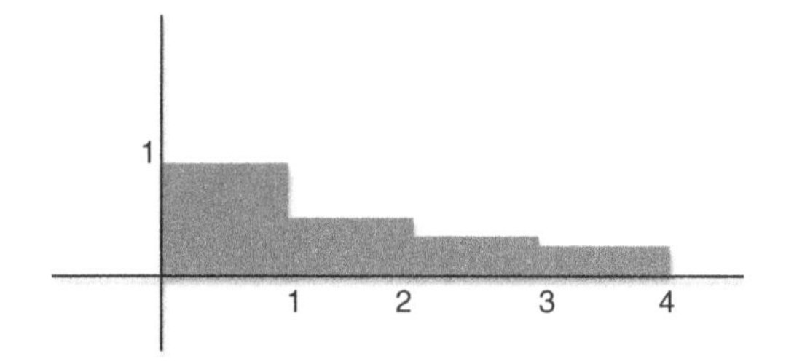

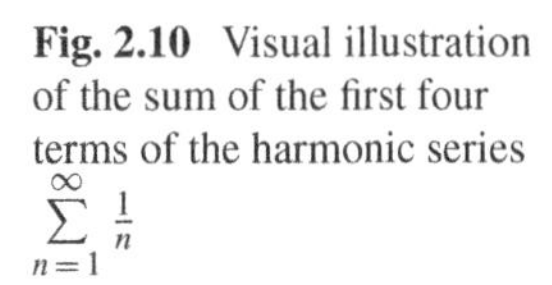

Fig. 2.10 Visual illustration of the sum of the first four terms of the harmonic series $\sum_{n=1}^{\infty} \frac{1}{n}$

Initially, the students used a centimeter graphing paper and constructed the picture shown in Fig. 2.10. Then they confirmed that each shaded rectangle had an area corresponding to a term in the given harmonic series. Next I asked them to imagine a train of shaded rectangles beyond the picture they drew. I had them estimating the amount of paint they would need to cover the entire shaded area. With the aid of a graphing calculator, they started adding the terms one by one. After verifying on the basis of a sufficient number of terms that the series would yield no finite sum, they began to realize the difference between everyday and mathematical ways of seeing.

The above experience allowed my Algebra 1 students to develop the view that visually drawn constructions of some mathematical objects, concepts, or processes despite being incomplete in form could also effectively assist in developing *structural awareness* (Mason, Stephens, & Watson, 2009) of the corresponding abstract knowledge. While their totality (i.e., essence) could not be fully apprehended through the sensory modalities alone, what is actually being sought is intellectual insight drawn from one's emerging structural awareness as mediated by the relevant visual representation. Certainly, having structural awareness drawn from visual forms would allow learners to construct inferences about the corresponding abstract object, concept, or process. This perspective should assist us in moving beyond dichotomous thinking relative to the symbolic/visual hierarchy in mathematics. Brown (1999) is certainly correct in pointing out that some mathematical knowledge could not be primarily derived from sense experiences, as follows:

> All measurement in the physical world works perfectly well with rational numbers. Letting the standard meter stick be our unit, we can measure any length with whatever desired accuracy our technical abilities will allow; but the accuracy will always be to some rational number (some fraction of a meter). In other words, we could not discover irrational numbers or incommensurable segments (i.e., lengths which are not ratios of integers) by physical measurement. It is sometimes said that we learn $2 + 2 = 4$ by counting apples and the like. Perhaps experience plays a role in grasping the elements of the natural numbers. But the discovery of the irrationality of $\sqrt{2}$ was an intellectual achievement, not at all connected to sense experience.
>
> (Brown, 1999, pp. 5–6)

His basic point is reflected via the dashed side of the operative triangle shown in Fig. 2.8, which symbolically conveys the treacherous path that characterizes the movement from the visual to the abstract or, more appropriately, from our everyday sense of the visual to the mathematically visual.

However, aiming for structural awareness allows us to engage in a more reconciliatory *interanimated* discussion of the perceived divide between the visual and the symbolic or, in Brown's terms, between sense experience and intellectual achievement. This move toward interanimation is derived from the work of Warren, Ogonowski, and Pothier (2005), who also saw the need to move past what they perceive to be dichotomies in modes of thinking in school science. They speak of interanimation in terms of some creative coordination, that is,

> [it] denotes a process whereby a person comes to regard one way of conceptualizing, representing, and evaluating the world through the eyes of another, each characterized by its own objects, meanings, and values. As such, it resists the strong temptation to dichotomize modes of thinking or being.
>
> (Warren et al., 2005, p. 142)

The move toward structural awareness is, thus, seen as being initially rooted in visual experience and, yet, serves as a route in gaining a better access of the corresponding abstract knowledge as well. Its essence, Mason, Stephens, and Watson (2009) note, lies in a learner's "experience of generality, not a reinforcement of particularities" (p. 17) that enables him or her to find meaning in manipulating both representational contexts in a simultaneous fashion.

3 General Notion of Visualization in this Book

In this book, I share the view that *visualization* is about "the many ways in which pictures, visual images, and spatial metaphors influence our thinking[1]" (Reed, 2010, p. 3). Hence, in its raw form, it could be either internal (mental images) or external (diagrammatic and spatial representations) (see Lohse, Biolsi, Walker, & Rueler, 1994 for an interesting classification of visual representations). Reed articulates it clearly when he insists that

> [l]anguage is a marvelous tool for communication, but it is greatly overrated as a tool for thought. . . . Much of comprehending language, for instance, depends on visual simulations of words or on spatial metaphors that provide a foundation for conceptual understanding. Sometimes we are conscious of visual thinking. At other times it works unconsciously behind the scenes.
>
> (Reed, 2010, p. 3)

[1]Bees provide a nonhuman example of species that appear capable of distinguishing between shapes and employing landmarks in tracing their route from some food source to their hive (Frisch, 1967).

However, as a tool for mathematical thinking, visualization has to take a much stronger sense in order to be significant and powerful. As I have used it over the past several years as a cognitive tool in helping my students learn mathematics better, I see visualization as a type of representation that employs "visuospatial relations in *making inferences* about corresponding conceptual relations" (Gattis & Holyoak, 1996, p. 231). Thus, I do not associate visual thinking in mathematics in static terms as merely about seeing images or pictures for the sake of having a visual or a sense experience in order to make mathematics learning fun. It is, more importantly, a concept- or process-driven seeing with the mind's eye.

Visualizing with the aid of, say, algeblocks, fraction strips, drawn pictures, and other manipulatives requires going beyond acknowledging that certain alphanumeric expressions could be mapped onto particular concrete images or diagrams. It also means employing it as a perceptual tool for exploring, sense making, constructing, and establishing the corresponding conceptual relations, "demonstrating and searching for a justification" (Malaty, 2008), and theorizing. Figure 2.11a–c provides examples of a bar representational approach in making sense of word problems involving percents. The examples have been drawn from several elementary school mathematics texts that are currently being used in classrooms in Singapore, the Netherlands, and Japan.[2] The rectangles shown are relation-based diagrams that convey particular meanings relevant to the problems being investigated. Another example is shown in Fig. 2.12. Parker's (2004) visual strategy in making sense of percent increase and percent decrease problems in prealgebra allows students to see relationships between various symbols (percent, numbers, terms associated with percent) involved in solving a particular type of percent problems. Kennedy's (2000) work with several groups of at-risk high school students emphasizes the need for them to see concrete-to-abstract and numerical-to-algebraic relationships between various manipulative models in integers (chips), fractions (strips and pattern blocks), percents and decimals (fraction strips), and the continuous number line. A fourth example is shown in Fig. 2.13 drawn from my second-grade class, which shows how they employed visual representations in making sense of, and establishing the conceptual relations involved in, adding, subtracting with regrouping, and multiplying whole numbers appropriate at their grade level.

The activity of theorizing in visualization means establishing conjectures and mathematical explanations, including "making speculative and intuitive work" (Jaffe & Quinn, 1993, p. 2) that in most cases might not be the same thing as establishing rigorous mathematics in which case the focus is on formal deductive proof. Visual theorizing shares many of the characteristics associated with the notion of *theoretical* – that is, according to Jaffe and Quinn (1993):

---

[2]See Ng and Lee (2009), Murata (2008), and van Galen, Feijs, Figueiredo, Gravemeijer, van Herpen, and Keijzer (2008) for an explanation of the role of rectangular, tape, and strip diagrams in Singapore, Japanese, and Dutch textbooks, respectively. I should note the interesting similarity in, and convergence toward, the use of visual approaches in teaching word problems in mathematics (more generally) in these countries.

a

Ali had $120. He spent 40% of the money on a watch and 25% of the remainder on a pen.

(a) What percentage of his money did he spend?

A shopkeeper had 4 handbags which were of the same cost price. He sold 3 of them at 40% more than the cost price. He sold the fourth handbag at cost price. He received $260 altogether. Find the cost price of each handbag.

140 × 3 + 100 = 520

520% ⟶ $260

1% ⟶ $\$\frac{260}{520}$

100% ⟶ $\$\frac{260}{520} \times 100 = \$50$

Cost price of each handbag = $50

b

The bill is $32.00. Calculate a 15% tip.

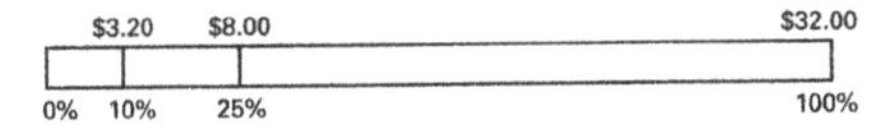

Here are two different solution strategies.

- Calculate 10% ($3.20), then 5% ($1.60), adding for 15% ($4.80).
- Calculate 25% ($8.00), then 10% ($3.20), subtracting for 15% ($4.80).

c

In Nishikawa City, each person produced 800g of waste each day last year. This is 125% of the amount produced 10 years ago.

What was the amount of waste produced 10 years ago?

Let's think about how to find the base quantity!

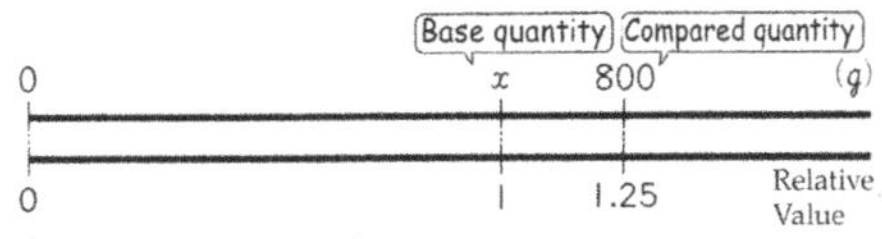

1 Make the amount 10 years ago xg and write a multiplication sentence. Then find x.

$x \times 1.25 = 800$

$x = 800 \div 1.25$

$x = \square$ Answer: $\square$ g

Fig. 2.11 (continued)

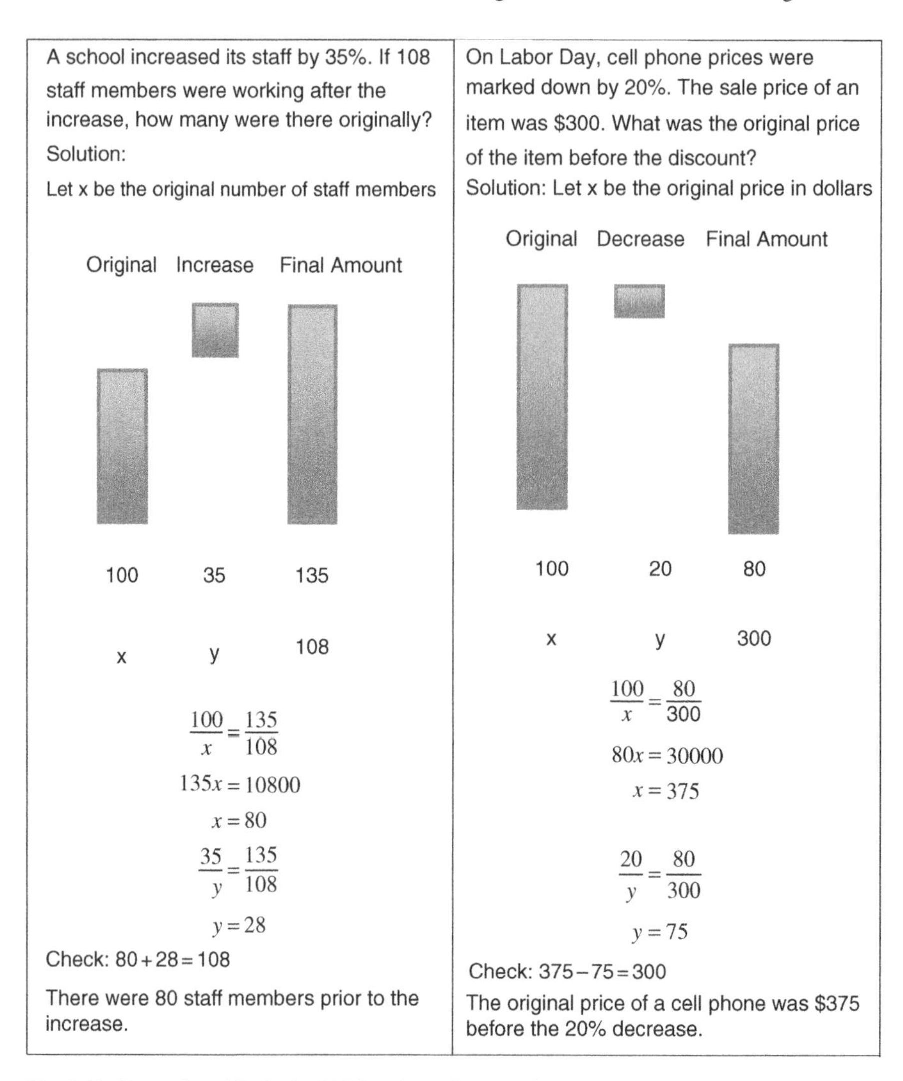

Fig. 2.12 Examples of Parker's (2004) unitary diagrams for solving percent increase and percent decrease problems

Fig. 2.11 a Visual approach to percent problems using unitary diagrams in a US version of a Singapore mathematics text (Curriculum Planning and Development Division, 2008, pp. 51 & 99; Copyright belongs to the government of the Republic of Singapore, c/o Ministry of Education, Singapore, and has been reproduced with their permission). **b** Visual approach to a percent problem using a bar diagram in a US version of a Netherlands mathematics text (Mathematics in Context, 2006, p. 23). **c** Visual approach to a percent problem using a tape diagram in a US version of a Japanese mathematics text (Tokyo Shoseki, 2006, p. 66)

Campos's Visual Subtraction with Regrouping Involving Two Two-Digit Whole Numbers

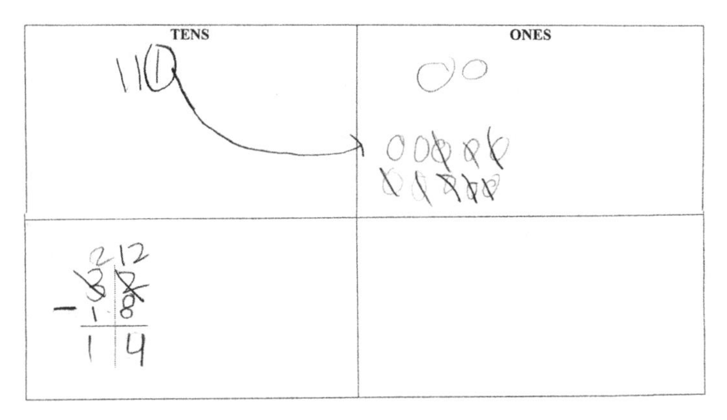

Lana's Visual Addition with Regrouping Involving Two Three-Digit Whole Numbers

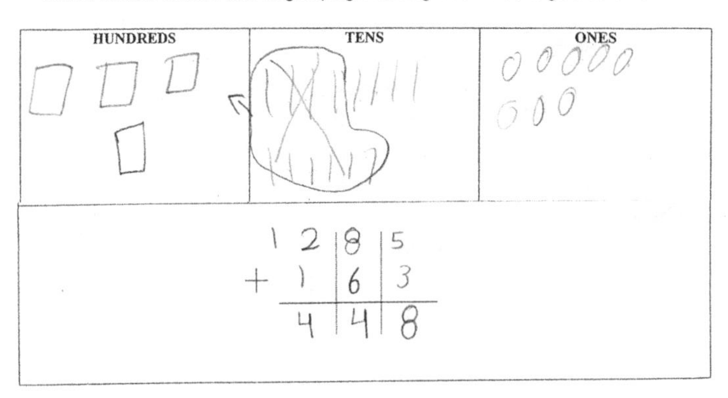

Eddie's Visual Multiplication Involving Two Single-Digit Whole Numbers

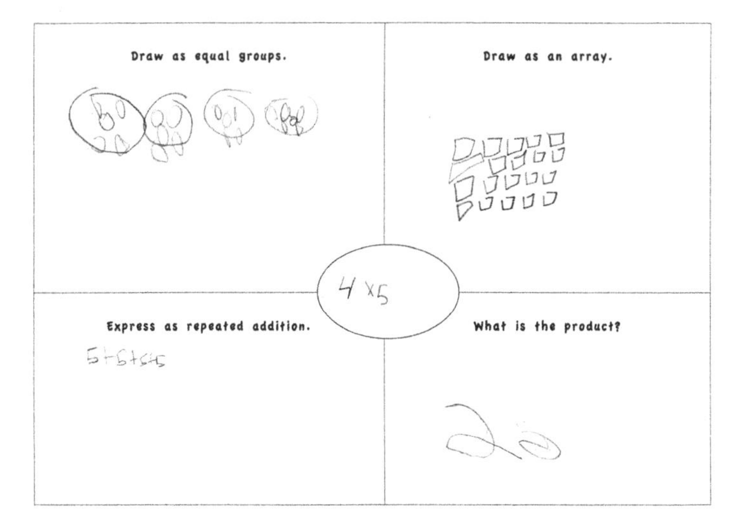

Fig. 2.13 Grade 2 students' visual approaches to adding, subtracting, and multiplying whole numbers

> The initial stages of mathematical discovery – namely, the intuitive and conjectural work, like theoretical work in the sciences – involves speculations on the nature of reality beyond established knowledge. Thus we borrow our name "theoretical" from this use in physics. There is an older use of the word "theoretical" in mathematics, namely, to identify "pure" rather than applied mathematics; this is a usage from the past which is no longer common and which we do not adopt.
>
> (Jaffe & Quinn, 1993, p. 2)

4 The Meaning of Representation in this Book

In using the term *representation*, I acknowledge the intricate relationship between its narrow and broad meanings [in Piaget's (1951) sense]. In a narrow sense, it refers to symbols that render an internal experience in some intelligible external form with the primary intent to communicate ("this expresses what I think I see"). In a broad sense, it conveys one's current worldview, that is, his or her interpretation of his or her experience of a phenomenon under investigation ("this is how I see"; cf. Pylyshyn, 1973) or, more concisely in Leyton's (2002) sense, "representation as explanation." For Vergnaud (2009), it is "a dynamic activity . . . a functional resource [that] organizes and regulates action and perception . . . [and is] a product of action and perception" (p. 93). Goldin (2002) offers a functional definition, that is, representation consists of inscriptions (or signs), rules for combining them, and a structure that allows the rules to make sense. Visual thinking, then, as a type of inference-based representation could be either personally drawn (i.e., based on individual subjective images) or externally mediated (i.e., based on what is acquired by the learner in a distributed context through acting with other learners and acting on institutional artifacts such as manipulatives and graphing calculators that, by their very nature, are loaded with intentional knowledge), or both. With more formal learning and socially structured experiences, it is likely the case that personal or subjective visualization and institutional ways of seeing will become integrated and more aligned than separate.

To illustrate, Fig. 2.14a shows the visual representations in written form of Jackie (Cohort 2, seventh grader) on a patterning task that I gave to my Algebra 1 class twice, before and after a month-long teaching experiment on linear patterning and generalization. With no external influence by way of formal instruction, she initially perceived and visually constructed an oscillating pattern of short and long shapes of V in which "every odd numbered stage . . . takes up 2 squares" and "every even numbered stage . . . takes up 4 squares." The pattern she visualized after the teaching experiment reflects an appropriation of an externally derived visual experience, which explains her choice of a linear pattern and her use of algebraic symbols in which she conveyed her algebraic generalization. Figure 2.14b shows the visual representations in written form of Arman (Grade 2) on two addition problems before and after a 3-week teaching experiment on addition of whole numbers of up to three digits long. Prior to the teaching experiment, he would employ a count-all visual

Fig. 2.14 (continued)

a

Before

B. From the following three figures below, pick at least two figures and create a pattern. Draw your figures below or use a grid paper.

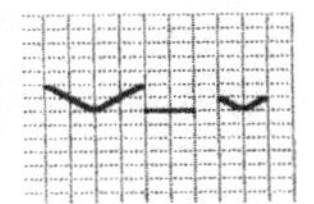

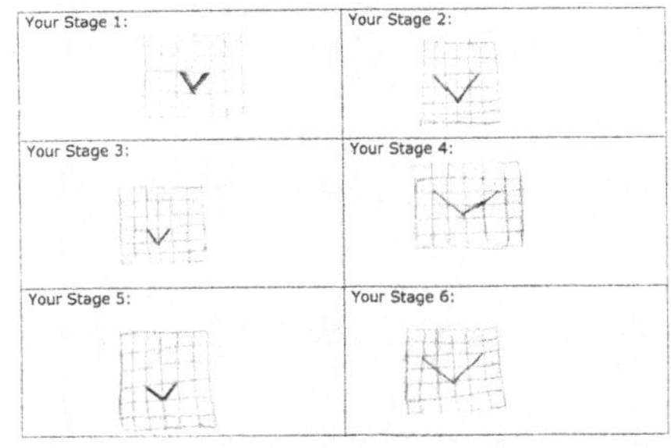

2. What stays the same and what changes in your pattern?

Every odd numbered stage is the same.
Every even numbered stage is the same.

3. Obtain a generalization for your pattern either by describing it in words or by constructing a formula. How do you know that your generalization works?

In every odd numbered stage, the figure takes up 2 squars.
In every even numbered stage, the figure takes up 4 squares.
It works because it increases every even numbered stage and decreases in every odd numbered stage.

After

B. From the following three figures below, pick at least two figures to create a pattern sequence of five stages. **Use the attached grid paper to draw your four additional stages.**

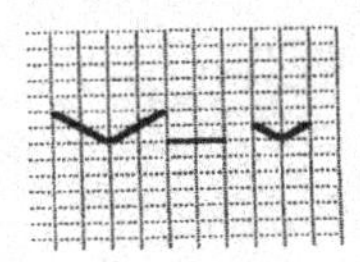

1. What stays the same and what changes in your pattern?

In the pattern, the way more is added stays the same, the stages changes, and the number of straight lines changes.

2. Obtain a generalization for your pattern either by describing it in words or by constructing a formula. How do you know that your generalization works?

L=3n
L=number of straight lines
n=stage number

Everytime the stage number increases, 3 more straight lines are added.

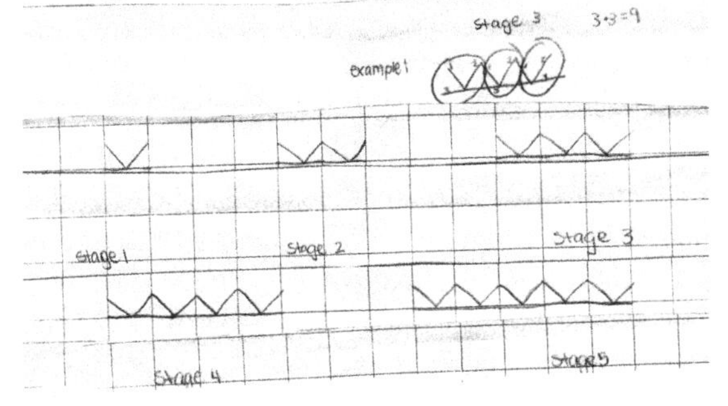

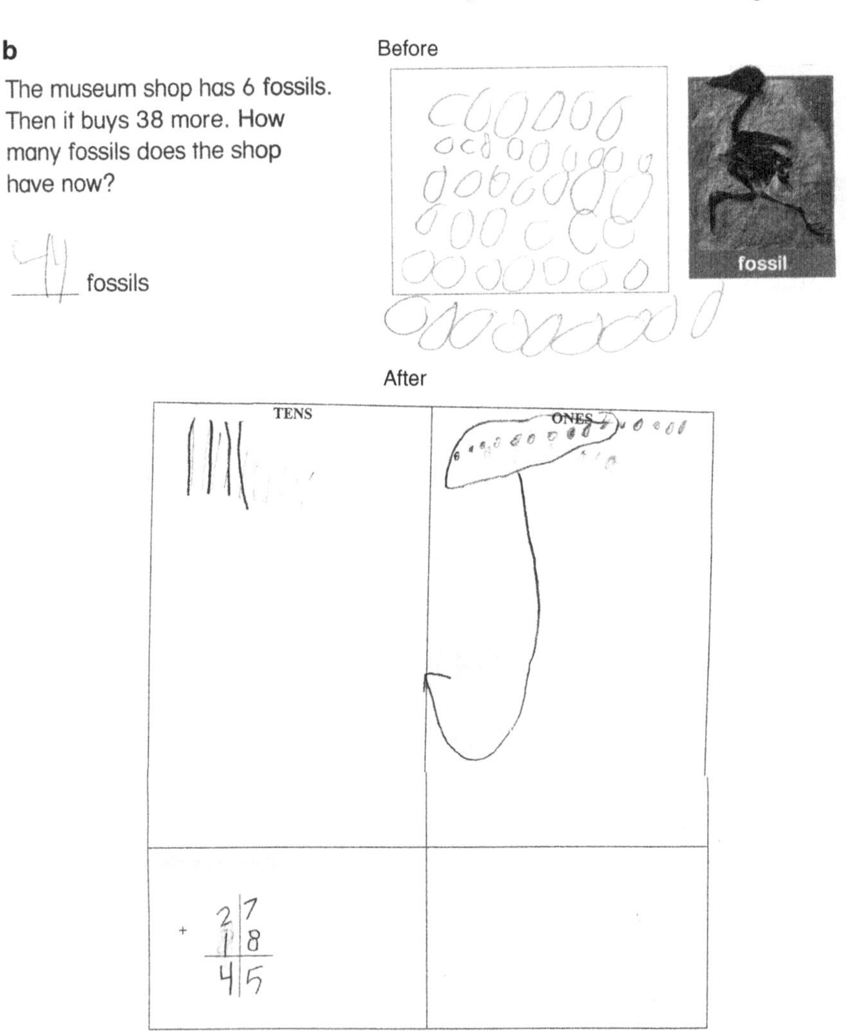

Fig. 2.14 **a** Jackie's written work on a free construction task after a teaching experiment on linear pattern generalization. **b** Arman's before and after visuals involving addition of whole numbers

strategy when adding whole numbers. He would initially draw all the required circles (or sticks) on the basis of the numbers that were presented to him and then would patiently count all of them together. After the teaching experiment, his combined visual/symbolic strategy reflected his understanding of the central role of place value and regrouping in the addition process. The solutions offered by Jackie and Arman exemplify what I consider in this book to be visual (and much later visuoalphanumeric) representations.

5 Three Fundamental Principles of Visualization

Visualization plays a fundamental role in any account of concept or process development, including problem solving. The manner in which it is addressed, however, depends on the role it plays in a particular theoretical paradigm, tradition, or orientation. For example, while the perceptual basis of knowledge is central to empiricist accounts, it is not so in the case of rationalist/nativist and historico-cultural traditions. Rationalist and nativist perspectives tend to have a taken-for-granted view of visual images or impressions since the primary interest lies in articulating innate structures that individuals are presumed to have at birth, which they use to acquire more knowledge about their world. In the historico-cultural tradition, any meaningful visualization, like all concepts, is cast in terms of its historically, socially, and culturally constructed and negotiated nature.

In this section, the fundamental principles of visualization that I suggest, which I have interpretively drawn from the general cognitive literature, are not aligned with any particular tradition since the jury is still out about whether a single paradigm could sufficiently characterize concept attainment within its stipulated rules and principles. Nevertheless, there are some common principles on visual representation that, albeit drawn from visual reasoning about everyday objects, are worth pursuing in a developing theory of visualization in school mathematics. This section is also meant for readers who are interested in how researchers outside of mathematics education have pursued, investigated, and analyzed the phenomenon of visualization with respect to objects and concepts that are not necessarily math related.

I should note that the last section in Chapter 3 revisits these same principles with the goal of situating them within work done on visualization in relation to mathematical cognitive activity. The need to have an extended discussion on these same principles in Chapter 3 deals with the fact that visualizing about mathematical objects, concepts, and processes is much harder than, and different from, visualizing about everyday objects, concepts, and processes. In the case of the former, they are theoretically derived (e.g., finding the zeros of a quadratic function by the quadratic formula), formal (e.g., variables and all definitions), and structurally complex. Also, the mathematical tasks in which visualization is employed necessitate a consideration of the intricate interconnectedness between and among the following constructs: (1) actions of a learner (i.e., to interpret or to construct by way of predicting, classifying, translating, or scaling); (2) situation (i.e., whether abstract or contextualized); (3) variables (i.e., the data, whether concrete or abstract and discrete or continuous, and the form, whether categorical, ordinal, interval, or ratio), and; (4) focus (i.e., the location of attention) (cf. Lienhardt, Zaslavsky, & Stein, 1990).

We begin Chapter 3 with strong claims about the visual roots of mathematical cognitive activity. Visual representations in mathematics are not simply personal images or visual images-in-the-wild but convey explicit knowledge structures that are constructed and negotiated in a lifeworld-dependent context. That is, the context of visual representations is seen to operate within shared rules, habits of seeing, and cultural practices. Further, the different kinds of visuals that are generated depend on the type of activity that is pursued, which could be imaginal (e.g., appeal to intuition), formational (e.g., concept or process development), or transformational (e.g., problem solving). Since all visual representations are conveyed through signs, we discuss different types of mathematical objects. Also, since all visual representations operate within rules and structures, we articulate the figural nature of the concepts that allow the objects to exist. Section 5 in Chapter 3 revisits the meaning of visual thinking but developed in the context of mathematics.

Principle of acquisition: Individuals abduce the intended meanings of visual representations involving everyday objects through the use of one strategy or a combination of two or more strategies. Some documented strategies include learned pairings; manipulating their corresponding iconic representations; associating with a relevant experience; and establishing relational structural similarities. Abduction refers to the logic that "covers all the operations by which theories and conceptions are engendered" (Peirce, 1957, p. 237). That is, it "consists in studying facts and devising a theory to explain them" and "its only justification is that if we are ever to understand things at all, it must be in that way" (Peirce, 1934, p. 90). Abduction, in fact, plays a central role in the development of mathematical concepts, processes, and representations since it is primarily concerned with developing and establishing inferences about rules (Thagard, 1978), rules and their extensions in simultaneity (Eco, 1983), and structures (Mason, Stephens, & Watson, 2009).

Gattis (2004) was particularly interested in how spatial representations acquire their meanings, but many of her claims also apply to visual representations in the general case. The meanings of symbols are oftentimes acquired through *learned pairings*, which involves explicitly learning the relationship between script and sound or between word and picture. *Iconic symbols* allow a mapping to take place between the elements of the image and the icon, especially if there is a good fit or an isomorphism between the icon or the physical model and the corresponding referent. However, no new information could be obtained beyond what icons represent because they represent an already existing object. Especially in the case of everyday objects, icons depend on physical resemblance and are not generally useful in representing abstract concepts such as "goodness."

Among abstract concepts, *associations with a relevant experience* significantly matter more than physical semblance. For example, children associate growth graphically with a vertical line or express feelings of warm and cold by imagining fire and ice, respectively. But association as a strategy itself could also be deceiving and limited and could possibly lead to misconceptions especially in cases when multiple mappings are possible or when counterintuitive concepts are involved.

Similarities of relational structure involve establishing relationships "between objects, relations between them, and relations between relations" (Gattis, 2004, p. 592). For example, interpreting a given spatial representation such as a map or a graph involves mapping "conceptual elements to spatial elements, conceptual relations to spatial relations, and higher-order conceptual relations to higher-order spatial relations" (Gattis, 2004, p. 592). Relational structure activates relevant cognitive processes such as metaphorical and analogical reasoning, which capitalize on similarities observed between and among objects, concepts, and relations. The novelty of Gattis's (2004) notion of relational structure stems from individuals' typical experiences with everyday diagrams in which only general meanings are available and follow-up work is needed, which involves constructing and reasoning about particular details. The details include establishing a mapping on three levels, that is, object to object, relation to relation, and higher order relation to higher order relation.

In Chapter 4, I discuss the mutually determining role of structured visual representations and alphanumeric symbolization in the construction of *visuoalphanumeric symbols*. We begin the chapter by considering at least three types of mathematical symbols – iconic, indexical, and symbolic – that appear to share many of the same characteristics above that Gattis (2004) inferred in the case of everyday objects and concepts. In the case of everyday and logic problems, Fig. 2.15 is a schematic model that situates the use of visual imagery in the beginning phase that is then carried through and progresses toward more verbal and other symbolic forms. Certainly, it is easier to state a rule or express a relationship in verbal or symbolic form. Further, the phenomenon of visual fading seems to occur in problem situations that are of the sequential reproduction type (i.e., sequential in the sense that general principles have already been established and reproduced because of the recognition of a familiar structure). However, the verbal or the symbolic phase does not necessarily imply the absence of visual images. Such images, in fact, basically support judgment, reasoning, and decision-making strategies. For example, among chess masters, their verbal articulations represent "judgments on more abstract representations of board positions" (Kaufmann, 1985, p. 62) without needing to repeatedly employ specific visual strategies. Among expert soccer players (and coaches, in particular, who communicate with their players using slates

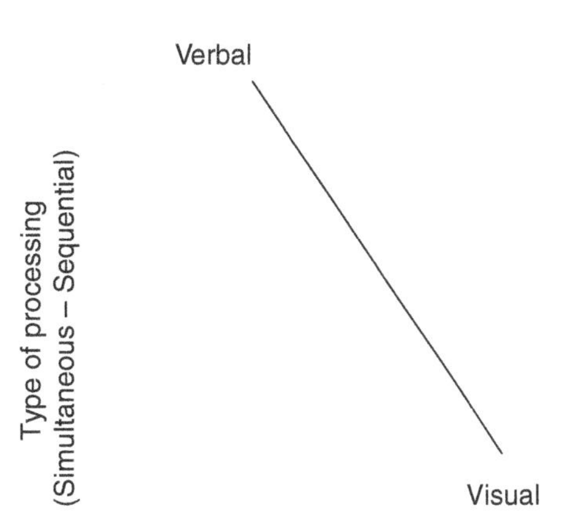

Fig. 2.15 Schematic model of visual–verbal representation (Kaufmann, 1985, p. 63)

or tablets), when presented with abstract representations of simulated visual soccer scenes (i.e., concrete details were removed and crosses were used instead of players), they have been documented to make quick and effective (verbal) decisions despite the absence of pertinent visual information (such as physical characteristics and actions of players). Their expertise could be explained in terms of the visual patterns and other relevant domain-specific knowledge that they have stored in long-term memory (Poplu, Ripoll, Mavromatis, & Baratgin, 2008).

We offer a similar argument in the case of visuoalphanumeric symbols. Chapter 4 explores the different mathematical symbols in some detail, especially the contexts in which they are used. Readers are introduced to the Wittgensteinian notion of modes of signification that highlights the use of the same symbol (say, fractions and variables) to convey different things. Throughout the chapter, we reinforce the notion of progressive schematization in both intra- and inter-semiotic resources and draw on many classroom instances in my middle and elementary school classes for empirical evidence of salient points. In particular, I make a strong claim about the necessity of transitioning from iconic and indexical representations and actions to symbolic via psychological distancing (DeLoache, 2005) as a way of supporting shifts that are needed in thinking from visual to abstract and from referential to operative (Fig. 2.8). One section provides an historical apercu of the progressive evolution of symbolic thinking in algebra. The last section provides the necessary context in understanding the ideas that are pursued in Chapter 5. In this section, readers obtain an empirical account of the development of visuoalphanumeric symbols among my middle school participants who were involved in a 3-year longitudinal design-driven research in pattern generalization activity. By *pattern generalization*, I refer to actions of construction *and* justification of direct, closed formulas relative to figural and numerical patterns (e.g., Figs. 2.6a, 2.7, and 2.14a, b).

In Chapter 5, readers are introduced to the meaning and significance of abductive reasoning in induction and generalization processes. While the notion of abduction is widely used in areas such as artificial intelligence, computer science, scientific discovery, and, more recently, philosophy of science, there is very little work done in cognitive science and mathematics education research that addresses abduction as a tool for (causal) reasoning and understanding.[3] Peirce writes:

> How is it that man every came by any correct theories about nature? We know by induction that man has correct theories; for they produce predictions that are fulfilled. But by what process of thought were they ever brought to his mind?
>
> (Peirce, 1995, p. 237)

In this chapter, we focus on patterns such as Fig. 2.6a and address the primary dilemma of how to develop reasonable abductive inferences (i.e., rules) and, consequently, algebraic generalizations on the basis of limited information (i.e., the

[3]Thagard's (2010) recent theorizing concerning this topic involves understanding the implications of *embodied abduction*, which involves "the use of multimodal representations" (p. 1). Embodied abduction is inferred in many sections of this book.

known stages in any pattern) that apply to the unknown stages (i.e., extensions). Since patterns are structures, in this chapter, we provide a visually drawn empirical account of progressive evolution of structural thinking and generalization involving patterns in both figural and numerical forms. We clarify what and how we mean by patterns and further explore Peirce's notion of abductive reasoning or, simply, abduction. We then justify a stronger claim in which accounts of progressive formalization, schematization, symbolization, and mathematization all involve accounts of progressive abductions. In two sections, we distinguish between entry-level abductions and later or mature abductions that produce visuoalphanumeric representations. We also discuss constraints and difficulties in making abductive transitions, which have implications in the content and quality of structural thinking and generalization. We close the chapter by considering how visuoalphanumeric representations in Algebra 1 and structured visual thinking at the elementary level could support hypostatic abstraction and growth in functional thinking and understanding.

Principle of reasoning: Many problem solvers employ imagistic reasoning in organizing their thoughts and as an alternative to purely symbolic or linguistic forms of reasoning. Imagistic reasoning oftentimes leads to meaningful associations, analogies, inferences, and relational structures. Yip (1991) observes that professional engineers, physicists, scientists, and mathematicians sometimes experience difficulty in verbally articulating their "intuitive grasp" of a problem situation but then still manage to convey their thoughts through graphs and visual, analogue, and diagrammatic representations. Nersessian (1994), Buchwald (1989), Kaufmann and Helstrup (1988), and Miller's (1997) engaging description of Einstein's propensity for visual thinking in various aspects of his scientific achievements all provide us with accounts of great scientists, mathematicians, and inventors who employed imagery (e.g., visual, nonvisual such as spatial, etc.) and visual analogies primarily in their process of discovery. For example, Hadamard, French mathematician, frequently employed visual imagery "when matters became too complex" (quoted in Kaufmann & Helstrup, 1988, p. 114). Einstein employed visual images in developing a thought experiment that played a central role in his special theory of relativity (Miller, 1997, p. 61). Buchwald's (1989) historical account of the wave theory of light in the 1830s includes a discussion of the significant role of drawings in Augustin Fresnel's polarization experiments. Inventors like Edison employed heuristics that were hinged on a visual process that "manipulate(d) both a device-like conception (mental model) and a set of physical artifacts (mechanical representations) in order to create a new object" (Carlson & Gorman, 1990, p. 417). Even in less dramatic cases, the comprehensive review of Kaufmann and Helstrup (1988) and the case studies developed by Nersessian (1994) show convergence in the use of imagery in problem solving among people of various abilities. Pinker (1990) also notes that in cases involving quantitative information, most people prefer to process them using graphic forms of representation than other nonpictorial means such as tables of numbers and lists of propositions.

Antonietti, Cerana, and Scafidi (1994) also share the above views on the general effectiveness of images in various contexts, especially the flexible power of images and their ability to assist people avoid the unnecessary use of procedures. However, they are also quick to point out that

> imagery is useful when subjects are aware of the goal, that is, when the free production of mental images is oriented toward a specific endpoint. Imagery flows produced without frames of reference ... can generalize in a wide range of directions so it is probable that subjects can follow an unproductive line of thinking.
>
> (Antonietti et al., 1994, p. 188)

In everyday life, we face choices about whether to use visual representations when, say, we are confronted with directions that have been expressed in words such as the one shown in Fig. 2.16. It is, as Reed (2010) points out, a spatial task that could be dealt with either verbally (by memorizing the steps) or visually (by forming a mental map image). The use of visual strategies in communication, discovery, problem solving, and interpretation exemplifies what I refer to as *imagistic reasoning*. Following the manner in which Gattis (2002) defines spatial reasoning, imagistic reasoning is the internal or external use of visual representations, such as impressions and diagrams, to reason. In the case of internally drawn images, individuals outsource them through diagrams and/or gestures that convey some intended meaning. Externally drawn images include physical models such as diagrams, pictures, and other concrete semiotic resources such as manipulatives and imaging software whose meanings have been constructed on the basis of rules that enabled their reification.

Imagistic reasoning can also be about the nature and content of the visual itself. However, it is the conceptual content of the visual representation that really matters. In employing imagistic reasoning, individuals need to establish a valid and consistent mapping between the visual representation and the corresponding concept and then use the mapping to construct and justify inferences about a possible conceptual relationship by manipulating what could be observed from the visual representation.

Four results in the general cognitive literature are worth pointing out in light of their implications to visualization and mathematics.

1: Start out going SOUTHWEST on E REED ST toward S 8TH ST.

2: Merge onto I-280 N via the ramp on the LEFT.

3: Take the BIRD AVE exit.

4: Take the BIRD AVE ramp.

5: Turn RIGHT onto BIRD AVE.

6: BIRD AVE becomes S MONTGOMERY ST/CA-82.

7: Turn LEFT onto PARK AVE.

8: Turn RIGHT onto LAUREL GROVE LN.

9: End on this street.

Fig. 2.16 Example of a mapquest direction to travel by car from location A to B

First, on matters involving nonspatial concepts such as time, age, and rate, Gattis (2002) observed that the spatial reasoning of young children (ages 6 and 7 years old) with little to no formal experiences in graphing appears to be "influenced by general constraints in reasoning" and "these constraints precede the learning of graphing [and more formal] conventions" (p. 1175).

Second, children's knowledge acquisition processes involving everyday objects involve marked transitions in orientation from (within and across) object properties to object relationships (see, e.g., Gentner, 1983; Markman and Gentner, 2000).

Third, Antonietti's (1999) investigation into college students' use of visual representations in various tasks (logical, mathematical, geometrical, and practical) illustrates the view in which usefulness of imagery appears to be task dependent. That is, an individual's choice of using a visual representation seems to depend on how he or she discerns the nature of the task. For example, if a task is perceived to be involving numbers, like Dina in my study, who consistently approached every pattern generalization task numerically, then he or she is likely to use a numerical method without "tak(ing) advantage of possible intuitions suggested by visualization" (Antonietti, 1999, p. 409). Antonietti also notes how students tend to perceive visuals as helpful in dealing with situations that involve concrete situations but not so in cases of abstract and conceptual problems.

Fourth, even when diagrams or pictures are shown with text, their level of complexity and realism influences the content and quality of knowledge and comprehension that are acquired by individuals (Butcher, 2006). For example, more detailed, schematic diagrams (e.g., Venn diagrams or the figural patterns in Chapter 5) are much harder to deal with than are iconic and simplified diagrams. What is at issue is the level of abstraction (i.e., transparency of relations) that is involved in both picture types.

In Chapter 6, I analyze the relationship between visual thinking and diagrammatic reasoning. Diagrammatic reasoning is viewed as a type of imagistic reasoning that individual learners employ in establishing necessary mathematical knowledge. It involves distributed actions of manipulating, discerning, interpreting, and inferring relations on diagrams. A typical example is shown in Fig. 2.17, which shows two diagrams in a school geometry text curriculum that have all the required information needed to establish their deductive proofs. Talking about ancient Greek mathematics and mathematicians, Netz (1999) refers to diagrams as "the metonym of [their] mathematics" (p. 12), where diagrams and their corresponding propositions and proofs all convey the same meaning. In Chapter 6, however, we pursue purposefully constructed diagrams in school algebra and number sense such as those shown in Figs. 2.11a-c and 2.12 in order to demonstrate their significant role in providing students with structured images of symbolic forms and perhaps, more importantly, in "set(ting) up a world of reference" (Netz, 1999, p. 31) that allows them to see the necessary conceptual relations. Readers are also referred back to examples in previous chapters to establish the view that progressive diagrammatization can support progressions in formalization, schematization, and mathematization of the corresponding alphanumeric forms. Diagrams are, thus, seen beyond their typical

1. Let M be the midpoint of the side AB of an equilateral triangle ABC. Let N be a point on BC such that MN $\perp$ BC. Prove that BC = 4BN.

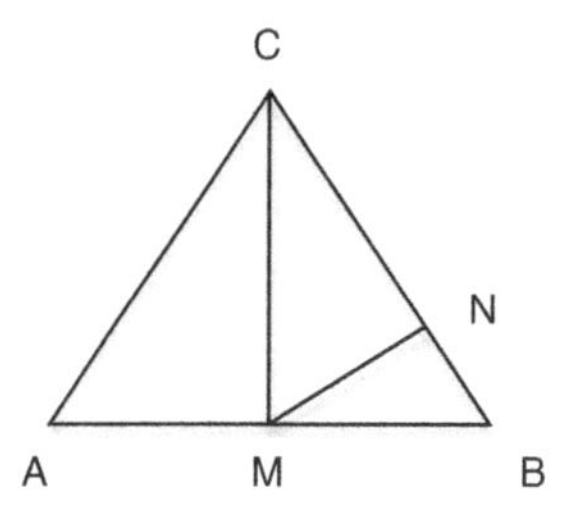

2. Let P be a point outside a circle Σ. Let A be a point on Σ such that $\overline{AP}$ is tangent to Σ. Let BC be a chord of Σ with C nearer to P than B such that BC II AP. Let the lines BP and CP meet Σ again at K and L, respectively. Let line KL meet AP at point M. Prove that M is the midpoint of $\overline{AP}$.

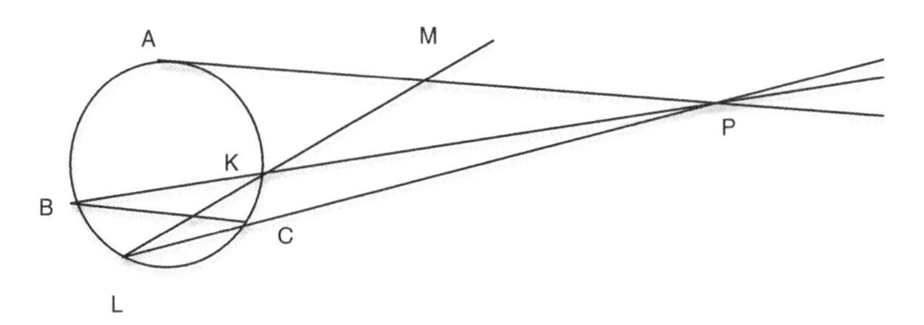

Fig. 2.17 Two examples of geometric diagrams (Bautista & Garces, 2010)

meaning as signs that externalize conceptual relations to fulfilling a mediating role in knowledge acquisition and development (cf. Stjernfelt, 2007).

Chapter 6 begins with a summary of cognitive issues relevant to the role of the human visual system in diagrammatic activity such as our ability to discern relationships in diagrams, the interpretive nature of seeing, and our predisposition to recall facts simply by recalling a diagram. Then we deal with fundamental issues of existence and universality of diagrams that are always surfaced in any account of diagrammatic activity in mathematics. We also discuss pedagogical issues surrounding presented versus generated diagrams. In the last two sections, we discuss different types of diagrams, address issues surrounding progressive diagrammatization, and explore the role of diagrammatic reasoning in mathematical reasoning.

Principle of individuation: While an individual's ability to visually represent is influenced by one's visual system, it is also influenced by socially constituted practices. Finke (1980) notes that while it is true that constructing images depend on the knowledge and expectations individuals have about a target object or an event, they also process them at varying levels of their visual system, some of which operate independently of those knowledge and expectations. But socially or culturally drawn visual practices exist, including recent visual-based technologies that cause particular groups to share similar visual representations and strategies.

Thus, the visual images individuals construct have physiological, social, and cultural origins, which "our adopted modes of representation guide our seeing itself" (Wartofsky, 1978, p. 22). An empirical demonstration of this multigroundedness of visualizing is seen in new and emerging technological tools that enable individuals to have shared visual experiences, in effect, becoming inseparably intertwined with their own visual images.

6 Context-Based Visual Representations

The principle of individuation has led me to consider the two issues in Chapter 7, which focuses on blind-specific issues and cultural influences relevant to visual attention, thinking, and performance. General, neuropsychological studies on non-mathematical tasks done by Farah (1988), Bavelier, Dye, and Hauser (2006), and Ward and Meijer (2010) empirically demonstrate some capacity for visual imagery in the case of blind individuals (and "both better and worse visual skills" among deaf individuals). There is, in fact, converging evidence drawn from a variety of neuroimaging studies that also support this finding (see, e.g., Cattaneo et al. 2008 for a comprehensive review). For example, Ward and Meijer document the visual-like experiences of their two blind participants when information was routed via an auditory sensory substitution device. Recent neuroimaging results also show that damages in particular visual pathways affect particular aspects of visual imagery (shape, color, location, etc.; cf. Jeannerod & Jacob, 2004; Kosslyn, Ganis, & Thompson, 2001). However, the production of (visual-like) images is still possible and could be seen as "the end product of a series of constructive processes using different sources of information rather than mere copies of a perceptual input" (Cattaneo et al., 2008, p. 1347). Thus, it is interesting to determine the nature and content of representations that learners with varying levels of visual impairments develop as they acquire knowledge of objects, concepts, and processes in light of constraints in their physical systems that affect the way they perceive and engage in imagistic reasoning. One important implication that could be drawn from these studies involves understanding the epistemic significance of a sensuous dimension to learning, which capitalizes on multisensory convergence.

Also, since much of learning mathematics involves a process of guided participation, it is interesting to determine how individual visual representations could be influenced by one's own cultural practices (cf. Solomon, 1989). Crafter and de Abreu (2010) note that every representation of (mathematical) knowledge "has a double character ... the representation of something (and therefore a cultural tool) and of someone (and therefore seen as associated to specific social groups" (p. 105). More generally, "conventions," Bishop (1979) writes, "are of course learnt, as are the reasons for needing them, *and* the relationship between the pictures and the reality that are conventionalizing" (p. 138). Miller (1990) makes similar claims relative to the more fundamental matter of assigning images to pictures. He notes the significance and robustness of "the convention of representation" – that is, "it's our understanding of the *convention*" that "allows us to assign the role of image to a given configuration of pigment" (p. 2). Mishra, Singh, and Dasen (2009) provide a nice example of the influence of conventionalizing in visual representing in the context of Indian folks who live in different geographical locations. The authors claim that "the choice of a particular frame of reference in language or cognition is encouraged by ecological conditions and is reinforced by cultural practices of the given populations" (p. 388). Folks in rural Nepal use the Himalayas as their reference point unlike their counterparts in the Indian villages of the Gangetic plains who rely on an absolute frame and some local landmarks and those in urban cities such as Varanasi who use a combined egocentric–geocentric frame. Gooding's reflections on "seeing biologically, seeing culturally, and seeing naturally" in scientific work might provide interesting inroads concerning the role of cultural constructions of visual representations in mathematical concepts, objects, and processes. Gooding writes:

> There is an important difference between what appears to us naturally on the basis of innate perceptual and cognitive functions (what humans can see naturally) and what is made to appear natural on the basis of conventions that engage our innate capacities and learned cultural preferences (what there is to see, according to science).
>
> (Gooding, 2004, p. 557)

The point, of course, is not to set up a pernicious dichotomy but to acknowledge the complicit functioning of both "common cognitive capacities" and "cultural conventions" in the construction of visual representations, whether internal or external, including the structures (syntactic aspect) and meanings (semantic aspect) that are associated with them and render the character of being objective. Further, Gooding poignantly articulates:

> The objectivity of the relationship between an image and what it depicts cannot be defined independently of the context in which that relationship is constructed and used. ... It is plausible to suppose that the effectiveness of a graphical convention depends on how well it promotes the engagement of cognitive capacities that are not culture-specific by cognitive skills that reflect local cultural conventions and preferences. This engagement is a cognitive process mediated by cultural resources and social conventions and expectations. The widespread use of a method of representation highlights the particular human cognitive capacities that cultural conventions make use of.
>
> (Gooding, 2004, p. 559)

While Tucker's (2006) interest is in understanding the role of visualization in scientific discourse, her suggestions below seem applicable in mathematics, which we pursue in some detail in Chapter 7:

> I suggest that one way to advance the study of the role of visualizations in science involves looking further at the social formation of communities of collecting and exchanging pictures and understanding the historical (and increasingly institutionalized) mechanisms that developed for framing some as "scientific" and some as "unscientific" objects. . . . Assessing the history of scientific enterprises and responses to them compels us to look closely at the social histories of individual images, including where they survive today and how and why they came to be there.
>
> (Tucker, 2006, p. 113)

The two complex issues that are pursued in Chapter 7 have much to inform us about cultural and neurophysiological factors that influence the performance of visual thinking in mathematical learning. Readers should be forewarned that the findings discussed in this chapter have been drawn from investigations conducted by experts in these fields, which certainly influenced various aspects of my own research study that I have reported in the previous chapters. We begin the chapter with a very interesting interpretive sociohistorical account that explains subtle differences in visual attention and construction of concepts among students in diverse classrooms. We also discuss the theory-laden and socialized nature of scientific knowledge construction. Then we talk about accounts of differences in the way mathematical objects and their images are perceived, captured, and interpreted as a consequence of sociocultural modes of seeing that influence the nature and type of mathematical content that is valued and produced. We also present a materialist view of objects in mathematics, which sees visual forms of mathematical knowledge as evocations of shared social and cultural feelings and practices. We then give a cultural account of the nature of mathematical proofs in terms of particular ontological perceptions of objects that influence the manner in which truths and the empirical world are constructed. The remaining sections limn findings from research conducted with individuals that have visual impediments at varying levels. We discuss perspectives drawn from a few studies that describe the nature of their image construction and processing. The main point that is addressed deals with the need to broaden our prevailing understanding of the sources and nature of visual representations to include all aspects of our sensory modalities (i.e., auditory, haptic, kinesthetic and, in some cases, olfactory) that provide redundant information and need to converge and overlap in knowledge acquisition processes (Millar, 1994). We close the chapter with a discussion on the significant role of multimodality learning in mathematics.

7 Forms and Levels of Visual Representations

Are visual representations necessarily visual? In the case of externally drawn visual representations such as manipulatives, diagrams, and images drawn from a graphing calculator, the answer is perhaps obvious. But the answer is not definite in the case of internally drawn images. In fact, cognitive science has not settled this issue and research investigations in this area address more fundamental, and certainly more difficult, questions such as the nature of representations that underlie mental imagery in neuroscientific terms. Despite the unsettling fact, I take the convergent view of Chambers (1993), who claims that images are neither propositional nor pictorial but both, that is, "images are meaningful representations, that is, descriptive information must accompany depictive information" (p. 79).

In mathematics, visual representations in the mind are seen as visual simulations of the relevant objects, (numerical) processes, and relationships among objects. This reminds of a situation in my Algebra 1 class when my students tried to understand and remember the process of deriving the quadratic formula $x = (-b \pm \sqrt{b^2 - 4ac})/2a$ to obtain the roots of the equation $ax^2+bx+c=0$, where a, b, and c are real numbers and $a \neq 0$. While they could recall the formula by singing, a mnemonic strategy they acquired from the Internet, some of them initially understood and eventually remembered the derivation process by visualizing a reenactment of the process as it is applied on a particular example (see Fig. 3.12). Hence, the modality of visual images in mathematics could also be seen as being meaningful in Chambers's (1993) sense, that is, they are neither figural/geometric nor numerical/algorithmic in form but could be both. More broadly, visual thinking in mathematics encompasses the use of all types of symbols from personally constructed images and signs to alphanumeric expressions to diagrams that all convey meanings that are used to reason.

In Chapter 8, I discuss a progressive modeling view on visualization – the traveling theory proposed in this book – in relation to the development and understanding of mathematical objects, concepts, and processes on the basis of the two-triangular structure shown in Fig. 2.8. The traveling theory articulates the necessity of progressions in visualization from the referential to the operative. Consequently, visual forms evolve as well from personally constructed images to visual representations to what I refer to as visuoalphanumeric representations. The structures of such representations, in fact, should convey the same power as their symbolic counterparts that are routed in solely alphanumeric terms. As I have noted in the Introduction, an effective visual process in mathematics enables personal or subjective visual images to evolve into more structured visual representations (in Goldin's (2002) sense) before fully transitioning into visuoalphanumeric representations. The term *visuoalphanumeric* conveys the interanimated symmetric reversal phenomenon of seeing the visual in the symbolic and the symbolic in the visual, a product of heterogenous inference drawn from several sources.

Also embedded in the progressive model is the notion of structural awareness that enables learners to use their visual representations in constructing and experiencing generality and (hypostasized) abstraction and in connecting intended mathematical properties in their own mathematical thinking. Figure 2.18 is a diagram of the progressive model that I extrapolate in further detail in this particular chapter. Briefly, visual thinking in mathematics takes places initially at the referential phase in which mathematical knowledge is seen as generalizations and (hypostasized) abstractions of arithmetical relationships. Visuals are employed that allow individuals to easily recognize relationships that are instantiated by and through the images. With more learning, the circle of individuation, acquisition, and reasoning

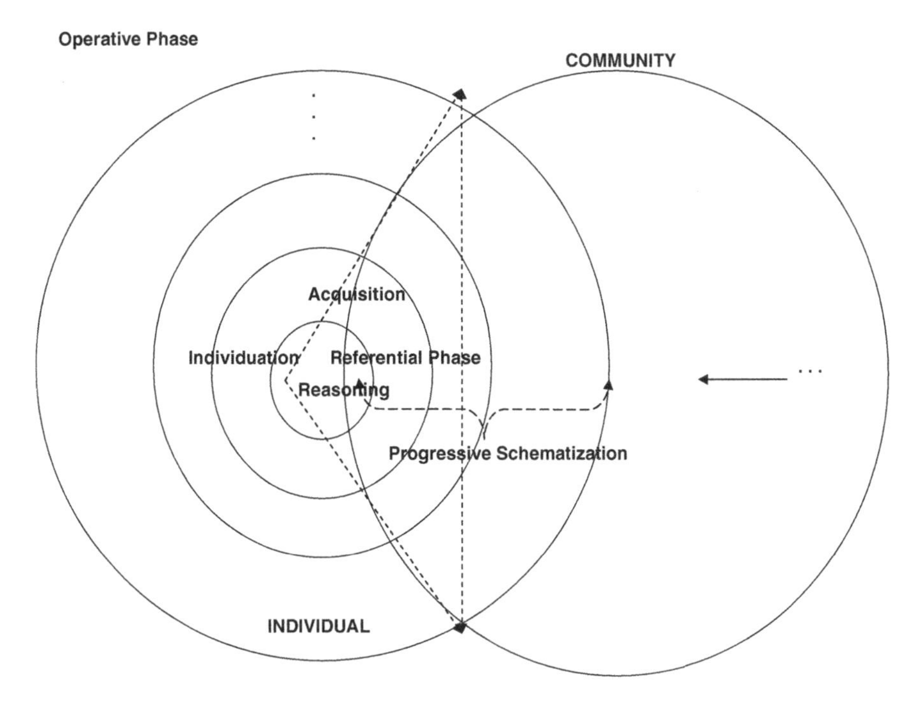

Fig. 2.18 A epistemological model of visual thinking in mathematics

grows in sophistication, leading to more formal types and specific forms of visual representations at the operative phase.

The concentric circles in Fig. 2.18 incorporate, and visually convey, progressive schematization in terms of local generalizations and (hypostasized) abstractions that eventually reach a global state at the operative level. That is, once something is generalized or abstracted from an object, an event, a process, or a relation, it then becomes the basis that supports a later generalization or (hypostasized) abstraction, and then again. Visual thinking is also seen initially as an individual affair through the construction of subjective images that later evolves as a result of frequent interactions with the community (in particular, the classroom). The community imposes its own practices at various levels of the epigenetic cycle, which is then appropriated and internalized in varying degrees. However, despite the social influence, individual learners *interpretively reproduce*[4] what they find meaningful to appropriate and internalize and so in some cases manage to retain some of

[4]Throughout this book, I emphasize the need to reconcile between individual construction and sociocultural practices. Corsaro's (1992) notion of interpretive reproduction is an approach that "stresses both the innovative and creative aspects of" every individual learner's "participation in society and the fact that" he or she "both contribute to and are affected by processes of social reproduction" (Corsaro, Molinari, & Brown Rosier, 2002, p. 323).

Fig. 2.19 Arman's count-all addition using circles only

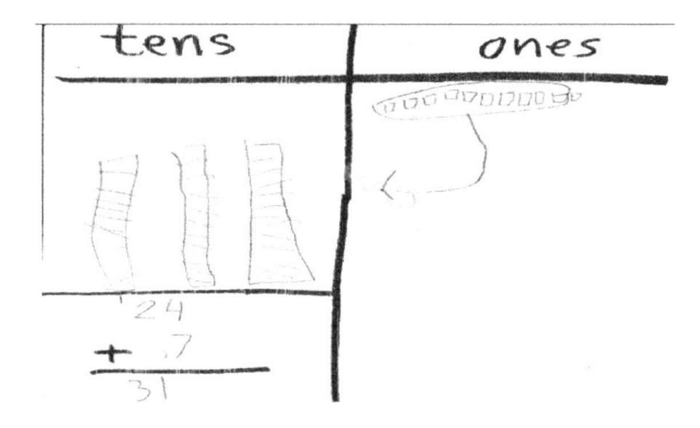

Fig. 2.20 Arman's visual addition approach using sticks and dots

their own impressions as a consequence of individuation. The overlapping regions are meant to convey negotiated visual practices that evolve over time, while the nonoverlapping regions represent practices that are kept at the individual and community levels. For example, it is common among children to be in possession of mnemonic visual strategies or unintended knowledge (misconceptions, misperceptions, etc.) that reflect variations in the established rules or personal constructions with distinct uses that are in most cases at odds with the institutional practices of the community.

8 Instructional Implications

Without a doubt, the hallmark of school mathematics is its symbol system so precise, economical, and universal that it bears little semblance to ordinary language systems in whatever cultural context. "The grammar of mathematical symbolism," O'Halloran (2005) points out, is "based on a range of condensatory strategies which facilitate rankshift for the rearrangement of relations" (p. 131). This explains why prevailing institutional practices and major stakeholders in school mathematics put premium in alphanumeric proficiency over visually oriented approaches to learning objects, concepts, and processes. Alphanumeric symbols capture the essence of an object, a concept, or a process in some dynamic and observable semiotic form that can withstand the critique of ambiguity and the messiness and arbitrariness of subjective impressions and experiences that are oftentimes foreshadowed in most visual representations.

In bridging internal and external phenomena, Bruner (1968) has suggested his widely accepted representational stages of enactive, iconic, and symbolic phases, which could be interpreted as a model of progressive schematization of visualization in mathematical learning. At the very least, visual representations exhibit all of Bruner's stages in a more dynamic manner at varying levels of use. Gestural activity, for example, is an enactive moment in which, say, a hand conveys a certain visual image that could be symbolic. The enactive stage is the domain of visible actions, while the iconic stage maps an internally constructed image to a relevant external representation. The symbolic stage is the phase of (hypostasized) abstraction that involves internalizing conceptual relations and formally naming them, which underlie all visually grounded sign systems (cf. Bruner, 1990).

While the intent in Chapter 8 is to explain Fig. 2.18 as an epistemological model that situates visualizing activity in mathematics in a much larger context of progressive schematization, instructional implications have also been embedded in the discussion. The final section in the chapter articulates the need for a more "pleasurable" approach to teaching and learning school mathematics, a kind of sensuous articulation of the "pleasures" that can be derived from developing visuoalphanumeric representations in school mathematics.

9 Overview of Chapter 3

In Chapter 3, I explain the contexts of important terms we use throughout the book. In doing so, readers obtain a more cohesive account of visual representations in school mathematics and gain insights into how progressive schematization could be achieved in a systematic manner. While such representations may initially emerge as personally constructed images and images-in-the-wild, performing visual actions within particular structures and customary ways of thinking allow such subjectively drawn images to transform into more meaningful structured representations. To take a case in point, Arman (Grade 2, 7 years old) initially added the two numbers supplied in the word problem in Fig. 2.14b visually by drawing circles and counting all. Later, when he learned to perceive whole numbers in terms of place value, his pictures transitioned formally that allowed him to visually add more efficiently using sticks and squares, as shown in Fig. 2.14b. How he developed such images and the corresponding visuoalphanumeric method is the subject of mathematical cognitive activity, which is rooted in explicit knowledge that is as much visual as it is symbolic.

Three contexts – imaginal, formational, and transformational – of mathematical cognitive activity are discussed in some detail. The basic purpose is to show the encompassing role of progressive schematizing in these contexts. Since we define representations in terms of the coordination between and among signs, rules, and structures, we make an analogous characterization of mathematical cognitive activity, that is, it basically consists of objects and the concepts that load them with meaning and value. Mathematical objects are categorized into several types with their visual essences explained in terms of their crucial role in depicting general representations, in hypothesis generation (i.e., visual abduction), and in developing structural awareness. Consequently, the nature of conceptualizing, and concepts for that matter, is rooted visually.

The last two sections address what I mean by, including principles that support, visual thinking in school mathematics. Both sections offer a refinement of the general ideas that have been raised and explored in Sections 3 and 4 in this chapter.

Chapter 3
Visual Roots of Mathematical Cognitive Activity

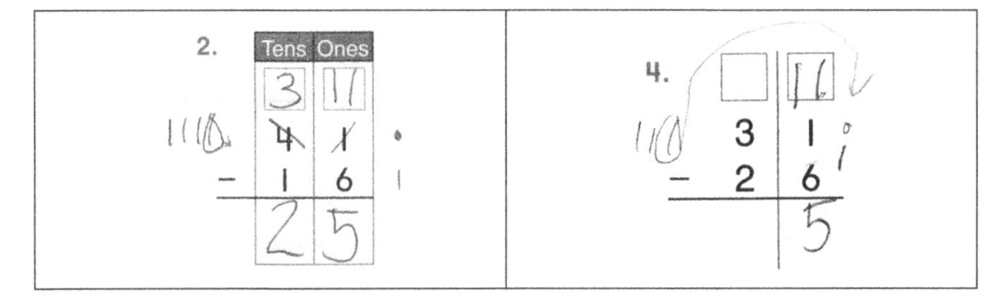

Fig. 3.1 Gemiliano's visual solutions on two textbook subtraction problems

I can do this!

(Gemiliano, Grade 2, 7 years old)

The human mind is inclined, naturally, to visualize facts. We think more easily with images because we are used to thinking of material objects.

(Fischbein, 1977, p. 155)

More generally, to see means to see in relation.

(Arnheim, 1971, p. 54)

Fischbein (1977) in the opening epigraph is certainly correct in pointing out our natural predisposition toward constructing images in order to make sense of some knowledge that appears to us perhaps initially in either linguistic or alphanumeric form. In the case of Gemiliano, he liked mathematics despite his many struggles with its symbolic aspect because in most cases he understood what was happening, at least visually. I should note that visualizing facts and images does not necessarily imply the use of pictures alone. They could also be routed propositionally, that is, in either linguistic or algebraic form. But whether those images take the shape of pictures or language, I underscore a basic problem some learners have in the case of school mathematical objects, concepts, and processes, that make sense despite the absence of any natural mapping with the real world. Papert (2002) expressly articulates in the following sentences this key problem in the context of algebra, albeit the argument could also be applied to school mathematics, more generally:

F.D. Rivera, *Toward a Visually-Oriented School Mathematics Curriculum*, Mathematics Education Library 49, DOI 10.1007/978-94-007-0014-7_3,
© Springer Science+Business Media B.V. 2011

> Language molded itself, as it developed, to genetic tools already there. The reason algebra is less well aligned with genetic tools is that it was *not allowed* to align itself; it was *made* by mathematicians for their own purposes while language *developed* without the intervention of linguists.
>
> (Papert, 2002, p. 582)

The basic goal in this chapter is to elucidate a unified account of the visual representational roots of mathematical cognitive activity. A necessary entry point involves understanding contexts or situations that engender visualizing activity to take place in mathematics. Perhaps Chambers' (1993) reconciliatory view of meaningfulness is a way of resolving the ongoing debate about whether images (or our modes of thinking, in general; cf. Pylyshyn, 2006, pp. 427–474) are pictorial or propositional. However, in school mathematics, meaningfulness of visual representations seems to depend on an understanding of the more complex relationship between the nature of the relevant mathematical content and the appropriate visual representations. Such visual representations are, in fact, constrained and mediated by content that also model (hypostasized) abstractions of some phenomenon. We dwell on this issue in Section 1, which focuses on three contexts of visualizing in mathematical activity.

In Section 2, we discuss the nature of school mathematical knowledge via Hoffmann's (2007) recent essay on knowledge types in mathematics epistemology. Hoffmann developed a useful categorization that allows us to distinguish between implicit and explicit knowledge and, consequently, helps situate the nature of mathematical knowledge in more realistic terms. In an oversimplified sense, we locate mathematical knowledge within a lifeworld-dependent context that also fits nicely within Radford's (2002) reconceptualization of the classical semiotic triangle (i.e., the triad of sign–object–*interpretant*) in terms of social-sign mediated activity. I think the matter of where and how to root mathematical knowledge needs to be settled in clear terms, at least provisionally in this book, so that the claims on visualization I present and discuss in various places would make better sense and could be analyzed and critiqued appropriately. We began Chapter 2 with reflections drawn from Peirce and Magnani in which thinking and inferring are matters that primarily involve the use of signs in a model-based activity that has a distributed nature. It is certainly an attractive thought. However, it needs further explanatory unpacking, especially in light of anticipated critiques from skeptics and, more importantly, Hoffman's (2007) view of "the distinction between 'implicit knowledge' and 'cognitive ability' " that "both can easily be mistaken in empirical research" (p. 188).

The ideas presented in the remaining sections hinge on the assumptions presented in Section 2 about the nature of mathematical knowledge in cognitive activity. Section 3 deals with different mathematical objects that exist within a lifeworld-dependent context that enables learners to develop necessary visual representations. In Section 4, the nature of mathematical concepts is also clarified and aligned with the manner in which mathematical objects have been characterized in the preceding section. In Section 5, I discuss an appropriate definition of visual thinking in mathematics that I assume throughout the remaining chapters in this book. In Section 6, we revisit the general principles of visualization pursued in the previous chapter and incorporate findings from studies in mathematical thinking and learning.

1 Three Contexts of Visualizing Activity in Mathematics

Kotsopoulos and Cordy (2009) implemented four teaching experiments involving soap bubble film over six 40-min periods in Cordy's Grade 7 class of 20 students. They used soap bubbles in assisting the students to develop their understanding of the mathematical properties relevant to three-dimensional objects such as volumes, surface areas, and minimal surface areas. In the beginning phase, the students initially identified their conceptions relative to their experiences with soap bubbles that allowed the authors to assess the extent to which their everyday senses reflected the mathematical sense that was needed in formally understanding soap bubbles. For example, none of the students identified important properties such as elasticity and tension. When Cordy used a bubble blower wand to produce bubbles, the students' verbal descriptions began to surface notions of stretching and elasticity. With more experiments, the conversation shifted to surface areas and the general shape of bubble films in at least two different contexts (one of which included determining the shape of the film inside a cubical-shaped bubble blower). For Kotsopoulos and Cordy, the activity enabled the students to come face-to-face with their internal embodied sense of an object drawn from their actual experiences with, and casual observations of, the object. In the authors' sense, visualizing already connotes the existence of a pre-existing image drawn from imagining whereas imagining tends to be novel, unpredictable, and a precursor to understanding some necessary relations.

In my Grade 2 class, Fig. 3.2 illustrates examples of the students' initial images of well-known unit fractions drawn from the activity of constructing fractions. Results of this activity show that young learners hold different conceptions of the same unit fraction despite the simplicity of the corresponding symbolic form and that their personally constructed images do not always align with the intended knowledge (i.e., the institutional version).

Many middle school teachers complain that students too frequently make the mistake of concluding that $a^m = a \cdot m$, where a and m are real numbers. Two popular errors are $2^3 = 6$ and $2^0 = 0$. Two recent studies on students' understanding of exponential expressions of the type a^m, where a is an integer and m a whole number, suggest that their preformal understanding of such expressions is rooted in a numerical view of exponents as repeated multiplication (Pitta-Pantazi, Christou, & Zachariades, 2007; Weber, 2002). In my Algebra 1 class, I used the paperfolding activity in Fig. 3.3 in assisting my students to deal with the common misconception that $a^m = a \cdot m$ for all values of a and m. For example, they initially acquired the image that 2^3 meant 8, and not 6, layers of paper and two trifolds, 3^2, produced 9, and not 6, layers. After empirically testing the proposition on a few specific instances, they felt convinced that $a^m \neq a \cdot m$ in most cases of a and m.

When I asked for a different explanation, Demetria (eighth grader, Cohort 1) used the example $2^3 \neq 2 \cdot 3$ and argued that $2 \cdot 3$ meant "two groups of three" while 2^3, which equals $2 \cdot 2 \cdot 2$, meant "two groups of two groups of two" or "two groups of four." When we tested the same visually drawn argument in several more examples, they began to view exponential expressions in a different way with most of them developing the particular sense that a^m and $a \cdot m$ would not behave the same way in most cases of a and m. Interestingly enough, young children as early

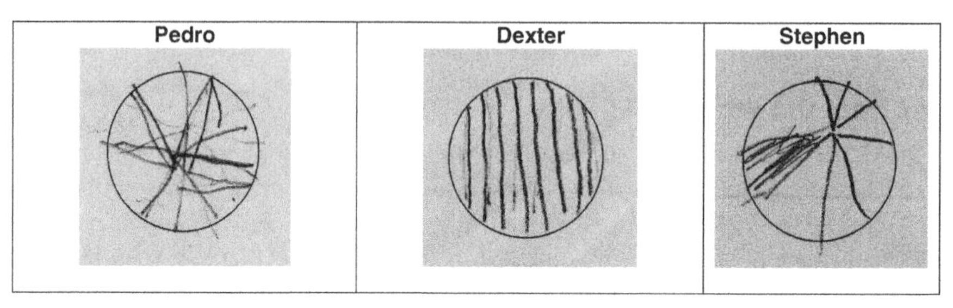

Different Versions of $\frac{1}{4}$ and $\frac{1}{10}$

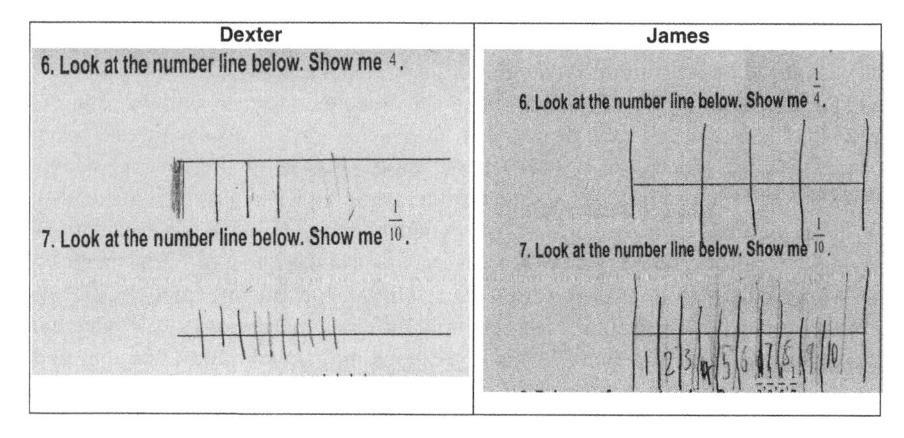

Fig. 3.2 Grade 2 students' initial images of several unit fractions

as 6 years of age have been documented to have an intuitive (or implicit) sense of the difference between linear and exponential situations at least in the context of everyday, informal growth tasks in which they tend to produce a higher forecast for exponential than for linear growth (Ebersbach, Van Dooren, Van den Noortgate, & Resing, 2008).

My Algebra 1 class also found it interesting when we associated the expression a^0, where $a \neq 0$, with an unfolded piece of paper, which meant one full layer with no fold. Hence, when they simplified algebraic expressions such as $(2x^2yz)^0$ and $(2xy/5x^2y^{-1})^0$, their typical response was "a piece of paper" that meant the value 1. In this situation, perhaps it could be easily argued that the initial visual representation was pictorial and that the final one was propositional, that is, the form of the visual transitioned from picture to a statement about number. However, there are other mathematical situations in which it is not quite simply possible to determine the actual visual in simple terms so that the issue of whether a visual is pictorial or propositional becomes secondary or is difficult to assess for meaningfulness. For

Fold a piece of paper into two equal parts once. Keep folding equally into two parts and record your results below.

Number of Folds	Number of Layers of Paper
0	
1	
2	
3	
4	
5	
6	

A. If you fold a paper for the seventh time, how many lay ers are there?

B. If you fold a paper for the ninth time, how many layers are there?

C. What does "row 0" mean in the table above?

D. What does 2^n mean in terms of the above activity?

E. Jaime claims that $2^n = 2 \cdot n$ for any whole number n. Is he correct? Why or why not?

F. How do we use paper folding to show 3^0, 3^1, 3^2, 3^3, etc.? True or False: $3^m = 3 \cdot m$.

G. How do we use paper folding to show 4^0, 4^1, 4^2, 4^3, etc.? True or False: $4^m = 4 \cdot m$.

Fig. 3.3 Paperfolding activity involving exponents

example, when my Algebra 1 students tried to explain how they would simplify a rational expression such as $(2/x) + (2/x^2)$, those who were unsuccessful said "I don't see it," which conveyed the difficulty of associating a corresponding image to the expression, whether in words or in pictures. Even with the aid of a graphing calculator, their reaction on the graphical relationship between the two addends and their sum never came close to their feeling of excitement when they used fraction strips in initially simplifying, say, (1/2) + (1/4), which actually shares a similar procedural structure with the given rational expression.

There are at least three kinds of visual activity relevant to school mathematical learning. *Level 0*, the first one, is *imaginal*. It is rooted in individual experiences and sense perception; it taps onto subjectively constructed or personally embodied images and intuitions (Kitcher, 1983; Schnepp & Chazan, 2004) that engender learners to "entertain possibilities for action[1]" (Nemirovsky & Ferrara, 2009, p. 159).

[1]Two points are worth noting. *First*, Miller (1990) spoke about visualizing that arises from "non-veridical perception ... simply the experience of visualizing something ... in the absence of an actual object, scene, or person" (p. 4). For example, in (pure) mathematics, it is common among mathematicians to visualize objects that have no corresponding physical existence. However, their reification rests on an axiomatic, deductive system, which then allows us to classify them under structured visual representations. *Second,* the manner in which we characterize Level 0 visualizing is different from Nemirovsky and Ferrara's (2009) notion of *mathematical imagination*.

The linguistic expressions used involve, at the very least, "indices, predicates, and metaphors" (Bakker, 2007, p. 23). Mathematical intuition for Kitcher (1983) "serves as a mode of basic knowledge" and is "a *prelude* to mathematical knowledge" (p. 61). In imaginal activity, individual learners are provided with a basis in developing *plausible reasoning*, the domain of guessing, and *creative mathematically founded reasoning* (Lithner, 2008). What is, thus, expected to occur among learners in Level 0 visualizing is for them to construct in imaginals tentative and, in some cases, surprising relations. Like Kitcher's (1983) mathematical intuition, imaginals do "not warrant belief, although [they] may play an important heuristic role and also serve as *part* of a warranting process" (p. 61).

For example, the initial responses of the students in the study of Kotsopoulos and Cordy (2009) reflect their imaginal ideas of soap bubbles in the absence of formal instruction. Another example is taken from my sixth-grade class. When my Cohort 1 students were learning to think about fractions in multiplicative terms (e.g., 1/4 is one-half of 1/2; 5/7 means 5 copies of 1/7) by paper folding, the class initially encountered a situation on halving that confused them. When I asked them to fold a rectangular piece of paper into two equal parts to convey 1/2, they inferred differences in the areas when the folding was done lengthwise, crosswise, and diagonalwise. (Interestingly, my Grade 2 students experienced the same conceptual difficulty.) This situation became a context for a Level 0 visualizing activity when the students began to simultaneously draw on their paperfolding experiences and intuition in establishing the same area for all three folds without entertaining any formal recommendation that in effect told them how the fraction ought to be visualized in the first place.

The second kind of visual activity is still basic but more structured than imaginals, that is, *formational*. Here, the "having (of) a visual experience" is a way of reifying or producing a visual analogue of mathematical objects, concepts, or processes in some structured image format. Hence, the "seeing" is loaded with *intentionality* (Rodd, 2000, p. 238). The imaginal aspect in constructing a formational image lies in how a structured (say, culturally constituted) visual conveys the impression of "at-homeness" that enforces itself "upon us as being true" (Kitcher, 1983, p. 61). Formational images could be drawn from any of the following contexts: prototypical; metaphorical, and; metonymic (Presmeg, 1992). For example, my Algebra 1 and Grade 2 students used fraction strips (Fig. 3.4) in developing one sense of fractions (i.e., as parts relative to a whole unit) that enabled them to develop a prototype model of unit fractions. In my Algebra 1 class, the students used

---

I share their view that imagination causes individual learners to "entertain possibilities for action" (p. 159), including the distinction they developed between pure and empirical possibilities with the latter referring to actions drawn from empirical evidence and the former to actions based on an unconditional acceptance of some assumptions. But then they went a step further in situating "mathematical imagination" under pure possibilities that have all the qualities of "logical necessity in all its deductive and inductive modalities" (Nemirovsky & Ferrara, 2009, p. 160). In my classification, what they consider to be mathematical imagination is a negotiated form of visualization, which could be either formational or transformational in context.

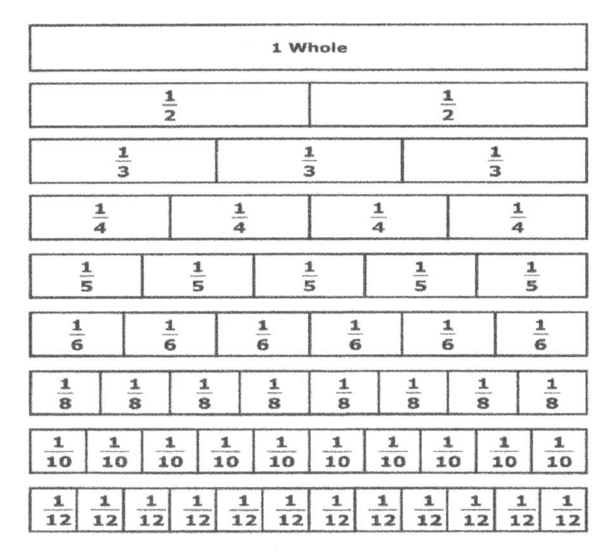

Fig. 3.4 Fraction strips (http://www.superteacherworksheets.com)

a graphing calculator to obtain the graphical behavior of linear and quadratic functions. In my Grade 2 class, the students drew sticks and dots to gain visual insight into complex addition and subtraction processes with regrouping (Figs. 1.1 and 3.1).

Students in other studies have also been documented to use visual metaphors in a formational sense. For example, Chiu (2001) lists images drawn from several studies that individuals employed in making sense of mathematical objects, concepts, and processes. For example, (1) sets were perceived as containers (object); (2) a polygon was viewed as a roundtrip path that starts and ends at the same location (concept); and (3) multiplying two positive integers $a \times b$ were seen in terms of the action of replacing the original a pieces with b copies (process). Cai (2000) documents American sixth-grade students' predisposition toward drawing formational images (versus their Chinese counterparts who were drawn to using alphanumeric expressions) that enabled them to successfully solve process-driven problems. A significant number of 151 Primary 5 children (mean age of 10.7 years) in a study by Ng and Lee (2009) effectively employed a rectangular modeling approach similar to the ones shown in Fig. 2.11a–c when they solved arithmetical problems that involve part–whole and comparison relationships. When the 34 graduate students in a study by Zahner and Corter (2010) were presented with six probability problems, they "spontaneously" (i.e., with no prompt) developed schematic, pictorial, and nondiagrammatic (e.g., listing) "external inscriptions" or images in solving them.

Hence, a visual activity of the formational kind in mathematics means having a visual experience in which the primary goal involves, following Arnheim (1971) in the opening epigraph, developing intentional conceptual relationships or a targeted structural awareness of the relevant elements in an object, a concept, or a process.

In formational activity, the imaginal forms relative to individual learners' personal or subjective experiences confront the visual representations that are valued by the mathematics community (cf. Duval, 2000). Pylyshyn's (2007) sense of formational visualizing below captures the essence of what we consider to be visual-images-in-the-wild:

> "What?" I hear you say. "How about imagining an object described by basonic string theory that has 25 dimensions?" "Okay," I reply, "as soon as you show me what it looks like I will imagine it." That's what imagining, in the sense of "visualizing," means – it means having a visual experience.
>
> (Pylyshyn, 2007, p. 56)

The purposeful action of "showing what it looks like" could then facilitate a structured visual experience, that is, a visual representation in Goldin's sense. For example, when my Cohort 1 students were in sixth grade, we used two-sided colored chips in making sense of the fact that $a \times -b = -ab$, where a and b are whole numbers, which then became our basis in establishing why $-b \times a = -ba = -ab$ (by commutativity) and $-b \times -a = ba = ab$ (by patterning). In seventh grade, Cohort 1 used a number line as a second approach in understanding why the product of two negative integers had to be positive, which also meant teaching them the appropriate way of applying the visual actions on the number line. In eighth grade, both Cohorts 1 and 2 used algeblocks in reestablishing the double-negatives rule for multiplication that also required them to model a correct sequence of concrete actions with the algeblocks.

Soto-Andrade (2008) provides an interesting account of a visual experiment that he conducted with his college-level students. Initially, he asked them to obtain the sum of powers of 1/4, that is,

$$\sum_{n=1}^{\infty}\left(\frac{1}{4}\right)^n = \frac{1}{4} + \left(\frac{1}{4}\right)^2 + \left(\frac{1}{4}\right)^3 + \left(\frac{1}{4}\right)^4 + \cdots$$

Using a calculator, they estimated the sum to be about 0.332. He then asked them what that value meant in visual terms. The visual images that the students produced and shared with one another (Fig. 3.5a) provided them with a visual experience that empirically illustrated why the series seemed to have a sum that is one-third the area of one original square. Cellucci (2008) offers another visual interpretation of the above series using an equilateral triangle (Fig. 3.5b). In all three figures, the visual experience is tied to developing a visual understanding of the necessary structural relationships that could not be immediately and conveniently accessed in symbolic form.

Presmeg's (1992) empirical work with 54 high school students and, more recently, Ng and Lee (2009) in the case of 151 middle school students, however, remind us of constraints in the use of formational images. When prototypical and metaphorical diagrams are used in metonymic terms, which involve manipulating a concrete figural case as an instance of a more general relationship, students might be misled into making false claims on the basis of the particular relations

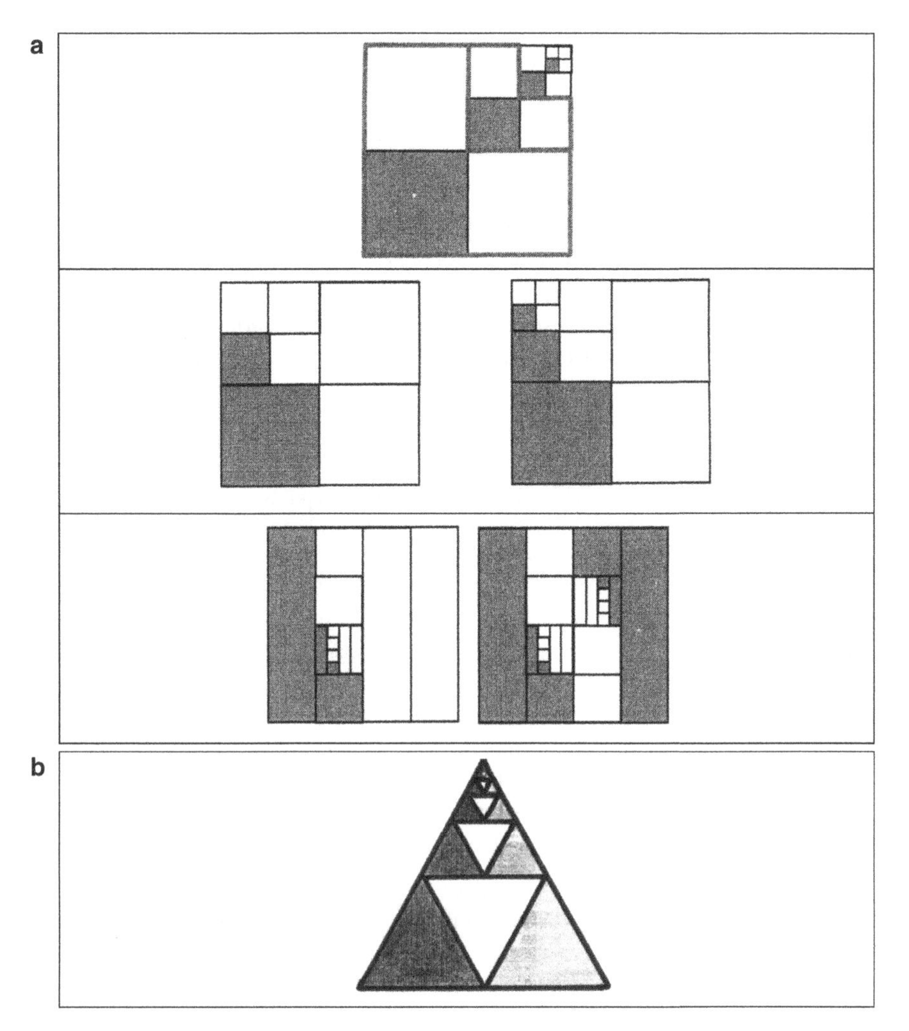

Fig. 3.5 **a** Students' visual interpretations of $\sum_{n=1}^{\infty}\left(\frac{1}{4}\right)^n$ using a square (Soto-Andrade, 2008). **b** A visual interpretation of $\sum_{n=1}^{\infty}\left(\frac{1}{4}\right)^n$ using an equilateral triangle (Cellucci, 2008)

they infer on the specific case. The most recent synthesis of research offered by Owens and Outhred (2006) in relation to geometric objects and concepts also indicate that even in relatively simple cases where students employ formational images to help them learn mathematics better, they need to (1) learn to attend to the more important aspects of such images; (2) overcome initial static perceptions in favor of abstract dynamic ones; and (3) acquire appropriate (mathematical) conventions in developing and labeling them both for practical ("better performance on problems"; efficient and systematic labeling) and conceptual ("recognize critical and noncritical features") purposes (pp. 90–97). Relative to the use of visual models in arithmetical problem solving, Ng and Lee (2009) recommend that teachers may need to "offer children a set of correct solutions" in order for them to "compare and contrast" with the incorrect ones so errors are articulated and that they are assisted in "improv[ing] upon their model drawings" (p. 311).

The third kind of visual activity is *transformational,* which may include either an imaginal or a formational aspect. The above situation with my Algebra 1 students and their experiences relative to the paperfolding activity in making sense of exponents and various situations involving exponents provide a good example of this transformational sense of visualizing in mathematical learning, which I classify under the category of *scenario visualization.* Presmeg's (1992) notion of pattern imagery, that is, image schemes that embody "the essence of structure without detail" (p. 603), falls under this category as well.

Arp (2008) writes:

> Scenario visualization is a form of conscious cognitive visual processing that enables one to select visual information while bracketing out irrelevant visual information. It also allows us to transform and project visual images into future scenarios, as well as coordinate and integrate visual information, so that the perceiver has a coherent picture of both the imagined and real worlds.
>
> (Arp, 2008, p. 165)

Arp (2008) provides a provocative telling of the history of advances in tool making beginning with the hand ax to the eventual construction of a javelin from a scenario visual perspective. He claims that the tools underwent several transformational phases because of changes in, and better knowledge of, the immediate environment, including anticipated possible scenarios that necessitated refinements in tool use and tool characteristics. In scenario visualization,

> (i)t is not the having of visual images that is important; it is what the mind does in terms of actively selecting and integrating visual information for the purpose of solving some problems relative to some environment that really matters.
>
> (Arp, 2008, p. 116)

Scenario visualizing in mathematics occurs when students develop a visual strategy in planning their succeeding steps toward solving a problem or when they are in the process of developing an inference. By inference, we appropriate Norman's (2006) view that defines it as "a type of mental act whose outcome is a possible change in belief" (p. 18). For example, prior to formal algebra, which utilizes alphanumeric symbols, Singapore and Japanese children learn to process word problems using scenario visualization in the form of diagrams such as the ones shown in Fig. 2.11a–c. In Fig. 2.12, Parker's students use scenario visualizing in solving problems relevant to percent increase and percent decrease, which are typically solved symbolically in all prealgebra textbooks with the use of variables.

Many metaphorical-based reasonings also reflect the use of scenario visualization. For example, Chiu (2001) notes the prevalent use of process-driven metaphors in the form of visual scenarios among the children and adults in her study when they tried to solve arithmetic problems involving positive and negative integers. My precalculus class of 10th- and 11th-grade students developed an abstract graphical process for solving simple and complex polynomial inequalities that used their knowledge of graphs of polynomial functions and an empty number line (Fig. 1.5c; Rivera, 2007a). In my Algebra 1 class, their visual experience with a^0, where $a \neq 0$, as conveying an unfolded piece of paper allowed them to infer the fact that a nonzero base that is raised to 0 would always yield the numerical value 1, which they used in follow-up tasks such as simplifying expressions like $(2xy)^0 - (3x)^0 + 2(1/3y)^0$. Also, when they used the transformation application program on their graphing calculators to scenario visualize transformations of graphs of linear and quadratic functions (Fig. 3.6) in a dynamic manner, the visual experience enabled them to make generalizations about the graph of $a \cdot f(x \pm b) + c$ relative to $f(x)$, where $f(x) = x$ and $f(x) = x^2$.

Suffice it to say, imaginal, formational, and transformational visualizing contexts allow students to use visual images "to see the unseen" (McCormick, DeFanti, & Brown, 1987, p. 1), in particular, "to see in relation" (Arnheim, 1971, p. 54). However, developing an image of the "unseen," whether internally derived or externally drawn, also depends on the content of the relevant mathematical knowledge and the value accorded by the community that supports its construction. Referring to Fig. 2.8, visualizing depends on whether the target object, concept, or process is referential or operative. Most reference-based visualizing activities appeal to imaginal and formational images since the relevant content could easily be represented in iconic terms, that is, the conceptual components of their structures have strong physical fidelity with the corresponding visual components of the pictorial representation (see, e.g., Chiu, 2001). But in many cases of operative content, they appear to defy simple categorizing in imaginal, formational, or transformational contexts since the "seeing" that is involved is oftentimes hinged on an abstract process. A classic extreme example of this situation is the case of irrational numbers.

In this section, we discussed in some detail the imaginal, formational, and transformational contexts of visual activity in school mathematics. The next important

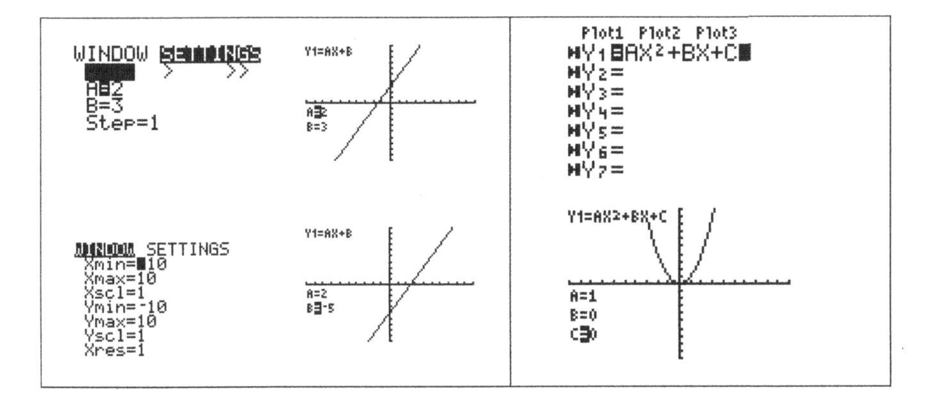

Fig. 3.6 Transformation graphs on the TI-84+

issue involves clarifying sources of visual activity, which we link to the complex nature of school mathematical knowledge.

2 Explicit and Implicit Knowledge in Mathematics

Hoffmann (2007) uses the following situation with a 4- or 5-year-old child in pointing out a distinction between implicit and explicit knowledge:

> When you ask [the] kid to count as far as [he or she] can, [he or she] might come to six before it gets difficult. But ask [him or her] to count wood blocks and [he or she] might come without any difficulties to 26 .... The disturbing question is: Does the child know the numbers from 1 to 26, or does [he or she] not? If [he or she] knows the numbers, why, then, is [he or she] not able to count them without having the wood blocks in front of [him or her]? And if [he or she] does not know these numbers, it should be impossible to count anything.
>
> (Hoffmann, 2007, p. 185)

Through many years of my own work with learners from different age- and grade-level groups on matters relevant to pattern generalization, I have on a number of occasions found myself in a similar situation where interviewees would claim they knew what was taking place in their patterns but were relatively unsuccessful in describing them in a representational form that Lee (1996) refers to as *algebraically useful*, that is, patterns whose structures could be conveyed by a direct formula. Even Radford's (2008) notion of algebraic generalization involving patterns necessitates closure in the form of a direct expression.

The "disturbing question" is resolved, which Hoffmann explains in the context of the counting child, once we make a distinction between implicit and explicit knowledge. That is, the counting child has implicit knowledge of counting numbers but could only demonstrate explicit knowledge – or cognitive ability – of counting with the wooden blocks. In the context of my research, some interviewees had implicit knowledge of patterns but harbored a constrained and less-developed explicit knowledge or cognitive ability for pattern generalization in either Lee's or Radford's sense.

Hence, a fundamental assumption in this book is that whenever we talk about visual representations and visual structures in the context of mathematical cognitive activity, we refer to the relevant explicit knowledge. Further, when we analyze an individual's cognitive ability or his or her cognitive activity, then we mean his or her relationship with the relevant explicit knowledge, which furnishes him or her with images that may or may not align with images he or she has in his or her implicit knowledge. Implicit knowledge is located in the mind, whereas cognitive ability or cognitive activity is situated in a cognitive system that has the nature of being dynamic and distributed (in Magnani's, 2004 sense). Two consequences are worth noting. *First,* the phenomenon of having a variety of cognitive abilities could be explained in terms of different environments in which cognition takes place. *Second,* the following view by Giere and Moffatt (2003) can help explain both the important role and positive status (i.e., the veracity) of explicit knowledge in mathematical cognitive activity:

> [T]he importance of distributed cognitive systems is simply that they make possible the acquisition of knowledge that no single person, or a group of people without instruments, could possibly acquire. And combining representations in ever simpler forms makes it possible for individuals with only the power of a pattern-matching brain to comprehend and appreciate what has been learned. This is a cognitive scientific explanation for how forms of external representation can provide scientific results with the power to persuade others of their veracity. . . . [R]esults do not come to be regarded as veridical because they are widely accepted; they come to be widely accepted because, in the context of an appropriate distributed cognitive system, their apparent veracity can be made evident to anyone with the capacity to understand the workings of the system.
>
> (Giere & Moffatt, 2003, p. 305)

Drawing on the empirical work of Reber (1989), Hoffmann also notes that much of implicit knowledge is far richer than, and is always ahead of, explicit knowledge. But he also sees the development of cognitive ability to be a precondition for developing implicit knowledge, which explains why "working with things – concrete objects or representations – is far more important for the development of knowledge than anything else" (p. 193). The transition from external to implicit knowledge takes place through a process of internalization that is driven by repeated experiences that are themselves rooted in cognitive activity.

Now that we have clarified the distinction between implicit knowledge and cognitive ability or explicit knowledge, it certainly makes sense to think of cognitive systems (1) in Peircean terms as semiotic systems whose explicit relationships can be known through their signs and representations and (2) in a Freudenthalian context as involving the progressive schematization of signs. In any cognitive system, objects and relationships among them are explicitly known. Also, cognitive activity involves processes that are either internal (mental) or external (social), or both. For example, in the case of the counting child, when he or she competently counts with the blocks, then he or she has a cognitive ability to count with blocks. When he or she is able to count numbers on his or her own without the blocks, then he or she has implicit knowledge of counting numbers. This particular situation exemplifies how cognitive ability can precede implicit knowledge, in which case the latter knowledge has clearly been abstracted from the concrete activity with the blocks. Certainly, we can have an implicit knowledge of something, but considering the manner in which we characterize the nature of cognitive ability, it depends on, and is constrained by, the relevant concrete activity. For example, counting numbers does not require a concrete activity, but counting with blocks does. Thus, we take it that learners' fundamental source of visualization is drawn in cognitive activity, which involves explicit knowledge.

When a child is able to count with blocks in cognitive activity, the knowledge that he or she uses is considered collateral. *Collateral knowledge* is a Peircean term that refers to those "forms of knowledge that remain hidden though being an essential condition for a cognitive activity" (Hoffmann, 2007, p. 188). Also when learners say they know something about an object, "the only thing we can know is that all knowledge about [the] object depends on the cognitive means that the involved persons have at their disposal" (Hoffmann, 2007, p. 189). Thus, objects are *lifeworld-dependent* in the sense that they make sense on the basis of hypotheses and the "customary ways" in which people "structure the activities that take place within" them (Hoffmann, 2007, p. 189).

Three important consequences are worth noting. *First,* we take a pragmatic standpoint on foundational issues relevant to neurophysiological sources of visual images. In Chapter 2, although I included internally drawn images as a source of visualization, I meant for them to have an interpretive quality that can undergo externalization in cognitive activity and, thus, analyzed as "thoughts in signs" (Peirce, 1958b, 5.251) within a lifeworld-dependent context. That is, they make sense and exist within a type of seeing that we shall refer to as *epistemically drawn images*. Hence, the following provocative thoughts by Jacob and Jeannerod (2003) should be interpreted as seeing from an organic point of view, which we oppose on pragmatic grounds:

> Many of the things human can see they can also think about. Many of the things they can think about, however, they cannot see. For example, they can think about, but they cannot see at all, prime numbers. Nor can they see atoms, molecules and cells without the aid of powerful instruments. Arguably, while atoms, molecules and cells are not visible to the naked eye, unlike numbers, they are not invisible altogether; with powerful microscopes, they become visible. Unlike numerals, however, numbers – whether prime or not – are simply not to be seen at all. Similarly, humans can entertain the thought, but they cannot see, that many of the things they can think about they cannot see.
>
> (Jacob & Jeannerod, 2003, p. ix)

Certainly, numerals are visual representations of numbers and, thus, could be seen as such. But, in a lifeworld-dependent context – perhaps an imagined reality (following Malone, Boase-Jelinek, Lamb, & Leong, 2008) – one can choose to "see" prime numbers, and all other types of numbers and numerical relationships for that matter, in which case the corresponding visual representations are considered interpretive signs based on some conceptual facts that perform a mediating role. That is, they, including our experiences with them, become intelligible through signs.

For example, in my Algebra 1 class, the students' visual representation of a prime number refers to a single rectangle whose only dimensions are 1 and the number itself. This particular concrete model of a prime number allowed them to see how it compared with a composite number that has at least two rectangles with differing dimensions of length and width. The Fibonacci sequence is another example of a mathematical concept that mathematicians and nonmathematicians alike have inferred, say, on the fruitlets of a pineapple, on a shell spiral, and in the family tree of honeybees. In such cases, the concrete objects represent useful (and approximate) images to individuals who perceive and sense the corresponding concepts either in terms of facts about them or as mediated in some form.

In Dretske's (1969) classification, visual representations in mathematics are secondary forms of epistemic seeing, which involve seeing facts relevant to the objects that can never be perceived in organic form. We manipulate, operate, and act on what we construct to be their visual forms in the same manner as we would on their corresponding conceptual content or facts. Doing this allows us to develop and further enrich the content of our explicit knowledge in cognitive activity. Grelland (2007) has a useful term for this kind of cognitive action, that is, "linguistic empiricism," which takes into account the mutually dependent relationship between intuition and formalism. In physics, especially, it is common practice to view the mathematical component of a theory to be providing the formal structures that assist in calculations and the physical component to what is sensed by the "the mind of the scientist (under the denotation of 'physical intuition')" (Grelland, 2007, p. 258). However, while some physical phenomena could be easily imagined such as fluid and classical mechanics, there are other phenomena that defy sense experiences such as energy in classical physics, including some that are simply "unphysical" such as aspects of electromagnetic field and quantum mechanics. In these latter cases, physicists – in particular, theoretical (versus experimental) physicists in the sense of Jaffe and Quinn (1993) – manipulate and perform experiments on the corresponding mathematical objects and their structures as a whole that then enable them to make observations in relation to a proposed theory.

Second, we are now in a better position to talk about *misperceptions* (misconstrued perceptions) or *aconceptions* (the absence of a particular conceptual content that matters to mathematics). Malone, Boase-Jelinek, Lamb, and Leong (2008) were disappointed when a majority of 720 Australian students "misperceived" the correct reflection of the vertical flag along the angled mirror shown in Fig. 3.7. The digital responses were all drawn in the same screen panel that shows a good percentage of the students producing an incorrect image. For Malone et al. (2008), the students who exhibited a misperception "perceived via a single sensory modality (e.g., seeing

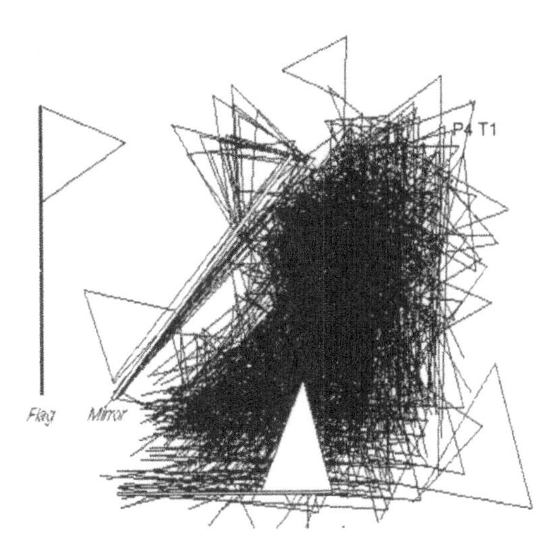

Fig. 3.7 720 Australian students' response on a reflection task (Malone et al., 2008)

in mathematics or hearing in music) something that (was) different from reality or an imagined reality (e.g., visualizing a rotated shape in mathematics)" (p. 1).

However, what constitutes "reality" is relative to one's position in this reality, whether real or imagined, which would then force us to dwell on irresolvable, philosophical ontological issues. In place of reality, we focus on cognitive activity in a lifeworld-dependent context, which addresses issues relevant to the content of visualization as it is routed in explicit knowledge. Hence, when students misperceive in cognitive activity, then we can assess the difference between their perceptions and imaginals and the customary habits, hypotheses, and shared collateral knowledge and (visual) representations that have been violated in the lifeworld context.

Pursuing an alternative analysis of those students who produced "incorrect" images in Fig. 3.7, perhaps the issue is not a simple matter of misperception but a lack of explicit knowledge or expectation in terms of some anticipated model of perception that students, as a matter of fact, acquire through more learning in cognitive activity. Oftentimes, the "reality" that Malone et al. (2008) refer to privileges institutional mathematical knowledge, which tends to assume notions that are counterintuitive to knowledge drawn from everyday objects and relations.

The reflection concerning lack of explicit knowledge came to me at least on two different occasions with two different grade-level groups whose thoughts on a few patterning tasks were assessed prior to formal instruction. *First*, in sixth grade, a good number of my Cohort 1 students extended the pattern in Fig. 2.6a in the same manner shown in Fig. 3.8. When a colleague and I reported our results, we received a concern from a reviewer who thought that patterns with three to four initial instances were not well defined and that our results were suspect on the basis of the students' misperceptions about such patterns. In other words, the reviewer's perception of Fig. 2.6a pattern involved several different possible extensions, which evidently was not in the repertoire of existing collateral knowledge of our Cohort 1 of sixth graders who were limited to just one way of seeing the pattern.

Second, Table 3.1 shows the results drawn from individual clinical interviews that I conducted with my Grade 2 class relative to the semi-free patterning task shown in Fig. 3.9 prior to formal instruction, indicating nine different extensions that were all flawed primarily because the students failed to see and articulate a reasonable structure for their constructed patterns. None of them paid attention to the shapes of the two initial stages in the semi-free pattern. Even when the interviewer asked them to determine what stayed the same and what changed from one step to another in their pattern, none actually saw any kind of change, which prevented them from establishing the correct value in stage 10 of their pattern.

The common pattern structure in Fig. 3.8 among my Cohort 1 students and the lack of any pattern structure in the case of the Grade 2 class in Table 3.1 could be interpreted *not* in terms of misperceiving patterns but a lack of more explicit knowledge about how such patterns could be dealt with. In fact, when I asked a subset of those sixth graders in eighth grade to construct and establish a pattern relative to the task shown in Fig. 3.9, they produced several different ones that were all consistent and valid (Fig. 3.10). Hence, visualizing images and representations

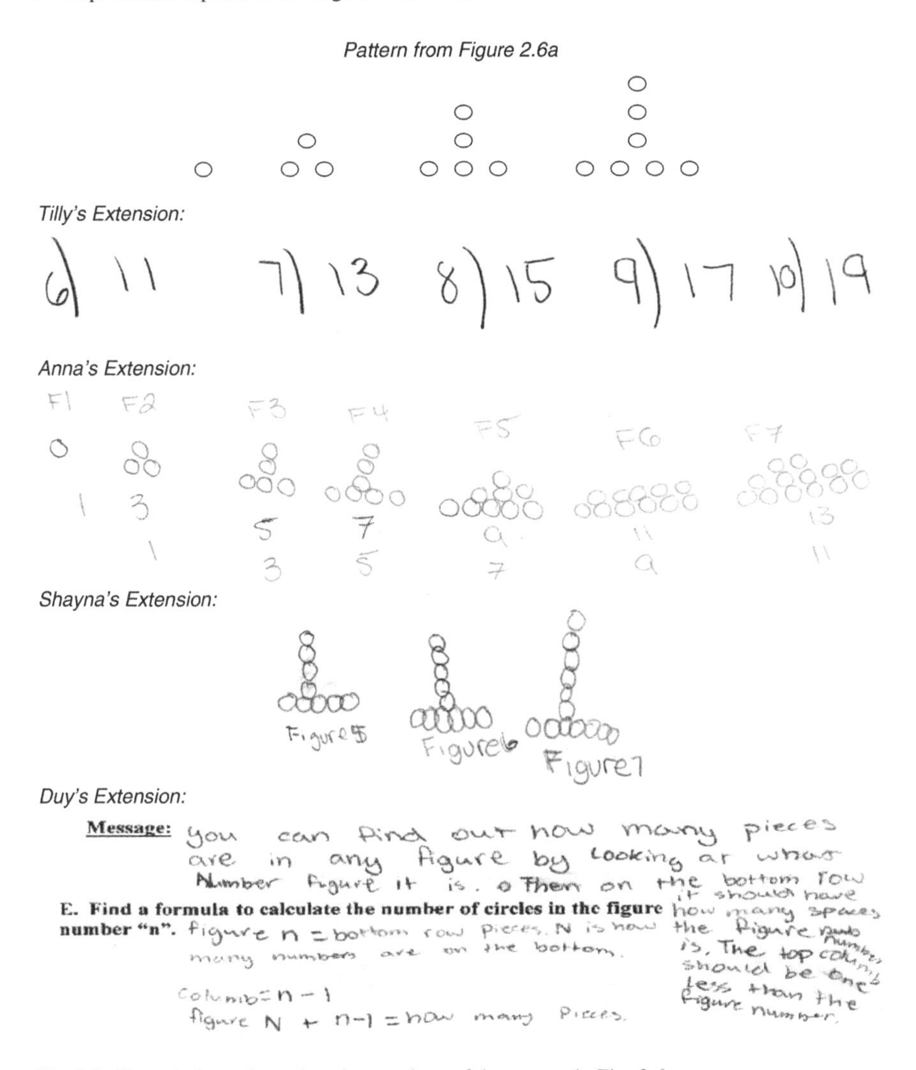

Fig. 3.8 Four sixth-grade students' extensions of the pattern in Fig. 2.6a

has to take into account the lifeworld-dependent context of objects, concepts, and processes that are learned and negotiated in cognitive activity.

The preceding situations with my Algebra 1 and Grade 2 classes lead us to our next point. *Third,* many reported studies in visualization and school mathematics learning describe different types of visualization and visual strategies as they occur in particular formational or transformational contexts (see, e.g., the different types of imagery identified in Brown & Presmeg, 1993; Dörfler, 1991; Presmeg, 1986; the different types of visual strategies in Campbell, Collis, & Watson, 1995; the local versus global kinds of visual reasoning by Hershkowitz, Friedlander, & Dreyfus, 1991; and the range of imagery forms in the context of a spatial problem-solving

Table 3.1 Grade 2 students' extensions relative to the semi-free task shown in Fig. 3.9 ($n = 21$)

Student Responses	Frequency
1, 3, 4, 5, 6, could not do stage 10	10
1, 3, 3, 4, 5, could not do stage 10	2
1, 3, 5, 7, 9, could not do stage 10	2
1, 3, 3, 4, 5, stage 10: 10	2
1, 3, 5, 6, 7, could not do stage 10	1
1, 3, 1, 1, could not do stage 10	1
1, 3, 6, 9, 19, could not do stage 10	1
1, 3, 6, 10, 12, could not do stage 10	1
1, 3, 1, 5, could not do stage 10	1

Below are the first two stages in a growing pattern of squares.

For Grade 8 Students

1. Continue the pattern until stage 5

2. Find a direct formula in two different ways. Justify each formula.

3. If none of your formulas above involve taking into account overlaps, find a direct formula that takes into account overlaps. Justify your formula.

4. How do you know for sure that your pattern will continue that way and not some other way?

5. Find a different way of continuing the pattern and obtain a direct formula for this pattern.

For Grade 2 Students:

Let us begin with a square and call it step 1.

Now suppose step 2 looks like as shown. How many squares do you see?

1. How might step 3 appear to you? Show me with the blocks.

2. Show me steps 4 and 5. How many squares do you see?

3. Pretend we don't have any more blocks and suppose we skip steps. If someone asks you how step 8 looks like, how might you respond? Can you describe or draw for me?

4. If someone asks you how step 10 looks like, how might you respond? Can you describe or draw for me?

Fig. 3.9 Semi-free pattern task

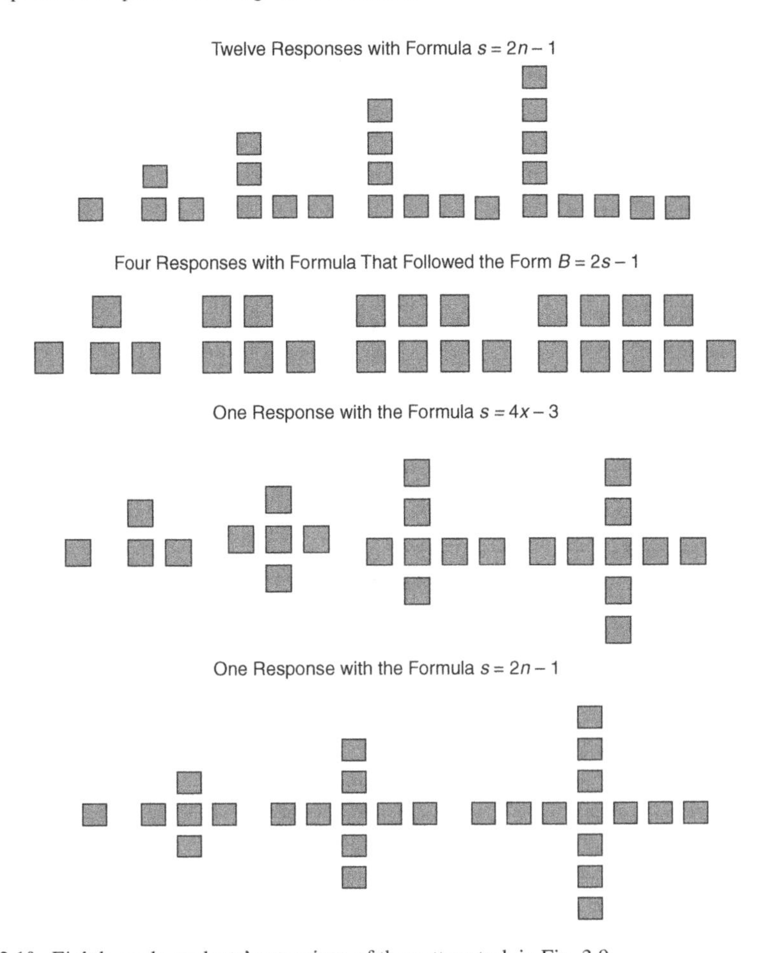

Fig. 3.10 Eighth-grade students' extensions of the pattern task in Fig. 3.9

task by Owens, 1999). While such studies provide important insights into the pervasiveness of visual actions in mathematics, the most crucial issue at this time is how to account for progressive growth in visual representation, say, in transition accounts from explicit to collateral to implicit knowledge. As students' thinking about mathematical objects, concepts, and processes evolves from the referential to the operative (Fig. 2.8) or from everyday ways to mathematical ways of seeing, the "centrality of visual reasoning" (Alcock & Simpson, 2004, p. 1) in such epistemically drawn images or secondary forms of seeing is highlighted and sustained in a developmental manner. Figure 3.11a–c, for example, illustrates how Gemiliano's formal knowledge of subtraction with regrouping evolved over a span of three phases, with the alphanumeric phase (Fig. 3.11c) drawing and relying on the earlier dominant visual phases (Fig. 3.11a, b). Figure 3.11a encouraged Gemiliano to merely engage in a count-all counting strategy in the units place (1, 2, 3, . . ., 10, 11). In Fig. 3.11b, he employed a count-on strategy instead (10 + 1) that allowed him to understand

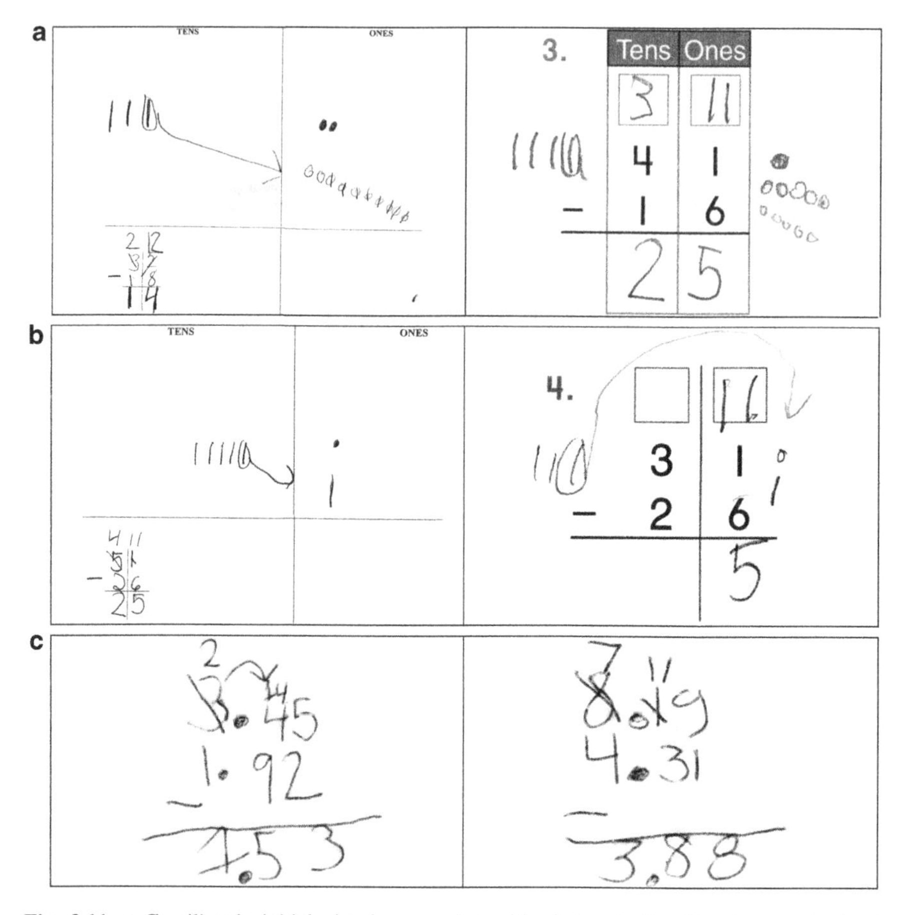

Fig. 3.11 **a** Gemiliano's initial visual regrouping with circles. **b** Gemiliano's second visual regrouping with a stick. **c** Gemiliano's numerical regrouping method

the standard numerical algorithm. We consider Fig. 3.11c to be an example of a visuoalphanumeric representation that involves seeing the visual in the numeric. Another evidence of this progressive view is shown in Fig. 2.4a–c, which shows my Algebra 1 students' diagrammatic method for factoring quadratic trinomials.

3 Mathematical Objects in Cognitive Activity

So, what are mathematical objects in cognitive activity? At the very least, they are all lifeworld-dependent objects – that is, they exist insofar as customary ways and hypotheses allow them to get reified in cognitive activity. In reifying, all objects are loaded with intentional content, which means a relationship exists between an object and its conceptually assigned meaning. Hence, within a lifeworld-dependent context, objects are symbols that stand for something. They offer, in a Gibsonian sense, *affordances* that cause those who use them to entertain, perform, or perceive some probable action.[2] However, we note that it is what individuals do with them through purposeful symbolic action in cognitive activity that they become (fully) symbolic (Rakoczy, Tomasello, & Striano, 2005; Uttal, Scudder, & DeLoache, 1997). We clarify the psychological dimension after we identify the kinds of mathematical objects that students deal with in cognitive activity.

There are at least five types of mathematical objects, namely: concrete objects; graphic or virtual objects; relational objects; epirelational objects; and alphanumeric objects.[3] At the very least, they exist as *models* that, in Fischbein's (1977) sense, have been purposefully constructed to "have an heuristic capacity as scientific models do" (p. 155) – that is, the "capacity to stimulate the process of reasoning and to permit its progress by their own means" (p. 164) – with roles that include "facilitat(ing) the interpretation of certain given facts" and providing "help [in] solv[ing] problems according to the original facts" (p. 155).

What is common across all the object types is the fact that they "are generative, internally consistent, and internally well-structured" (Fischbein, 1977, p. 155); hence, they are useful visual representations (following Goldin, 2002). Fischbein (1993), in a later paper, gives these models or objects the character of being *general representations* in the sense that, while they may appear to refer to particular concrete objects or mental images, they really are not. For example, when we manipulate a certain geometric object and make a claim about its shape, we are, as a matter of fact, referring to the relevant infinite class of objects (p. 141), that is, what Norman (2006) considers to be the *representational scope* of the object (p. 30). Also, the general representational quality of the objects enables learners to develop structural awareness of the corresponding concepts or processes.

Concrete objects are those physical objects and tactile manipulatives that are found in almost all mathematics classrooms. They are primarily used to assist students gain access to abstract concepts and, in Dickson's (2002) words, "test their honesty" (p. 221) by routing the students to familiar and interesting objects. For

---

[2]Gibson (1986/1979), of course, was referring to visual perception involving everyday objects in a person's (or an animal's) environment. Land, for example, naturally affords particular kinds of animals to physically move or pose in ways that adapt to their intentions. In the case of mathematical objects, visual affordances of objects require an interpretive act of discerning and constructing some intention relevant to the objects.

[3]Begle's (1979) notion of mathematical objects had him classifying various kinds of mathematical ideas, as follows: (1) facts (notation-based and deduced ones); (2) concepts; (3) operations (assignment of meaning from object to object); and (4) principles (relationships). Leushina (1991) also wrote about mathematical objects but in the context of visual aids. Leushina's classification of aids depends on "how the surrounding reality is reflected," as follows: (1) natural; (2) representational; and (3) graphic. My notion of mathematical objects does not conflate object and the meanings, associations, and ontological sources we derive from them. I leave aside conceptual meanings and intentions in favor of what I consider to be the raw dimension in which inscriptions and physical or material manifestations generally are reified forms of their ideal referents.

example, in my Algebra 1 class, the students used algeblocks to learn simple polynomial expressions and operations (Fig. 2.2). In my Algebra 1 and Grade 2 classes, the students used fraction strips (Fig. 3.4) to learn part–whole relationships and establish rules for combining fractions. Binary chips are also popular in middle school classrooms and are used to represent positive and negative integers. In geometry, three-dimensional solids and expertly constructed plaster models are used to describe parts of solids and surfaces and relationships between and among them. In my Grade 2 class, the students initially worked with pattern blocks to help them perform composition and decomposition of shapes. Overall, the tactile dimension in all concrete objects enables learners to "experience" and "apprehend" the "integrated idea" of their total nature. Further, their physical character, as Dickson (2002) has eloquently pointed out, "relates to us as physical beings. Only the physical object has life. Only the physical object has the power to resonate with our lives" (p. 221).

Graphic and virtual objects are objects that are generated with the use of technological tools such as applets, animations, virtual-based manipulatives, graphs of algebraic functions on a graphing calculator, and geometric objects drawn on computer graphics software like curves and surfaces in single- and multi-object graphics worlds. Palais (1999) points out the significance of computer-generated images in not only allowing computers to "produce such static displays (i.e., the concrete objects) quickly and easily, but in addition it then becomes straightforward to create rotation and morphing animations that can bring the known mathematical landscape to life in unprecedented ways" and help "obtain fresh insights concerning complex and poorly understood mathematical objects" (pp. 647–648). Also, compared with static objects, Taylor, Pountney, and Malabar (2007) point out that the "dynamic, evolving process" in most animations helps students "construct mental models of various 'processes' ... as well as the details of calculations" (p. 251). However, they also stress the need for an effective combination of animation and interaction in mathematical learning.

Relational objects are diagrams or inscriptions that model a structure, display relationships, express concepts and (hypostasized) abstractions in some detail, and facilitate problem solving, reasoning, and discovery (Dörfler, 2007; Glasgow, Narayanan, & Chandrasekaran, 1995; Tversky, 2001). For example, Ng and Lee (2009) characterize the rectangular models used in the Singapore mathematics curriculum for problem solving (e.g., Fig. 2.11a–c) as "visual analogues that capture all the information" relevant to the problems and whose structures are "made overt" through the rectangles that explicitly indicate the necessary relationships (pp. 284–285). Relational objects appear in schematic form with labeled components and clearly defined spatial relations among the components. Additional exemplary examples include those figures that accompany geometric theorems, which express relationships between vertices and segments in a particular way (e.g., Fig. 2.17). Abstract algebra, statistics and data analysis, discrete mathematics, design theory, graph theory, and real analysis use a significant amount of relational objects.

Epirelational objects pertain to objects that are known through special relationships that are inferred among their parts. For example, π is an epirelational object whose approximate value could be visualized by physically measuring and calculating the ratio of the circumference and diameter of a circle. The whole number part in π can be visually verified by initially measuring the length of the diameter of a circle with a piece of string and then showing that its circumference is slightly more than three times the diameter by tracing three copies of the string on the circumference (see Fig. 3.13a). Phi, ϕ, whose exact value is $(1 + \sqrt{5})/2$, is another example of an irrational number that could be visualized by obtaining the ratio of the length and width of a Golden Rectangle (see Fig. 3.13b). Visualizing such ratios requires inferring particular relationships between parts unlike, say, every square root of a nonperfect square number that could be easily depicted by taking the length of the segment corresponding to the hypotenuse of the relevant right triangle. Many analytic concepts such as functions (try $f(x) = 1$ if x is rational and $f(x) = 0$ otherwise), the integral, and continuous and differentiable functions have no corresponding canonical images that could capture their complex nature as a whole in sufficient form (Giaquinto, 1994).

Alphanumeric objects refer to variables, numerals, and combinations of variables and numerals that are used to describe any of the above objects. For example, algeblocks (Fig. 2.2) as concrete objects could be expressed alternatively as polynomials. Most algebraic and transcendental functions are relational and/or graphic or virtual objects that have corresponding alphanumeric descriptions in the form of equations. Kirshner and Awtry (2004) note the "effects of visual salience" (versus the associated conceptual content) of alphanumeric forms and rules in mathematics learning, which could be one way of explaining students' errors when they manipulate symbols in algebra. "Visually salient rules," Kirshner and Awtry point out, "have a visual coherence that makes the left- and right-hand sides of the equations appear naturally related to one another" (p. 229). For example, the identities $a(b+c) = ab+ac$, $(xy)^m = x^m y^m$, and $(a/b)(c/d)=ac/bd$ $(b, d \neq 0)$ appear to be visually salient for many students, while the identities $(a + b)^2 = a^2 + 2ab + b^2$, $x^0 = 1$ $(x \neq 0)$, and $(a/b)/(c/d)=ad/bc$ $(b, d \neq 0)$ are not. Drawing on their empirical work with 114 Grade 7 students (mean age of 12 years) who have not had any formal instruction in algebraic rules, Kirshner and Awtry found that students tend to initially engage in "visual pattern matching" rather "immediately and spontaneously" (p. 243). Further, they tend to overgeneralize the "context of application" (p. 246) of visually salient rules (e.g., extend the power rule $(xy)^2 = x^2y^2$ to $(x+y)^2 = x^2+y^2$), which does not appear to be the case with less and nonvisually salient ones because the latter type "forces students to attend more fully to the structural descriptions" (p. 246).

The quadratic formula, $x = \left(-b \pm \sqrt{b^2 - 4ac}\right)/2a$, which assists in obtaining the roots of a quadratic equation of the type $ax^2 + bx + c = 0$ $(a \neq 0)$, exemplifies a type of object that is both alphanumeric and epirelational. When the ancient Babylonians sometime in 1700 B.C. dealt with quadratic equations of the simple type $x^2 + bx = c$, they, in fact, considered it to be a visual (i.e., geometric) problem (Katz, 1999, pp. 27–28). The epirelational aspect arises in the visual demonstration of the solution, which involves the following steps below (Fig. 3.12a is the

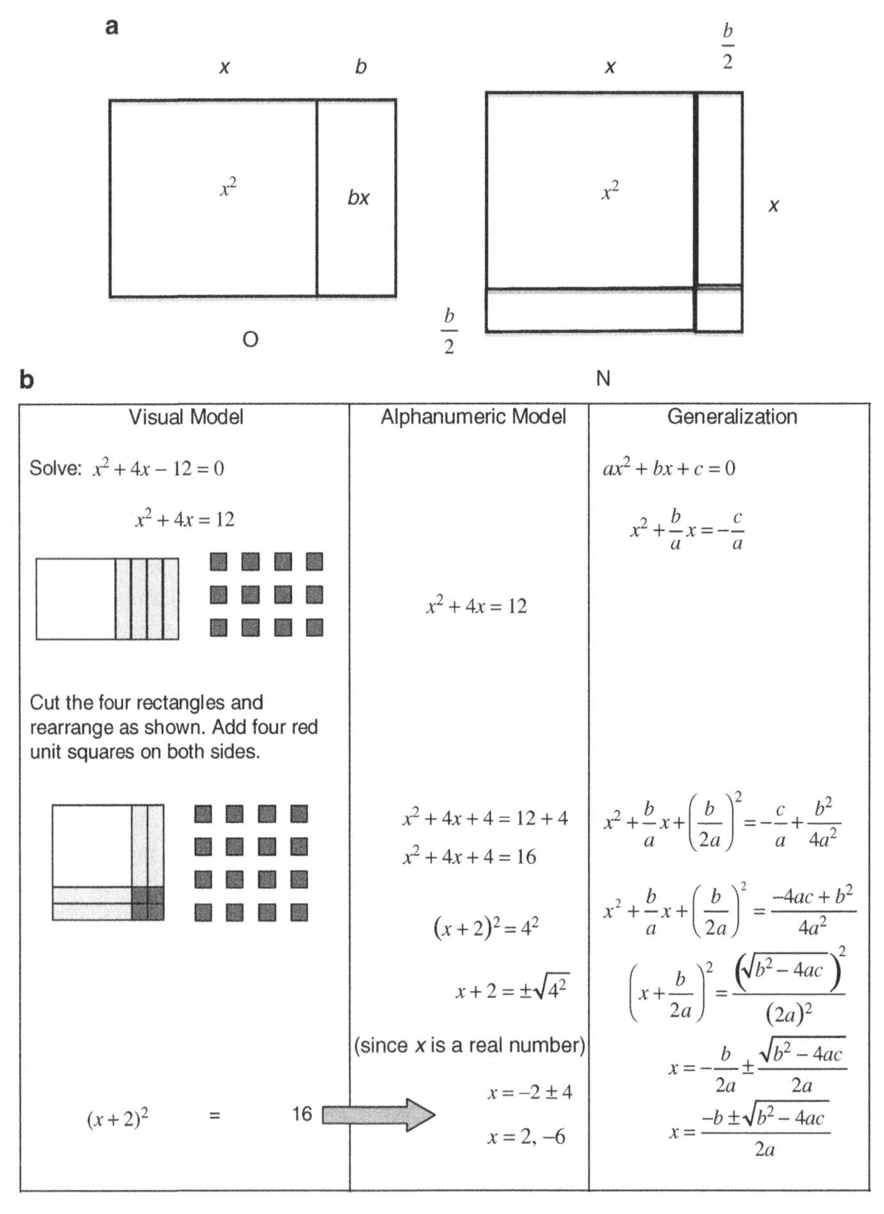

Fig. 3.12 **a** Visual demonstration of the quadratic formula (Katz, 1999, p. 28). **b** Visual and alphanumeric models for solving simple quadratic equations

corresponding generalized geometric model and Fig. 3.12b is a particular instance of a visual solution to a quadratic equation):

(1) The equation in geometric form means we have an original figure O that consists of a square having side x and a rectangle having dimensions b and x with a combined area of c units.
(2) Take half of b, $b/2$, and then reconfigure O so that the two smaller rectangles are on two adjacent sides of the square.
(3) Add a small square to O having side $b/2$ and, thus, area $b^2/4$ to complete the new square N with dimension $(x + (b/2))$.
(4) Add $b^2/4$ to c.
(5) Take the square root of the value obtained in step 3 in order to obtain the length of the side of square N and do the same process in step 4.
(6) Subtract $b/2$ to obtain the correct value of x.

The visual power of the geometric demonstration is actually in steps 2, 3, and 5. In taking half of b and reconfiguring O so that the two halved rectangles now appear on two adjacent sides of the original square, we actually obtain N with side $(x + (b/2))$ by *necessarily* adding a small square with area $b^2/4$. Further, applying the square root process *necessarily* helps in determining the length of the side of N.

Drawing on my work with elementary and middle school children and undergraduate students, something more actually needs to take place between learner and object. Uttal et al. (1997) and Rakoczy et al. (2005) point out that comprehending objects "as symbols rather than as substitutes for symbols" (Uttal et al. 1997, p. 37) and symbolic action, respectively, are necessary components in any interaction between learner and object.

Uttal et al. (1997) used a scale model activity to assess whether children between 2.5 and 3 years of age would see a relationship between a scale model of a room and the actual room in the context of a retrieval task. In the activity, a child is shown a situation in which an experimenter hid a small version of a toy in the scale model. The child is then brought to the actual room to locate the larger version of the toy. Next, the child goes back to the scale model to retrieve the hidden toy. For Uttal et al. (1997), the scale model is a symbol for the actual room in the sense of the former "standing for" the latter, and they interpret children's success in the retrieval task as an indication that the children are able to comprehend the symbolic and abstract relationship between the scale model and the actual room.

For Rakoczy et al. (2005), when children perform symbolic action on an object, it means they are acquiring "processes" relevant to "cultural learning" on the basis of their "nascent understanding of intentional action and on cultural scaffolding" (p. 69). Beyond the view of Uttal et al. (1997) in which objects are symbols, Rakoczy et al. (2005) point out that "it is people who mean and refer, using symbols as their instruments for doing so" (p. 70). This reminds me of the Peircean concept of *interpretant* that oftentimes gets lost in semiotic accounts of sign or symbol formation that tend to dwell exclusively on relationships between a signifier (object) and signified (referent). The notion of *interpretant* conveys the importance accorded to an individual's personal meaning, but Rakoczy et al. (2005) situate such meanings in cultural activity and load objects with "intentional affordances"; what children do with objects is reflective of what they intentionally read and imitate from others in social activity. Further, while the objects above, except in most cases of virtual and epirelational objects, may have natural meanings associated with them by virtue of "causal and covariational regularities in the external world" (p. 70), all symbolic actions are nonnatural in the sense that

a

Exploring Pi

A. Draw a big circle on a construction pad. Construct a diameter and use a piece of string to measure its length.

1. How many copies of the string do you need to measure the "perimeter" of the circle? Fill in the blank: The circumference of a circle is slightly more than ______ times its diameter.

2. Now draw two other circles having different diameters and follow the same procedure above. What can you infer about the circumference of a circle relative to its diameter?

B. You will need to measure at least 10 different circular objects. Carefully measure the circumference of each object with a piece of string. Then use a metric ruler to measure the length of string in millimeters. Also measure the diameter of each object in millimeters.

Create a spreadsheet with three columns. In column A, enter the circumferences of the items you measured. In column B, enter the diameter. In column C, create a formula for the value of π using the circumference and diameter. (= A1/B1)

Your spreadsheet should look like this:

	A	B	C
1	27	8	3.375
2	13	4	3.25
3			

Use your spreadsheet to answer the following questions.

1. What is the value of π used by your spreadsheet?

2. How close are your calculations to the actual value of π? To how many decimal places are your calculations correct?

3. What could cause your calculations of π not to be exact?

4. What could make your calculations more exact?

5. Suppose you created another spreadsheet with the same columns as this one. The new spreadsheet calculates the diameter when given the circumference and the value of π. What formula would you enter in column B?

6. If you create another spreadsheet that calculates the circumference when given the diameter and the value of π, what formula would you enter in column A?

Fig. 3.13 (continued)

b

1. In rectangle ACDF below, AF = CD = 2 units. Show that rectangle ACDF is a Golden Rectangle by following the steps below.

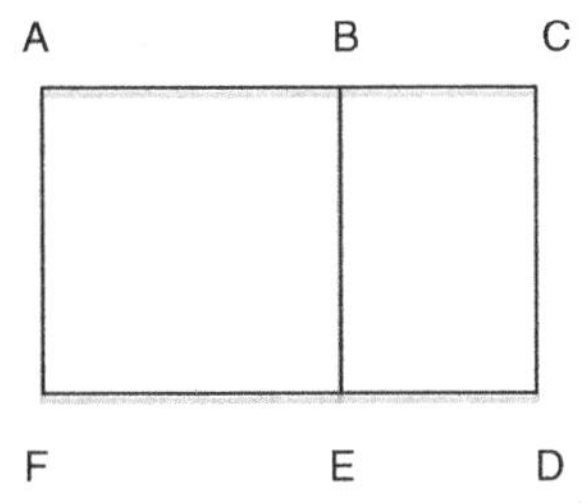

a. Consider the 2 x 2 square ABEF. Locate the midpoint of FE and label this point M.

b. Draw $\overline{MB}$ and obtain its length.

c. Verify with a ruler that $\overline{MB}$ and $\overline{MD}$ have the same length. Note that another way of verifying this is to see that diagonal $\overline{MB}$ has been projected onto $\overline{MD}$.

d. Obtain the ratio of the length $\overline{FD}$ to the width $\overline{DC}$. This verifies that rectangle ACDF is a Golden Rectangle.

2. Using the same figure above, show that rectangle EDCB is also a Golden Rectangle.

3. Draw a 4 x 4 square S. Construct a rectangle *R* that will be joined to S on one of its sides so that the resulting (bigger) rectangle *B* is a Golden Rectangle.

b. Verify that *B* is indeed a Golden Rectangle.

4. Generalize. Begin with a square whose side has a length of *x* units. Show how we can construct a rectangle on one side of the square so that the resulting rectangle is a Golden Rectangle. Then verify that the resulting rectangle is indeed a Golden Rectangle.

Fig. 3.13 **a** Middle school level activity exploring π (Curriculum and Development Division (2008) and http://stamford.region14.net/webs/greesonl/upload/pi_handout.pdf). **b** Algebra 1 activity exploring ϕ

> they are not out there in the world, but rather are socially constituted through the way persons use and interpret them. Symbols are objects with derived intentionality that is conferred on them through the attitudes, actions, and practices of persons that possess intrinsic or original intentionality.
>
> (Rakoczy et al., 2005, p. 70)

What should be clear at this stage is that mathematical objects are visual representations in some way or another. Concrete, graphic and virtual, and relational objects are externalized as figures in some format, while alphanumeric objects utilize everyday numerical and alphabets. Most epirelational objects involve a constructive process in which case the visual experience is felt through a complex synergy among elements in several dimensions. For example, the two activities in Fig. 3.13a are often used in middle school mathematics classrooms that allow students to visually construct the idea that π exists as a constant ratio for any number of circles with a known circumference and a given diameter. Figure 3.13b is an interesting Algebra 1 activity that provides students with a visual experience of simultaneously constructing Golden Rectangles and verifying that ϕ, which is approximately equal to 1.618, is indeed the length-to-width ratio of such rectangles. In both activities, the elements involve the use of alphanumeric and concrete objects. Further, the reification process involved in the development of the corresponding visual representations requires students to pay attention to the associated conceptual meanings that are purposefully constructed through instrumental action. In other words, they need to act on them in ways that share the intentions of, say, the lifeworld-dependent context that allow those objects to exist in the first place.

4 Mathematical Concepts in Cognitive Activity

If all mathematical objects convey conceptual content, then what is a mathematical concept? Mathematics education researchers have talked about the nature of a mathematical concept in several different ways; however, in light of the overall goal in this book, we focus on the definitions suggested by Davydov (1990) and Fischbein (1993). We begin with two powerful perspectives on concepts that we assume, that is, the notions of (1) *concept image and concept definition* by Tall and Vinner (1981), which Fischbein (1993) uses in constructing his version of a figural concept, and (2) *conceptual fields by* Vergnaud (1996, 2009), which focus on the triad of scheme, situation, and linguistic/symbolic representations that characterize concept development.

For Tall and Vinner (1981), a conceptual image refers to "the total cognitive structure that is associated with the concept, which includes all the mental pictures and associated properties and processes. It is built up over the years through experiences of all kinds, changing as the individual meets new stimuli and matures" (p. 152), while a concept definition takes the shape of "words used to specify that concept" (p. 152). Concept definitions produce concept images and are acquired in several different ways (through rote work or in some meaningful context) or evolve as a personal reconstruction that may differ from a formal concept definition.

For Vergnaud (1996), his conceptual field refers to "a set of situations, the mastering of which requires several interconnected concepts. It is at the same time a set of concepts, with different properties, the meaning of which is drawn from this variety of situations" (p. 225). Thus, concepts take meaning in a complex domain such as space that involves knowing and understanding the relevant qualitative and quantitative dimensions.

While the above definitions of concepts in mathematics acknowledge the crucial role of structures or situations in giving meaning and form to the relevant concepts, Davydov and Fischbein offer perspectives that address the relationship between object and concept in more explicit terms.

Davydov (1990) initially analyzed the meaning of concepts from an empirical perspective in which generalizations and (hypostasized) abstractions take place through a process that surfaces essential common attributes among objects. He then critiqued the empirical model because it favored the sequence of perception–conception-critique that he found problematic in many aspects such as the phases being viewed separately with no guarantee of a successful transition or passage from one phase to the next. In the sequence, perception provides "the raw material" that is drawn from "isolated, sensorially perceived objects and phenomena in the world around us" (p. 17), while conception signals the stage in which visual images are perceived, compared, and captured in verbal form. A conception is marked by the acquisition of important, primary, fundamental, and general attributes drawn from a set of objects and carefully delineated from, and compared with, secondary, individual attributes. Expectedly, a concept evolves after all the attributes have been ordered and the important ones have "become genuinely abstract – abstracted from any particular forms of their existence" (p. 18). A concept, in its final form, is thus seen as a nonvisual, abstract universal, and rational generality (p. 246). A typical example of this empirical approach is the manner in which children's concepts of numbers, properties, and operations are oftentimes depicted, that is, they initially occur when the children begin to count discrete objects, which they then combine in ways that allow them to induce the relevant properties and operations.

Dayvdov (1990) developed a dialectical model as a counterstance to the observation that language emerges from empirical thought as a result of the "method of obtaining and using sensory data" (p. 248). For him, thought is not nakedly empirical but "is *rational* cognition" (p. 248). He writes:

> Consequently, with respect to the activity of social man in general it is impossible to apply the category of "sensory cognition" as a particular and special level that precedes "rational cognition." Cognition on the part of socialized humanity acquired a rational form from the very beginning.
>
> (Davydov, 1990, p. 248)

In light of the above assumptions, Davydov does not distinguish by level between the sensory and the rational. This explains why he considers a concept as a concrete (hypostasized) abstraction, that is, it is an instance that has properties of being a particular case and containing a universal property. He writes:

> A concept functions here as a form of mental activity by means of which an idealized object and the system of its connections, which reflect in their unit the generality or essence of movement of the material object, are reproduced.
>
> (Davydov, 1990, p. 249)

Further, we come to know the relevant properties and attributes of this ideal object and relationships among ideal objects through "cunning" that allows us to

> reveal the properties of objects and recreate relationships and connections with one another through them. One thing becomes a means of embodying the properties of other things, functioning as their standard and measure. ...Different symbol systems (materials ones, graphic ones) are means of "standardizing" and, thus, idealizing material objects, means of translating them to a mental level. ... The functional existence of a symbol involves its ... being a means, a tool for portraying the essence of other sensorially perceived things – that is, of their universal [essence] .... Disclosure and expression in symbols of the mediated being of things, of their universality, is a transition to the theoretical reproduction of reality.
>
> (Davydov, 1990, p. 251)

Davydov's (1990) view of concept as concrete abstraction shares views offered by Mandler (2007, 2008) in relation to the "highly global, rather sketchy and abstract" nature of object-concepts observed in infants. That is, when infants initially confront an object, they tend to pay less attention to details since they appear to be more interested in what the object does in some context in which it is seen. Through more experiences and over time, a detailed characterization of the object is constructed.

Davydov's notion of concrete abstraction was, in fact, the basis of an experimental curriculum he developed with Russian colleagues that focused on demonstrating the possibility of an "ascent from the abstract to the concrete" (p. 355) or, in Mandler's (2008) sense, a "top-down" approach to concept formation. In the first semester of first-grade students' mathematical experiences, for example, they did not deal with the concept of numbers at all but came to know facts about quantity by "singl(ing) it out in physical quantities and becom(ing) familiar with its basic properties" (p. 359). They also learned to compare things (i.e., "more" to "less") in relation to an identified unit of measure and eventually transitioned to the use of letters and (relational) symbols. Thus, formal mathematical concepts emerged on the basis of the students' informal, everyday experiences that were rooted in the visual and the concrete.

Davydov's (1990) beginning approach to numbers among young children echoes basic findings in current psychological research that confirm the initial structure of preschool children's number sense as comprising of protoquantitative concepts that are not numerically precise. For example, they "express quantity judgments in the form of absolute size labels such as big, small, lots, and little" (Resnick, 1989, p. 162), which they also use on tasks that involve making comparisons that they accomplish at a perceptual level.

A similar approach to concept acquisition, development, and formalization can, in fact, be done in the upper grades. For example, in my Algebra 1 class, the students initially explored simple polynomial expressions as volume quantities through algeblocks. Next, they compared quantities and saw differences between and among cubes and right rectangular prisms of various dimensions. The quantitative experience allowed them to see differences, say, between expressions x^2y and xy^2 that later became their basis in understanding the significance of a *unit* – that is, the ideal object – in adding and subtracting polynomials (see Sophian, 2007, pp. 64–83 for an interesting interpretive analysis of the concept of a unit based on the El'konin–Davydov curriculum). Hence, for them, it made sense to combine, say, $3xy^2 - 5xy^2$ on the basis of seeing xy^2 as a common unit unlike the case of, say, $3xy^2 + 3x^2y$ that involves two different units. Further, it was their clear concept of a unit drawn from their experiences with the algeblocks that enabled them to make sense of adding and subtracting rational and radical algebraic expressions later in the year. To illustrate, when they simplified rational expressions in both arithmetical (e.g., 8/3 – 2/3) and algebraic contexts (e.g., $(1/x) + (1/x^2)$), they initially parsed the expressions by identifying a common unit, which also explained the significance of finding a least common denominator (see pp. 218–220 of this book). The activity in Fig. 2.9b uses $\sqrt{2}$ as the basic unit that enabled them to perform operations involving $\sqrt{2}$. They also used the unit concept in making sense of basic transformations of graphs involving linear and quadratic functions. With the use of a graphing calculator, they compared the graphs of $g(x) = a \cdot f(x \pm b) \pm c$ with the basic "unit" graphs of $f(x) = x$ and $f(x) = x^2$.

Fischbein's (1993) notion of a *figural concept* fuses both object and concept in a way that is compatible with Davydov's (1990) view of ideal objects and his critique of the empirical sequence. In mathematics, we do not simply conceptualize from a perceptual image and then develop a concept in words since "images and concepts interact intimately" (Fischbein, 1993, p. 145), especially in geometrical situations. Fischbein (1993) writes:

> Is the course of the reasoning process determined essentially by conceptual constructions (symbolized or mediated by imaginary means) or vice versa: is it the play of images which pushes forward the reasoning process in its creative attempts? The most plausible hypothesis seems to be that we deal in fact with *one game* in which active conceptual networks interact with imaginative sources. Moreover, we have reasons to admit that, in the course of that interplay, meanings shift from one category to the other, images getting more generalized significance and concepts largely enriching their connotations and their combinational power.
>
> (Fischbein, 1993, p. 145)

Figural concepts are, thus, "simultaneously conceptual and figural" (Fischbein, 1993, p. 160); they are "abstract, general, ideal, pure, logically determinable entities, though they still reflect and manipulate mental representations of spatial properties (like shape, position, metrically expressed magnitudes)" (Fischbein, 1993, p. 160). Further, we work with an image on the basis of what is allowed by the corresponding concept. Fischbein (1993) also points out that the development of figural concepts, like Davydov's idealized objects, "is not a natural process" (1990, p. 161) and, by nonnatural, it conveys the same meaning as Rakoczy et al. (2005) who trace such processes as "not out there in the world, but rather are socially constituted through the way persons use and interpret them" (p. 70) in a lifeworld-dependent cognitive activity.

5 Visual Thinking in Mathematics

For Peirce, the German mathematician Karl Weierstrass was basically responsible for the jettisoning of imagery in mathematics: "The whole Weierstrassian mathematics is characterized by a *distrust of intuition*." The Weierstrassian perspective, Peirce points out, "betrays ignorance of a principle of logic of the utmost practical importance; namely that every deductive inference is performed, and can only be performed, by imagining an instance in which the premises are true and observing by contemplation of the image that the conclusion is true" (Peirce quoted in Houser, 1987, p. 18). In 1872, Weierstrass was the first to construct and justify a nonvisual example of a continuous function that was nowhere differentiable at any point.

Davis (1993) traces the "historical decline of the image" in geometry beginning with the successful algebraization of geometry in the seventeenth century through Descartes. What followed next were the powerful conceptual inventions in the nineteenth century of objects at infinity and complex geometries in projective geometry, which could not be represented visually, and the full acceptance of abstract deductive systems as a result of non-Euclidean geometries that, in effect, celebrated the demise of a "unitary view of truth" (p. 334). Also, the disappointment with the geometric image was felt with the discovery of analytic concepts such as continuous, nondifferential functions whose complex nature defied any suitable canonical graph that contained all the required properties. Hence, the adage "appearances can be deceiving" took on more meaning and prestige that further supported the "devisualization" and "despatialization" of geometry. Despite the negative history, Davis (1993) remains hopeful that "vision can yield something deeper than formulaic-deductive mathematics and hence can contribute to a wider view of mathematics" (p. 336).

For Katz (2007), algebra in an historical context could be interpreted primarily as solving for unknowns. Its beginnings could be traced to the Babylonians and Greeks who developed it in the context of geometric problems that were visual in nature. Later, the Islamic influence transformed algebraic practices symbolically, which favored the use of variables to solve numeric problems. The third phase in the development of algebra was motivated by concerns in physics in which algebraic methods were used to solve for unknowns that represented curves of geometric figures. In this phase, functions and their graphical representations became fully developed concepts that seem to have been a driving force behind our recent journey back to the fold of the image.

"The limits of deductivism are at last," Rival (1987) writes, "dawning on mathematicians, thanks largely to computers [and] combinatorics" (p. 44). The history of the cultural present is characterized by revolutionary advances in technological tools that also contribute significantly to new emerging epistemic practices that see visual thinking in mathematics in a central role.

We now define visual thinking in mathematics, as follows:

> *Visualization in mathematics is a situation-specific epistemo-ontological capacity of forming and transforming figural images in a lifeworld-dependent activity.*

While the above definition should be clear on the basis of the preceding discussions, a summary explanation for ostensive purposes is instructive.

First, the italicized definition is reflective of the conceptual constraints in, and assumptions about, visualization that are specific to mathematics as discussed above. At the very least, it incorporates the following characterization offered by Arcavi (2003):

> Visualization is the ability, the process, and the product of creation, interpretation, use of and reflection upon pictures, images, diagrams, in our minds, on paper or with technological tools, with the purpose of depicting and communicating information, thinking about and developing previously unknown ideas and advancing understandings.
>
> (Arcavi, 2003, p. 217)

Second, the italicized definition does not make any claim that privileges internal constructs in either mentally or explicitly constructing a relevant object, which seems to be the suggestion of Zazkis, Dubinsky, and Dautermann (1996) in their definition below:

> Visualization is an act in which an individual establishes a strong connection between an internal construct and something to which access is gained through the senses. Such a connection can be made in either of two directions. An act of visualization may consist of any mental construction of objects or processes that an individual associates with objects or events [i.e., some dynamic, external phenomena] perceived by her or him as external [i.e., anything that an individual accesses through sensory/motor experiences]. Alternatively, an act of visualization may consist of the construction, on some external medium such as paper, chalkboard, or computer screen, of objects or events that the individual identifies with object(s) or process(es) in her or his mind.
>
> (Zazkis, Dubinsky, & Dautermann, 1996, p. 441)

What our definition above does emphasize is a capacity that is *not* merely an *act* that is added to our repertoire of knowing but conveys a mutually determining phenomenon of acting and knowing. This shares Magnani's (2004) view of distributed cognition as a "simultaneous co-evolution of the states of mind, body, and external environment" (p. 520). Thus, visualizing involves knowing that something occurs as a result of performing action(s) on a target object, concept, or process. For example, my Algebra 1 students acquired their understanding of the proof of the Pythagorean theorem in two ways as shown in Fig. 3.14. First, they laid down

A. Use Four Copies of the Same Right Triangle (Classic Approach)

B. Use Two Copies of the Same Right Triangle (Garfield's Approach)

Fig. 3.14 Two visual explanations of the Pythagorean theorem

four copies of the same right triangle in a particular manner one at a time on a table, which then allowed them to establish an algebraic proof by equating the sum of the individual areas and the area of the larger square. Second, they laid down two copies of the same right triangle on the table and established the theorem by relating the sum of the areas of the three right triangles and the area of the trapezoid. The mathematical activity in this case exemplifies an instance of visualization in which both knowing and acting in a distributed context enabled successful knowledge construction. O'Regan (2001) describes this way of visualizing as a "practical kind of knowledge," of "knowing that certain laws of co-variance apply between our actions and the resulting changes in the sensory input" (p. 278). He makes the following observation about "seeing," which summarily explains what we mean by our capacity to visualize in mathematics in a distributed context:

> It's like the feel of driving a car – you're not able to describe verbally every single aspect of the experience. Nevertheless all the things you can do, like press on the accelerator and know the car will whoosh forward, constitute the "what-it-is-likeness" of driving a car.
>
> (O'Regan, 2001, p. 278)

Third, Giaquinto's (2007) definition below leaves out psychosocial factors that matter in visual thinking about figural concepts:

> Visual thinking [is] thinking that involves visual imagination or visual perception of external diagrams.
>
> (Giaquinto, 2007, p. 1)

However, he points out two important aspects of visualizing in mathematics, that is, *discovery* (without necessarily having to prove) and *explanation* (such as "help(ing) make a theorem that one already knows more intuitive" (Giaquinto, 2005, p. 78)). He writes:

> In general, visualizing provides an almost effortless way of acquiring new information, and its results often come with a degree of immediacy, clarity, and forces that makes visualization apt as a means of discovery and explanation. Thus the epistemic value of visual thinking in mathematics becomes obvious, once we give due importance to discovery and explanation.
>
> (Giaquinto, 2007, p. 264)

Both aspects of discovery and explanation fall under the general description of theoretical mathematics that is concerned with conjectural and intuitive work and, thus, are incomplete. But their value is seen in terms of providing the conceptual elements that engender the emergence of rigorous mathematical work. Rigor could take the form of a formal deductive proof (cf. Jaffe & Quinn, 1993). For example, Fig. 3.14 provides illustrations in visual form that provide support leading to a rigorous proof. In my Algebra 1 class, we initially used the visual activity in Fig. 3.15 to discover a relationship between the hypotenuse of a right triangle and its two sides. We then followed it up with the activity in Fig. 3.14 that provides an effective visual explanation of the famous theorem in two interesting ways, which then led to an algebraic proof.

Rigor at the elementary level could mean conceptually establishing the validity of some procedural knowledge. For example, in my Grade 2 class, the visual

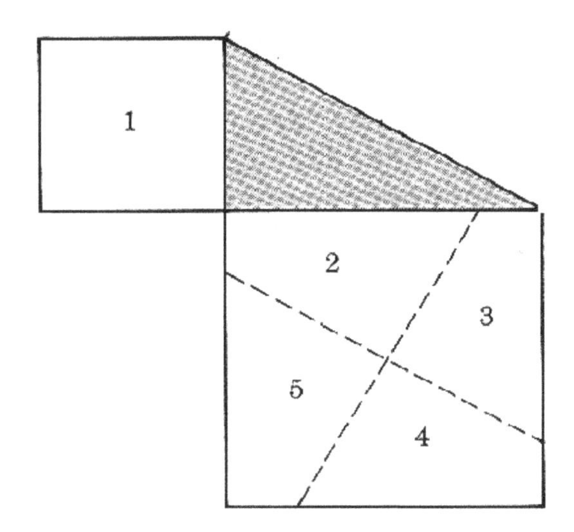

Fig. 3.15 A visual discovery activity involving the Pythagorean theorem (O'Daffer & Clemens, 1992)

experience with sticks and dots shown in Fig. 3.11a helped the students obtain a more systematic algorithm for subtraction when regrouping objects became a necessity. However, compared with Fig. 3.11a, the visual explanation in Fig. 3.11b assisted the students to see the connection between the visual and the corresponding alphanumeric algorithm shown in Fig. 3.11c.

Fourth, we acknowledge the significance of a *domain of situations* that oftentimes goes without saying in constructing mathematical knowledge. For example, Vergnaud (1996) characterizes children's mathematics as pertaining to a domain of cognitive processes that includes object identification and properties, construction of relations and transformations between and among objects, and comparison and combinations of the relevant quantities and magnitudes (p. 224).

Fifth, visualizing as an acquired *epistemo-ontological habit* that learners draw in *a lifeworld-dependent activity* is made in light of Duval's (2006) analysis of the differences between common and mathematical ways of comprehending figures (p. 115), which also makes sense in the case of visualizing. For example, there is a difference between seeing a roof corresponding to the top part of a table as a representation of a quadrilateral and seeing parts of the quadrilateral as consisting of four sides represented by four line segments. In the former, one apprehends perceptually (i.e., a figure as a single gestalt), while in the latter, discursively (i.e., the figure either as consisting of several constituent gestalts or as subconfigurations). Epistemo-ontological habits refer to matters involving *how we see* and *what we see*. My sense of epistemology is Foucauldian, that is, it pertains to both senses of *savoir* and *connaissance* with the latter focusing on the relationship of the individual to an object and the rules governing that relationship and the former to the historical, necessary conditions that make the latter possible (Foucault, 1982).

Thus, epistemo-ontological habits of forming and transforming figural images are subject to a set of situations (following Vergnaud) that is reified in local (classroom and teacher practices) and institutional rules. *Forming* and *transforming* a figural image refer to the having of a visual experience and the inferring in scenario visualization, respectively, in the context of cognitive activity. Clarifying the source of cognitive activity allows us to deal with the predicament pointed out by Nemirovsky and Noble (1997) about the difficulty of categorizing a visual representation in terms of whether it is an internal or an external phenomenon by situating the sources of construction and justification of figural images in a lifeworld-dependent context that focuses on explicit knowledge. In other words, what is constructed in the mind and outside the mind are mutually determining actions in cognitive activity.

6 Revisiting Principles of Visualization from a Mathematical Point of View

In Chapter 2, we discussed in some detail the following three principles that matter in visualization involving everyday objects: acquisition; reasoning; and individuation. In this section, we revisit the three principles and highlight matters relevant to visualizing in mathematical cognitive activity.

Principle of Individuation: Much of one's visual processes in relation to school mathematical content is derived in a lifeworld-dependent context that focuses on exact mathematical knowledge. At least two decades of research in numerical cognition have provided interesting foundational accounts of the existence of biologically and behaviorally determined core representational structures in mathematics among preverbal infants and preschool children (see, e.g., Cordes, Williams, & Meck, 2007; Dehaene, 2001; Gelman & Gallistel, 1978; Pica, Lemer, Izard, & Dehaene, 2004; Simon, 1997; Spelke, Breinlinger, Macomber, & Jacobson, 1992). Dehaene (1997), for example, has empirically demonstrated *the number sense* claim in which preverbal infants seem to be in possession of a mental number line that enables them to make approximate sense of numbers nonverbally. He also noted that increased facility with acquired language and explicit representations extends the core structures that produce explicit mathematical knowledge involving exact numbers and numerical understanding. In a synthesis of research on children's mathematics, Resnick (1989) also echoes the shared view that the "culturally transmitted formal system" of counting provides "the first step in making quantitative judgments exact" (p. 163).

The school mathematics curriculum consists of mathematical objects, concepts, processes, and explicit knowledge whose figural images have an exact nature. The principle of individuation addresses the concern in which the aspect of acquisition entails exact mathematical understanding, which can be "hard" to learn in the same sense that Feigenson, Dehaene, and Spelke (2004) described why numbers are "hard," as follows:

> (N)umber is *hard* when it goes beyond the limits of these [core representational] systems. When one attempts to represent an exact, large cardinal value, one must engage in a process of verbal counting and symbolic representation that children take many years to learn, that adults in different cultures perform in different ways, and that people in some remote cultures lack altogether. When humans push number representations further to embrace fractions, square roots, negative numbers and complex numbers, they move even further from the intuitive sense of number provided by the core systems.
>
> (Feigenson et al., 2004, p. 313)

Hence, there should be no confusion in interpreting the context in which the principle of individuation has been conceptualized in relation to visualizing in mathematics. In any lifeworld-dependent activity, there are particular sociocultural forms of symbol systems that provide the foundation for exact mathematical practices, which also influence the nature and content of the corresponding figural images whether in an imaginal or a formational or transformational context.

Principle of Acquisition: Visual representations of mathematical objects, concepts, and processes acquire their meanings in a lifeworld-dependent context. While most everyday objects are acquired naturally (in the sense of Rakoczy et al., 2005), all mathematical objects are acquired purposefully in a nonnatural manner. They do not merely convey signs that reflect "internal cognitive processes" but evolve "as tools or prostheses of the mind to accomplish actions as required by the contextual activities in which the individuals engage" (Radford, 2002, p. 241). Further, visually forming and transforming mathematical objects is not so much about capturing what they "represent" as it is more about "what they enable us to do" (p. 240). This insight has been drawn from Radford (2002) who situates the emergence of signs and thinking in a social-sign mediated context. Certainly, there have been documented cases in which some learners in classroom situations have produced individual images (i.e., imaginals) that in many cases have resulted in the construction of inventive concepts and strategies (cf. Resnick, 1989). However, meaningful figural images are oftentimes shared in cognitive activity in relation to some mathematical content.

The task, as it were, involves finding ways in which personal, subjective imaginals and images-in-the-wild could be "progressively transformed by the students" (Radford, 2002, p. 240) as visual representations that share the complex conceptual content of the corresponding institutional context (i.e., as figural images). This transformation is likely to take place through activities that (1) enable students to engage in explicit knowledge construction and cognitive praxis and (2) are framed "by social meanings and rules of use" that provide them with "social means of semiotic objectification" (Radford, 2002, p. 241). Certainly, the social practice of acquisition should not be interpreted as a simple matter of reducing images to convey "direct imprints of the external environment" but more in terms of a "[cognitive] encounter between the individual's subjectivity and the social means of semiotic objectification" (Radford, 2002, p. 260), that is, an act of "meshing" the "personal and interpersonal tones within the limits of the contextual possibilities" (Radford, 2002, p. 260).

Principle of Reasoning: The nature and content of imagistic reasoning depend on the type of mathematical object, concept, or process that is pursued. Imagistic and symbolic reasoning co-emerge in cognitive activity. Imagistic reasoning is used in imaginal, formational, or transformational contexts. It is useful in developing initial relationships, hence, imaginal. It is useful in illustrating a mathematical definition or concept, hence, formational. It is helpful in establishing visual inferences and justifications and in problem-solving contexts through scenario visualizing, hence, transformational. By reasoning, we mean "a personal-level psychological process consisting of inferences" (Norman, 2006, p. 18) with inference tied to the epistemic value of justifying belief (ibid., p. 11).

A good example of imagistic reasoning that infers at the elementary level is shown in Fig. 3.11b. Compared with Fig. 3.11a, Fig. 3.11b enabled my Grade 2 students to infer the actual steps in the subtraction algorithm with regrouping without getting distracted by the large number of circles that are drawn in the case of Fig. 3.11a. Another good example is shown in Fig. 3.16, which shows a pebble arithmetic demonstration that the sum of any number of even numbers is an even number (Livingston, 1999, p. 869). Step 1 in the pebble proof is formational in which every even number can be represented by an equal number of white and black circles. Step 2 visually demonstrates an alternative way of looking at even numbers, that is, by regrouping the circles and forming a one-to-one map between the white and black circles. In bringing the circles together in step 3, the one-to-one map between the white and black circles establishes the even parity of the sum.

Figure 3.16 is a visual inference whose primary epistemic value is to justify a belief that would eventually be transformed in symbolic form corresponding to a "formal" proof of the proposition. In a related sense, Polya (1988) notes: "When you have satisfied yourself that the theorem is true, you start proving it" (p. 181). Thus, when individuals use imagistic reasoning, the visual route provides them with feelings of accessibility, certainty, clarity (Norman, 2006, p. 25), correctness, confidence, and satisfaction (Alcock & Simpson, 2004).

Also, the effectiveness of imagistic reasoning lies in providing a gestalt-like experience in understanding the relevant figural image. Livingston (1999) articulates this condition in relation to Fig. 3.16. He points out that the "the reasoning involved

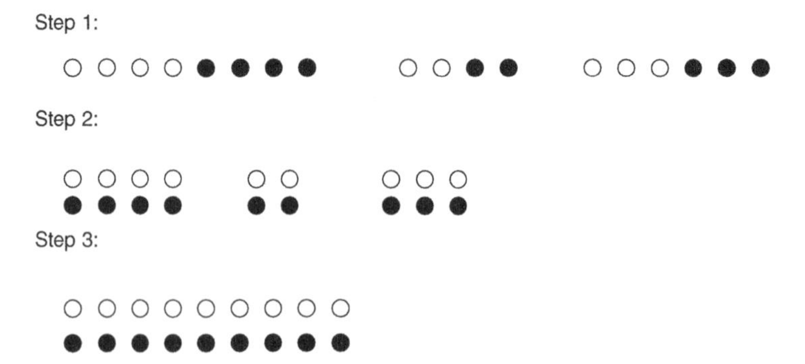

Fig. 3.16 A pebble arithmetic demonstration that the sum of any number of even numbers is even (Livingston, 1999, p. 869)

in the [visual] demonstration is neither 'in' nor not 'in' the proof figure" (p. 869) as it is more about developing its meaning via simultaneously coordinating the "materially definite writings (in this case, arrangements of dots)," "see(ing) through the notational particulars (the dots) to what they represent," and "organiz(ing), rearrang(ing) and rework(ing) such displays to find gestalts of reasoning and practice adequate to a stated theorem" (p. 869).

7 Overview of Chapter 4

In Chapter 4, we establish the visual roots of mathematical symbols. We begin by understanding the symbols that are used in school mathematics and claim that at least three kinds exist depending on how they are interpreted and used in context. While almost all symbols are, indeed, symbolic, other symbols could also be iconic, and still others indexical. Despite the impression that their meanings differ, they could all be grounded visually. Progressive mathematization, however, is a theory that anticipates transitions and growth in symbol use, from iconic and/to indexical and then to symbolic, leading to visuoalphanumeric representations. We illustrate such transitions in the number sense and algebra strands of the school mathematics curricula.

Chapter 4
Visual Roots of Mathematical Symbols

A symbol is symbolic if it describes or expresses or stands for an idea but has not yet become an enactive element. It can only become an enactive element if it has meaning, in other words, if there is associated with it at least one icon, an image, metaphor, picture or sense with itself [that] captures a pattern or relationship. Thus the state of being symbolic is highly relative.
(Mason, 1980, p. 11)

It is one of the essential advantages of the sign ... that it serves not only to represent, but above all to discover certain logical relations – that it not only offers a symbolic abbreviation for what is already known, but opens up new roads into the unknown.
(Cassirer quoted in Perkins, 1997, p. 50)

The differences between sign-types are matters of use, habit, and convention. The boundary line between texts and images, pictures and paragraphs, is drawn by a history of practical differences in the use of different sorts of symbolic marks, not by a metaphysical divide. And the differences that give rise to meaning within a symbol system are similarly dictated by use; we need to ask of a medium, not what "message" it dictates by virtue of its essential character, but what sort of functional features it employs in a particular context.
(Mitchell, 1986, p. 69)

In my Algebra 1 class, solving for the unknown in a linear equation occurred when they had to deal with reversal tasks in patterning situations such as item 4 in Fig. 4.1. Figure 4.2 shows the written work of Dung (eighth grader, Cohort 1), who understood the process of solving for the unknown in the context of finding a particular stage number p whose total number of objects is known. As shown in the figure, he initially took 1 away from 73 and then divided the result by 3. For Dung and his classmates, this particular process of undoing has been drawn from their everyday experiences in which doing and undoing form a natural and intuitive action pair. When my Cohort 1 participated in a teaching experiment on patterning and generalization in sixth grade, the first time I saw them use the undoing strategy occurred in the context of the patterning activity shown in Fig. 4.3. When they were confronted

F.D. Rivera, *Toward a Visually-Oriented School Mathematics Curriculum*, Mathematics Education Library 49, DOI 10.1007/978-94-007-0014-7_4,

© Springer Science+Business Media B.V. 2011

Consider the following pattern sequence below.

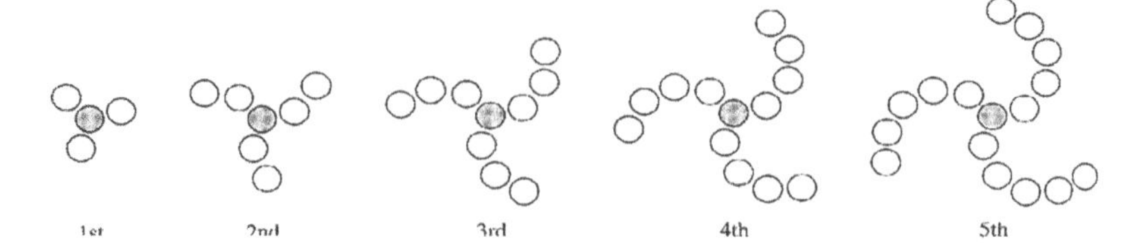

1. What stays the same? What changes?

2. Find a direct formula that allows you to obtain the total number of circles *C* at any stage *n* in the sequence. Explain why you think your formula makes sense.

3. How many circles are there in stage 37? Explain.

4. What stage number contains 73 circles altogether? How do you know for sure?

Fig. 4.1 Circle fan pattern task (pattern drawn from Beckmann, 2008, p. 496)

with the situation in item 21, the first thought that came to them was to take away the height of the original cup hold and then divide the result by 3. At this stage in their learning, they interpreted their constructed formula as a recording of their actions on a problem that they then conveyed in mathematical form. Hence, because a formula signified a process of doing, which mathematically meant performing a sequence of arithmetical operations leading to a total value or an output, reversing the steps in the formula or undoing by taking the opposite enabled them to determine the relevant input.

In the same Algebra 1 class, I wanted to determine the extent to which the students were capable of generalizing the undoing process above in the following two situations: (1) $Ax \pm B = \pm C$ and (2) $A(Bx \pm C) = \pm D$. However, instead of using a patterning context, I asked them to use algeblocks (Fig. 2.2). While they considered the task to be trivial in the case of situation (1), in which case they did not even use algeblocks, they found situation (2) interesting. Using algeblocks, they saw the significance of the distributive property of multiplication over addition in seeing the equivalence of $A(Bx \pm C) = ABx \pm AC$ that then allowed them to treat the resulting equation as a case similar to situation (1). In the third phase of the activity, I asked them to find a way of solving for x when the situation involved case (3), that is, $Ax \pm B = Cx \pm D$. We initially focused our attention on the equation $5x + 4 = 3x + 8$. Working in pairs, the students first gathered five yellow squares and four green cubes and set them aside. Then they gathered three yellow squares and eight green cubes and pulled them aside. In thinking about a possible strategy, several pairs saw that since there were three common yellow squares in both piles, they argued that taking them away from either pile would not affect the equation. As they were demonstrating with algeblocks, I was recording on the white board what they were saying, as follows:

A. Consider the following pattern sequence below.

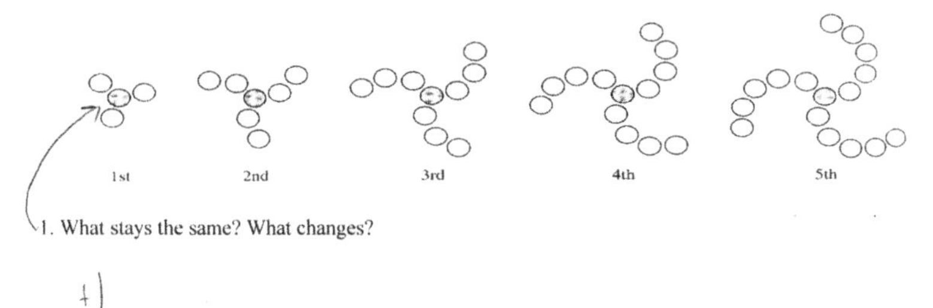

1. What stays the same? What changes?

+1

2. Find a direct formula that allows you to obtain the total number of circles C at any stage n in the sequence. Explain why you think your formula makes sense.

3. How many circles are there in stage 37? Explain.

37 × 3 = 111

111 + 1 = 112

4. What stage number contains 73 circles altogether? How do you know for sure?

reverse formula

c = 3p+1 = p(-1):2

73 - 1 = 72

3 ⟌72 = 24

Fig. 4.2 Dung's written work on the pattern task in Fig. 4.1

$$
\begin{aligned}
5x + 4 &= 3x + 8, \\
-\,3x \quad &\quad -\,3x, \\
2x + 4 &= \quad 8.
\end{aligned}
$$

Another pair immediately argued that since the resulting equation resembled situation (1), then solving for x in the final stage would involve demonstrating the undoing process. We tried the same strategy over several more examples until

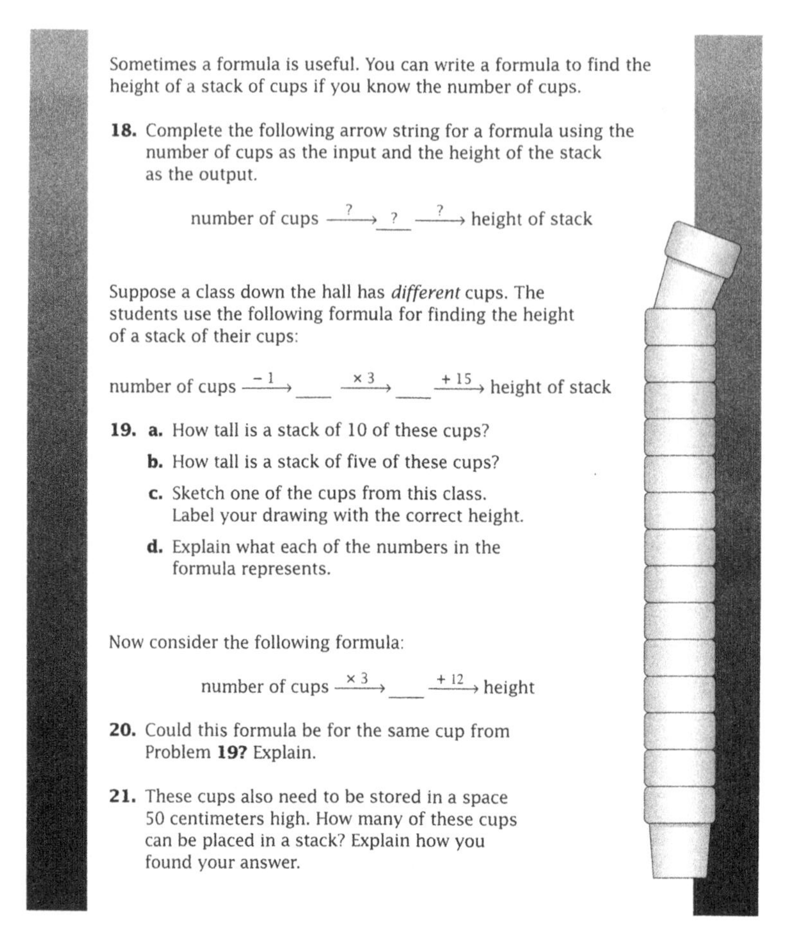

Sometimes a formula is useful. You can write a formula to find the height of a stack of cups if you know the number of cups.

18. Complete the following arrow string for a formula using the number of cups as the input and the height of the stack as the output.

number of cups $\xrightarrow{?}$? $\xrightarrow{?}$ height of stack

Suppose a class down the hall has *different* cups. The students use the following formula for finding the height of a stack of their cups:

number of cups $\xrightarrow{-1}$ ___ $\xrightarrow{\times 3}$ ___ $\xrightarrow{+15}$ height of stack

19. a. How tall is a stack of 10 of these cups?

b. How tall is a stack of five of these cups?

c. Sketch one of the cups from this class. Label your drawing with the correct height.

d. Explain what each of the numbers in the formula represents.

Now consider the following formula:

number of cups $\xrightarrow{\times 3}$ ___ $\xrightarrow{+12}$ height

20. Could this formula be for the same cup from Problem **19?** Explain.

21. These cups also need to be stored in a space 50 centimeters high. How many of these cups can be placed in a stack? Explain how you found your answer.

Fig. 4.3 Stacking cup task (MiC, 2003, p. 20)

it became evident that the algeblocks-induced process worked in all cases of situation (3).

The examples we used were also planned carefully. Following Harel (2007), I wanted them to transition to a nonreferential symbolic way of thinking about linear equation solving. In the sequence shown in Fig. 4.4, the coefficients changed in complexity and magnitude. The basic intent was to assist them to shift their attention from performing actions on and with the blocks to manipulating the coefficients numerically using the addition property of equality. This concern addresses the issue of "operating on unknowns" that Filloy and Rojano (1989) have pointed out as crucial when students solve linear equations in a single variable involving case (3). In phase 2 of Fig. 4.4, they needed to see, for example, how taking away -2 green cubes (or -3 cubes) and adding 2 green cubes (or 3 green cubes) in the top equation were equivalent actions. The examples in phase 3 of Fig. 4.4 focused on two cases,

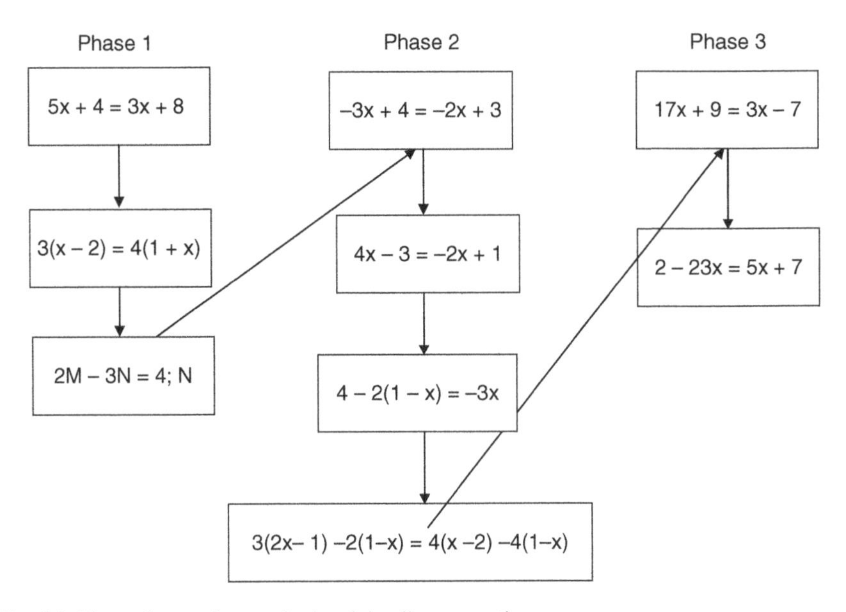

Fig. 4.4 Three phases of examples involving linear equations

namely: (1) linear equations with large coefficients that could not be easily represented and accomplished with algeblocks and (2) linear equations that necessitated the application of the distributive property rule.

Unlike the students in the study of Filloy and Rojano (1989) who initially approached situation (3) by trial-and-error methods, my students used a visual model (algeblocks) in making sense of an emerging formal strategy. The formal strategy, as a matter of fact, could be characterized by the co-emergence of visual and alphanumeric strategies that led to the intended nonreferential algebraic process for solving linear equations involving one unknown. There were two apparent visual strategies that my students employed in cognitive activity, that is, externally and physically manipulating the algeblocks and internally performing actions of taking away and undoing.

For Wittgenstein (1961), every symbol consists of a sign and a mode of signification. A sign refers to sounds, marks, or inscriptions that are perceived from a symbol. A *mode of signification* means that every symbol derives its content in relation to some *situation* that allows it to be *used* in a particular way. That is, "(i)n order to recognize a symbol by its sign we must observe how it is used with a sense" (Wittgenstein, 1961, 3.326), which means understanding how it is "*used* significantly in propositions" (Wittgenstein, 1973, p. 59) with propositions referring to "situations" (Wittgenstein, 1961, 4.03) or "pictures of reality" (Wittgenstein, 1961, 4.01). His famous example involves the two uses of the word "green" in the sentence "Green is green" in which the first word is a (proper) noun, while the last word is an adjective; "these words do not merely have different meanings: they are different symbols" (3.323). Another example is taken from my Algebra 1 class: "Zero is not the zero of the function $f(x) = x + 1$."

Among my Algebra 1 students, the equation $Ax + B = C$ was a mathematical proposition that depicted at least two different situations. The first situation was in the context of patterning activity, say, Dung's work on item 4 in Fig. 4.2, in which the variable p (sign) referred to stage number (symbol) and the numerals 3, 1, and 73 to various aspects (symbols) in the given patterning task. The second situation was in the context of solving for x (sign) that represented an unknown value to be determined (symbol) by an arithmetical process of undoing with A and B (signs) standing for coefficients (symbols) and C (sign) the constant term (symbol) in the equation. The above two situations demonstrate Wittgenstein's (1961) point about signs as possibly referring to different symbols (3.321).

Four points are worth noting at this stage. *First*, different symbols that have the same sign are prevalent in the school mathematics curriculum. For example, a fraction is associated with the sign (a/b), but it could mean different things depending on the sense or the context of use. It could be parts of a whole unit; parts of a discrete group of objects (in a set context); a measurement value (as a point on a number line); a ratio; and a division problem. In my Algebra 1 class, we interpreted nonunit fractions as copies of their respective unit fractions. For example, (1/4) was seen as a half copy of (1/2), (1/9) was a one-third copy of (1/3), and (3/4) was three copies of (1/4). This multiplicative view of fractions emphasizes the general concept of a unit that we talked about in some detail in Chapter 3. It is also the basis of an empty fraction strip approach that my students used in making sense of rational algebraic expressions and their operations. Another example involves letters or variables in school mathematics. Parker and Baldridge (2004) point out that their meaning could be interpreted in any of the following contexts: naming an object (e.g., A and B as points on a line in geometry); abbreviating words (e.g., t stands for time in hours in word problem solving); representing a frequently used number (e.g., π); representing specific but unknown numbers (e.g., solutions in conditional equations); and developing statements and identities that are true within their domain (e.g., direct formulas in patterns, algebraic identities, and variable-based theorems such as the Pythagorean theorem).

Second, we emphasize the inter- and intra-semiotic emergence and complementary nature of visual thinking and alphanumeric symbol representation in the development of mathematical symbols. While the title of this chapter focuses on visual roots of mathematical symbols, it does not in any way suggest preferring the visual and neglecting the alphanumeric. In fact, visually drawn symbols have as many limitations (and successes) as their corresponding alphanumeric symbol representations. For example, in my Grade 2 class, adding and subtracting whole numbers by primarily drawing sticks and dots (e.g., Figs. 1.1 and 3.11a, b) was not efficient in the long haul and confused some of them who could not draw clearly and count correctly. Especially in the case of subtraction of two two-digit whole numbers, in comparison with Fig. 3.11b, the visual strategy in Fig. 3.11a could not easily support the transition to the alphanumeric model since the students were too busy counting all the dots in the unit place and taking away. Various research studies have also consistently shown that overemphasizing alphanumeric acquisition translates to mere procedural knowledge among students.

In my Algebra 1 class, while the cognitive peg of their knowledge involving solutions of linear equations was visual, which they used to easily transition into alphanumeric form, I was aware that the visual process alone would not work for more complicated cases of linear equations. Also, the basic process of undoing in a simple linear equation, which is appealing in visual form, would not work in all kinds of equations. For example, some of my Algebra 1 students near the end of the school year tried to solve a quadratic equation by an undoing process. Figure 4.5 shows the work of Emma, who, in a clinical interview, tried to reverse the operations in her quadratic equation (see item 4) as her way of determining the values of the unknown.

The foregoing situation with Emma leads us to our next point. *Third*, Filloy and Rojano (1989) raise an interesting issue about *cut points* when students transition from one mode of thought to the next. In their study, in particular, they suggested a cut in the arithmetic-to-algebraic thinking among their students who solved linear

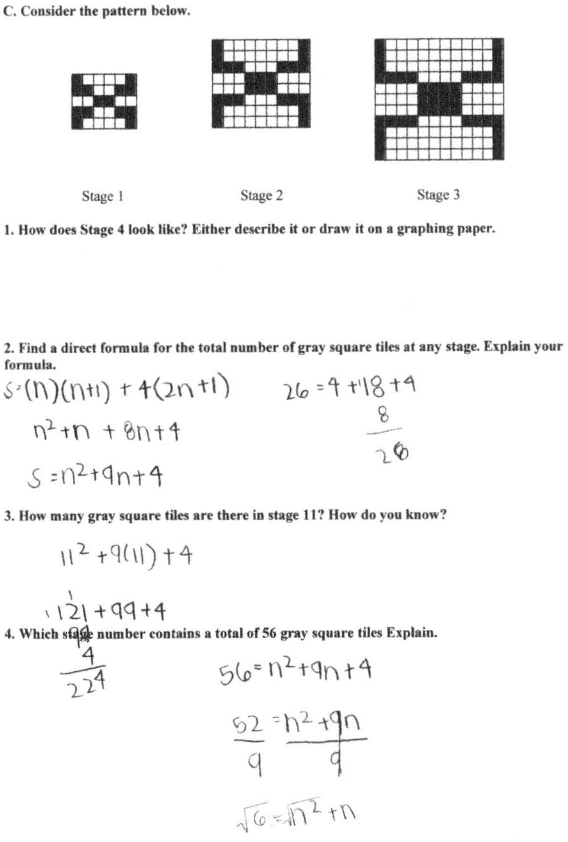

C. Consider the pattern below.

Stage 1 Stage 2 Stage 3

1. How does Stage 4 look like? Either describe it or draw it on a graphing paper.

2. Find a direct formula for the total number of gray square tiles at any stage. Explain your formula.

$S=(n)(n+1)+4(2n+1)$ $26=4+18+4$

$n^2+n+8n+4$

$S=n^2+9n+4$

3. How many gray square tiles are there in stage 11? How do you know?

$11^2+9(11)+4$

$121+99+4$

4. Which stage number contains a total of 56 gray square tiles Explain.

$56=n^2+9n+4$

$\frac{52}{9}=\frac{n^2+9n}{9}$

$\sqrt{6}=\sqrt{n^2}+n$

Fig. 4.5 Emma's reversal strategy in solving a quadratic equation

equations involving case (3). That is, while solving case (1) equations by undoing could be accomplished by applying basic arithmetical processes, solving case (3) equations would initially involve "operating on the unknowns" – the terms Ax and Cx – that fall "outside the domain of arithmetic" (p. 19).

Filloy and Rojano's notion of abscission points also applies in some cases of visual-to-alphanumeric symbol transitions such as the above situation with Emma or when students deal with epirelational objects. The cut in this case could be explained in terms of Berkeley's (2008) notion of *representational power*, that is, "any set of symbols [should] have the necessary flexibility to make manifest all the things that it needs to for the task at hand" (p. 97). Berkeley (2008) uses the representational power of Arabic over Roman numerals as an example. Unlike Roman numerals, Arabic numerals could deal with situations that involve 0 and place value in which positions matter.

In the case of solving quadratic equations, in particular, using the visual approach shown in Fig. 3.12b might have sufficient representational power in dealing with "friendly" coefficients; however, it could not be sustained in most other cases of $ax^2 + bx + c = 0$ in which case the quadratic formula yields the most representational power. Also, prior to solving quadratic equations, when my Algebra 1 students were learning to factor general quadratic trinomials, the visual activity with the algeblocks was helpful in dealing with simple trinomials (Fig. 4.6a) whose factored form is of the type $(ax + b)(cx + d)$, where (a,b,c,d) are positive integers since all the blocks could be reconfigured in the first quadrant. However, the visual activity became cumbersome to use and thus less effective when they had to deal with most other cases of trinomials such as the ones shown in Fig 4.6b, c since algeblocks had to be added and subtracted in different quadrants, which required a clever strategy in most cases. In these latter cases, the representational power that comes with manipulating algeblocks was diminished because the visual activity contained too much detail that the students had to acquire, which distracted them from obtaining a general process. Consequently, they refined the knowledge they acquired from the visual process involving simple trinomials (Fig. 4.6a) so that what eventually emerged was the rectangular tic-tac-toe method shown in Fig. 2.4a–c, a visuoalphanumeric method that still retained the basic elements of the visual approach (in terms of the overall structure of the diagram) but was sufficiently flexible at the alphanumeric level in dealing with most trinomials. Thus, it suffices to say that in many simple cases, the students used algeblocks to conceptualize the factoring process involved. However, in most cases, the visuoalphanumeric approach provided the most representational power.

Fourth, consistent with our proposed traveling theory of progressive schematization, intra-semiotic changes or transitions in symbolic notation and use (in Wittgenstein's sense of the symbol) are most likely to occur in various stages of learning and understanding mathematical concepts, objects, and processes. For example, Bialystok and Codd (1996) provide an interesting summary of empirical studies conducted with children that shows (1) transitions in their notation use (numerals) with more knowledge of a number system (counting) and (2) a developmental pattern in terms of the notations they use to represent quantity from

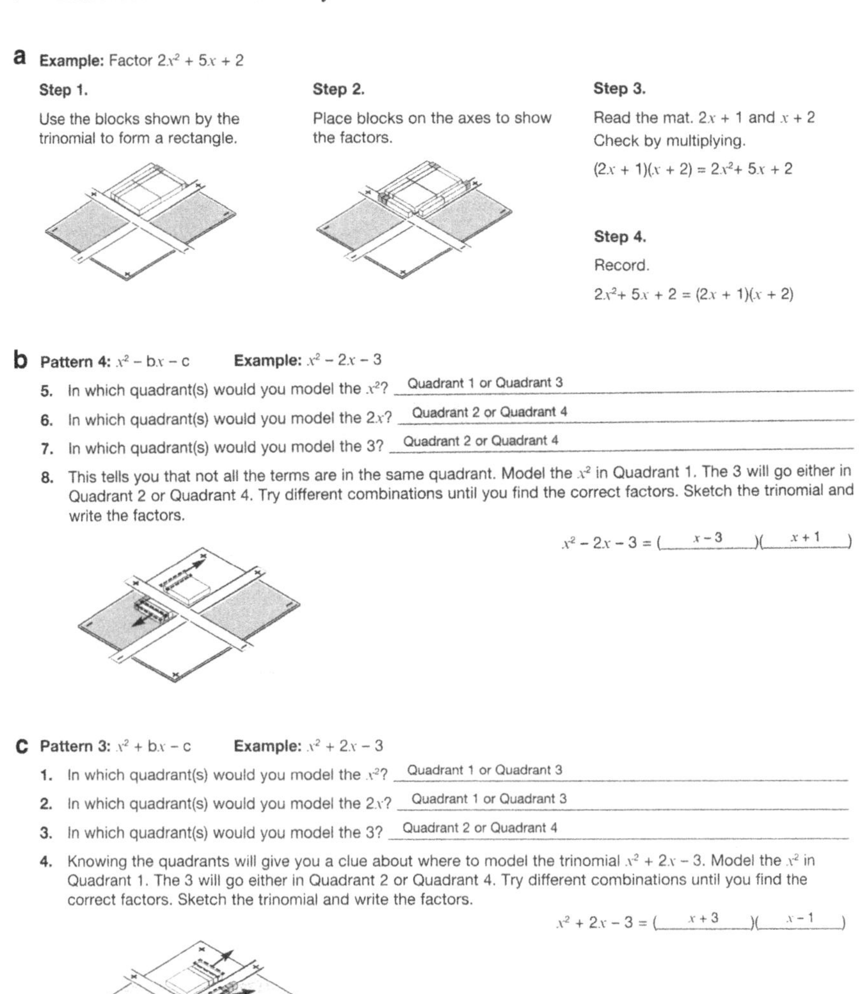

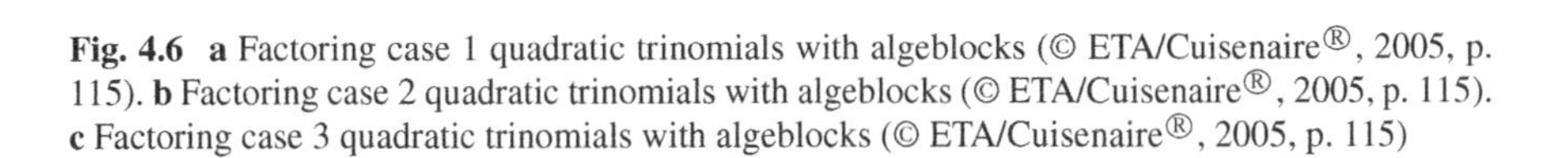

Fig. 4.6 **a** Factoring case 1 quadratic trinomials with algeblocks (© ETA/Cuisenaire®, 2005, p. 115). **b** Factoring case 2 quadratic trinomials with algeblocks (© ETA/Cuisenaire®, 2005, p. 115). **c** Factoring case 3 quadratic trinomials with algeblocks (© ETA/Cuisenaire®, 2005, p. 115)

idiosyncratic (personal) to analogical (dashes; marks) to the conventional symbolic form of numerals. But we emphasize repeatedly that a meaningful progressive account for learners has to consider changes, transitions, and connections between and among intra- and inter-semiotic resources.

This chapter is divided into three sections. In Section 1, we discuss the visual roots of mathematical symbols in relation to at least three modes of signification, namely Peirce's icons, indices, and symbols. In Section 2, we further explore visuoalphanumeric symbols in relation to objects in school algebra. In Section 3, we raise some issues concerning intra- and inter-semiotic transitions. In Section 4, we illustrate the formation of visuoalphanumeric symbol sense that my Cohort 1 students developed in the course of learning about pattern generalization in purposeful cognitive activity over 3 years of design-driven experiments.

1 Mathematical Symbols and Their Signs and Modes of Signification

Wittgenstein's (1961) notion of mode of signification in relation to symbols and signs provides the necessary link between a sign and its object. While Peirce was instrumental in transforming what was once considered a triangular view of signs – that is, signs in relation to a signifier and signified (Fig. 4.7a) – to what is now accepted as the pyramid of sign, signifier, signified, and *interpretant* (i.e., effect on the knower; Fig. 4.7b), Wittgenstein's mode of signification further refines our understanding of the semiotic pyramid.

There are different modes of signification. Peirce identifies three important modes: iconic; indexical; and symbolic. Symbols in an iconic mode are determined by their physical similarity or resemblance with a situation. They are what they are "quite regardless of anything else" (Peirce, 1958b, 5.44) and, hence, are not subject

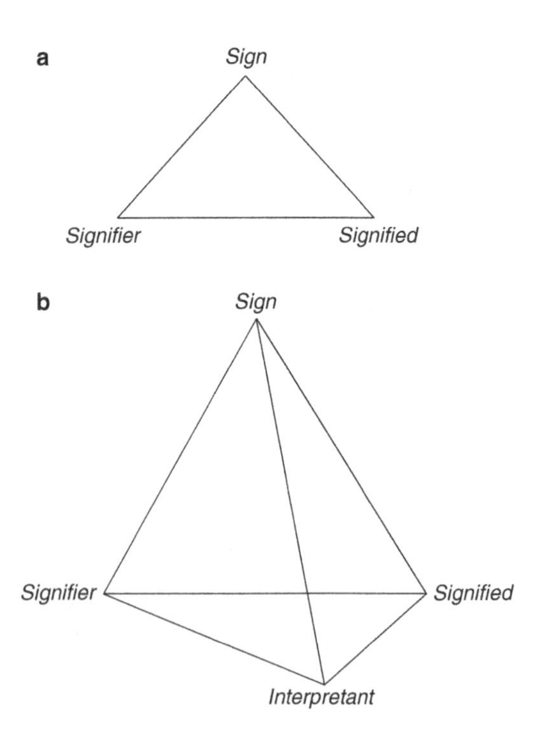

Fig. 4.7 **a** Classic signifier-signified relationship. **b** Peircean semiotic triangle

to reflection. Symbols in an indexical mode derive their use or meaning through association with a situation or one's experiences with them. They are products of our "experiences," where their "effects on us" exceed "our effects on them" (Peirce, 1958b, 5.45). Symbols in a symbolic mode are produced through conventions and rules that are shared in a situation. While they have no need to be similar (iconic) or have an analogical relation with their referents (indexical), they assert their nature and purpose by fueling their "sanction to action" (Peirce, 1958b, 8.256). How to choose which mode is appropriate for a given object, concept, or process in fact depends on what is negotiated among students in cognitive activity, which implies a collective sense or shared intentionality of "seeing as." This is a sociocultural response and one that shares and extends the perspective offered by Legg (2008) in an interesting comment below.

> As Wittgenstein said, all seeing is seeing-as. And of course every "seeing-as" presupposes the possibility of a "seeing-not-as." In this way, then, a seeing *is* a doing. As Wittgenstein also noted: "An aspect is subject to the will. If something appears blue to me, I cannot see it red, and it makes no sense to say 'See it red;' whereas it does make sense to say, 'See it as . . ." And that the aspect is voluntary (at least to a certain extent) seems essential to it.
> (Legg, 2008, p. 226)

Many concrete and relational objects are iconic. They resemble the objects they signify. Some icons have *structural resemblance* with their objects, while others have *pictorial resemblance* (Legg, 2008, p. 208). Fraction strips (Fig. 3.4), for example, are pictorial symbols of the part–whole notion of fractions. In structural resemblance, only relationships among the parts in an icon need to behave in the same manner as the parts in the actual object. "Every pictorial resemblance is," Legg points out, "a structural resemblance, so structural is a generalization of pictorial resemblance" (2008). In my Grade 2 class, squares, sticks, and dots (e.g., Figs. 1.1, 2.20, and 3.11a, b) conveyed to my students structural relationships involving place value relative to a decimal numeral system with one square representing a hundred or 10 sticks, one stick representing a 10 or 10 dots, and one dot representing a unit. In school discrete mathematics, a good example of the use of an icon that has only structural resemblance is the topological transformation of the famous Seven Bridges of Koenigsberg (Fig. 4.8a) in graph-theoretic terms. In Fig. 4.8b, vertices and edges represent bridges and connections between and among bridges, respectively. In school geometry, the monotonic structure in which theorems and definitions are oftentimes stated enables the construction and reification of the corresponding pictorial symbols. There are also *simple iconic symbols* such as using color to convey a shared quality of an object. Such icons satisfy Peirce's view in which "an icon is a sign fit to be used as such because it possesses the quality signified" (Peirce quoted in Legg, 2008, p. 208). Binary or two-colored chips that are used to model integer operations, for example, are simple icons with one colored side representing positive integers and the other colored side the negative integers.

The minus sign is an example of an indexical symbol. The minus sign as an operation is an index that conveys a complete turnaround on a number line. For example, in Fig. 4.9a, the expression $7 - 3$ means the turtle starts at 7 facing the positive axis, "turns around," and then moves 3 units forward facing the negative axis. The minus

a

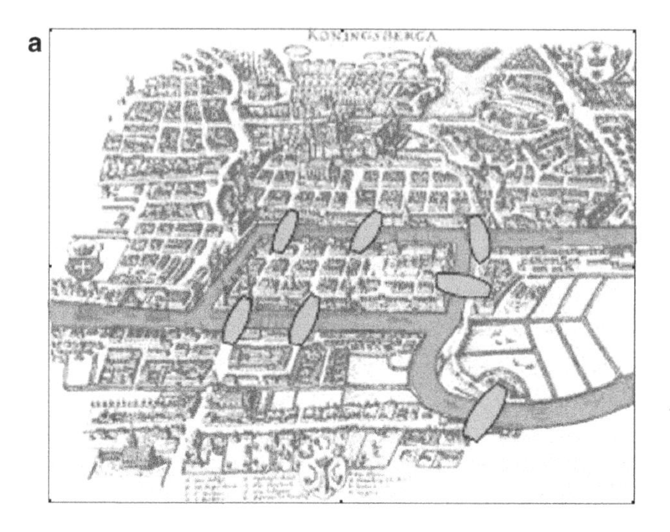

b

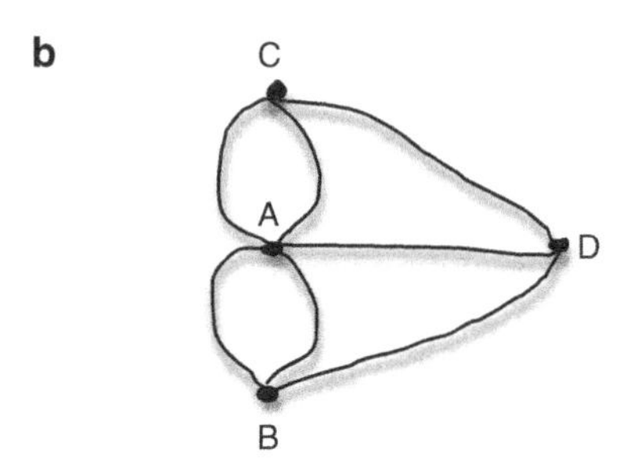

Fig. 4.8 **a** A map of the Seven Bridges of Koenigsberg (Permission granted from http://en.wikipedia.org/wiki/File:Konigsberg_bridges.png) **b** A topological graph of the map in Fig. 4.8a

sign that is associated with negative integers is an index that conveys moving backward on a number line. For example, in Fig. 4.9b, the expression $7 - -3$ means the turtle starts at 7 facing the positive axis, "turns around," and then "moves backward" 3 units still along the positive axis. In the context of binary chips, the minus sign as an operation conveys a "taking away" action. For example, in Fig. 4.10a, $5 - 2$ means "take away" 2 white circles from 5 white circles. The negative sign associated with negative integers conveys the use of the colored side of the binary chip. For example, in Fig. 4.10b, $-8 - -3$ means "take away" 3 black chips from 8 red chips.

With indexical symbols, Legg (2008) writes, the "brutely [i.e., unmediated] dyadic sign-object relationships" (p. 208) are known through "direct reference," say, through demonstrative and ostensive actions on the corresponding signs. Unlike icons that are monads, that is, the property (or properties) of their signs could exist with or without their objects, indices require both sign and object to exist in order for the relationship to make sense. Atkin (2005) also notes that indices stand for objects "through some existential or physical fact" (p. 163) that causally connects the former with the latter. Indices "assert nothing" and merely "show or indicate

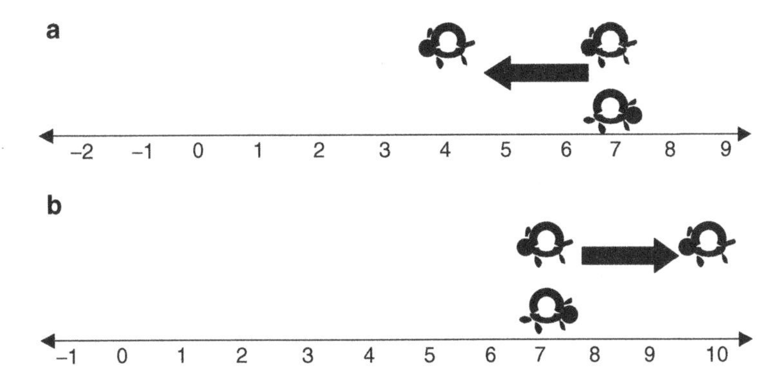

Fig. 4.9 **a** Number line model for $7 - 3 = 4$. **b** Number line model for $7 - -3 = 10$

their objects rather than describe them" (p. 165). They are not icons and symbols and, thus, have only a "secondness" quality (unlike icons, which are classified as having "firstness" or symbols, which have "thirdness"). In other words, while some indices may have some iconic or symbolic elements that could describe them more fully, they do not matter since it is the causal relationship that matters more than anything else. In my Algebra 1 class, for example, one index that the students found useful was the *n*th-root radical symbol $\sqrt[n]{\ }$, which conveyed to them the action of "dividing every exponent by *n*" on the basis of the horizontal bar that gave them a cue to divide the exponent(s) of the radicand by its index. To illustrate: $\sqrt{2^4}$ means initially dividing the exponent 4 by the index 2 so that $\sqrt{2^4} = 2^2$; $\sqrt[3]{2^3 \cdot 5^6} = 2 \cdot 5^2$; $\sqrt{3 \cdot 2^2} = 2\sqrt{3}$; etc.

Most mathematical objects (i.e., concrete, graphic and virtual, relational, epirelational, and alphanumeric) are symbolic and triadic. That is, any knowledge that is acquired between a sign and its object is mediated by rules and conventions that govern the relationship and should be learned correctly (Legg, 2008, p. 207). In Chapter 3, all mathematical objects are lifeworld dependent; hence, the possibility of their reification rests on the relevant intentional content and action that shape their images, meanings, and particular significance in some mathematical content. Further, we situate students' symbolic action on those objects within a much larger

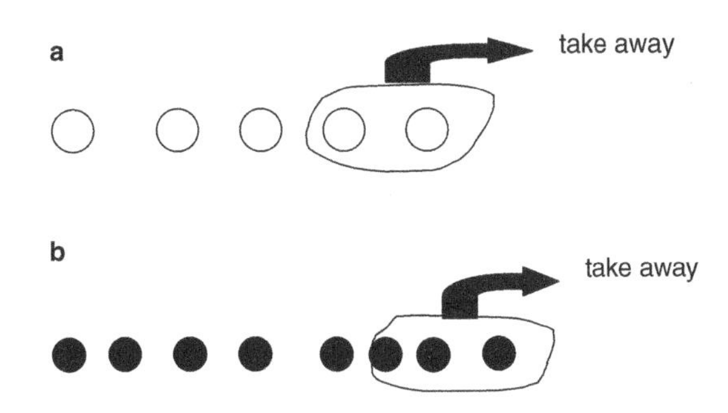

Fig. 4.10 **a** Binary chip model for $5 - 2 = 3$. **b** Binary chip model for $-8 - -3 = -5$

goal of cultural learning that the students internalize as they acquire knowledge about the objects in cognitive activity. In other words, the lifeworld-dependent nature of the objects loads them with intentional action and cultural scaffolding that purposefully lead to the acquisition of the relevant mathematical knowledge.

DeLoache (2005) points out, including Vygotskian scholars and practitioners, how our experiences are routed symbolically and how any account of human development is also about how individuals come to understand and use objects in a symbolic manner. The aspect of rules and conventions in characterizing mathematical objects as symbolic could either be individually constructed or be institutionally drawn. However, in mathematical cognitive activity, the most productive individual intentions are those that share, or are aligned, with institutional intentions (Radford, 2002). For Rakoczy, Tomasello, and Striano (2005), symbols are "objects with derived intentionality that is conferred on them through the attitudes, actions, and practices of persons that possess intrinsic or original intentionality" (p. 70). Hence, what is symbolic is seen as "an act that assumes a background of shared rules and practices as an interpretive framework" (2005).

Symbols, whether iconic, indexical, or symbolic, have a *dual representational* nature (DeLoache, 2005, p. 51), that is, the corresponding sign possesses concrete characteristics and bears an abstract relation to its object depending on the mode of signification. Issues with dual representation have been documented among young children who deal with iconic symbols such as large-scale photograph copies of their feet or pictures of objects. When they fail to distinguish between a symbol and its object (e.g., putting their feet on a photocopy or gesturing to indicate an attempt to pick up an object from a photograph), they tend to either focus exclusively on the object or the symbol or treat the symbol and the object to be one and the same thing. Beginning algebra learners go through a similar phase. For example, in my Algebra 1 class, the irrationality of $\sqrt{2}$ became an issue because some wanted to "see" it in its precise physical form or some others wanted to know where in the corresponding sign would they be able to get a glimpse of its expanded representation.

DeLoache points out the need for individuals to *psychologically distance* themselves in order to achieve dual representation and, thus, overcome *symbol realism*, "the tendency to treat entities which are patently symbolic . . . as if they possessed the properties and the causal efficacy of the objects which they denoted" (Werner and Kaplan quoted in DeLoache, 2005, p. 58). In the case of the irrationality of $\sqrt{2}$, the distancing occurred among my Algebra 1 students as soon as they began to reflect on Brown's (1999) point in relation to the activity involving the harmonic series (see pp. 32–34 of this book).

In some cases, students manifest *symbol attachment*, a tendency to rely on the entities themselves at the expense of acquiring their abstract relation to the objects they denote. For example, during the first quarter of the school year in my Grade 2 class, many students could not name the correct whole number after "xty-nine" on their own because they relied too much on the number line, a concrete object that they used to help them count quickly. When my Cohort 1 students were in sixth and seventh grade and participated in a teaching experiment on integers, some of

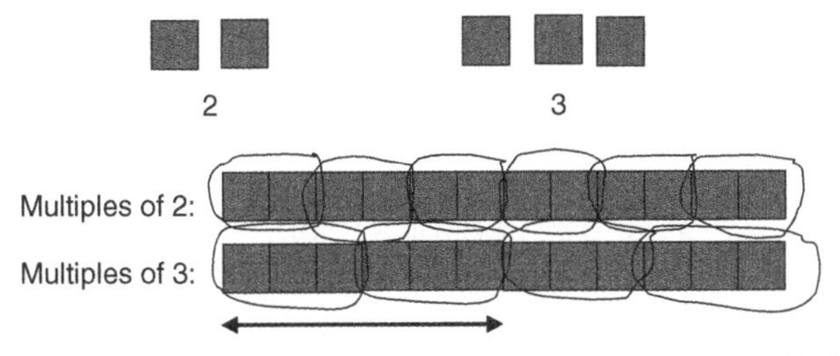

Fig. 4.11 Least common multiple (LCM) of 2 and 3 is 6

them consistently added and subtracted integers with either the binary chips (in sixth grade) or the number line (in seventh grade) like the child counting with the blocks in Hoffmann's (2007) essay, which prevented them from fully understanding, and appreciating the power of the symbolic rules. In my Algebra 1 class, when the students acquired a way of obtaining the least common multiple of two positive integers by building chains of unit cubes (Fig. 4.11) and the greatest common factors of two positive integers by algeblocks (Fig. 4.12), some of them had difficulty transitioning to the abstract phase because their images were pegged to either the cubes or the blocks. For example, some students were successful in obtaining the least common multiple of $(2^a, 2^b)$, where $a < b$ and a and b are whole numbers, because they used the unit cubes to model specific cases but failed on more complex cases such as $(2^a \cdot 3^b \cdot 5^c, 2^d \cdot 3^e \cdot 5^f)$, where *a, b, c, d, e,* and *f* are whole numbers and $a<d$, $b>d$, $e \leq f$. When asked to explain their difficulty, they expressed their disappointment at not being able to form chains of unit cubes because they found the numbers too large to model with unit cubes.

Psychological distancing in mathematical cognitive activity is crucial in situations when students employ, say, manipulatives in either iconic or indexical context to explore, generate hypotheses and pattern-imagery schemes, and develop

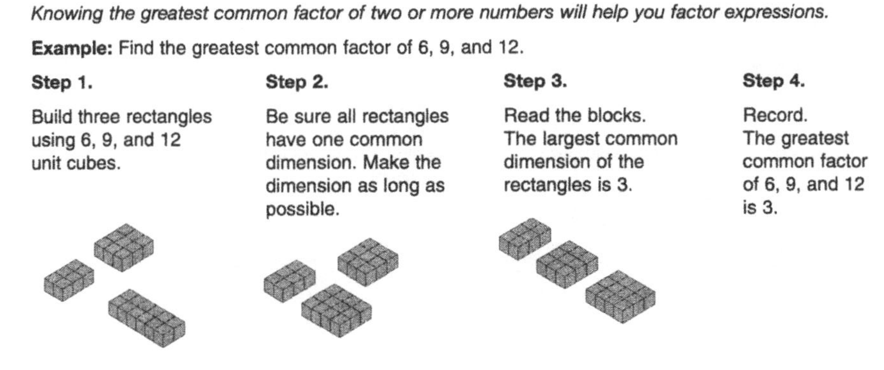

Fig. 4.12 Greatest common factor (GCF) of 6, 9, and 12 using algeblocks (© ETA/Cuisenaire®, 2005, p. 76)

structural awareness in order to progressively move toward more symbolic or formal concepts and processes. Uttal, Scudder, and DeLoache's synthesis of existing research on the use of manipulatives in mathematics classrooms paints a bleak picture, which is as follows:

> Unfortunately, however, research on the effectiveness of manipulatives has failed to demonstrate a clear, consistent advantage for manipulatives over more traditional methods of instruction. Meta-analyses of the literature have shown at best inconsistent and rather limited effects. In addition, several intensive, longitudinal studies of the use of manipulatives in individual classrooms have shown that children do not readily acquire new mathematical concepts from using manipulatives. Extensive instruction and practice may be required before manipulatives become effective. In some cases, manipulatives seem to do as much harm as good.
>
> (Uttal et al., 1997, p. 38)

On the basis of their work with various groups of young children, Uttal et al. (1997) infer that the "relation between manipulatives and their intended referents may not be transparent to children" (p. 44). Further, they note that the concreteness factor does not favorably translate into better learning for children; "it is the concreteness of the model that makes it difficult to use as a symbol" and "making symbols concrete and interesting as objects in themselves may decrease the chances" that they will "treat them as representations" (p. 45). They also note that multiple kinds of manipulatives for the same referent might be more distracting than helpful to children.

In light of the above constraints, Uttal et al. (1997) see teachers' instruction as instrumental in helping students move progressively toward the intended relation between manipulatives (signs) and referent (object). Ploetzner, Lippitsch, Galmbacher, Heuer, and Scherrer (2009) offer a similar recommendation in relation to their study that investigated 111 eleventh graders as they acquired their understanding of virtual objects, in particular, line graphs involving motion phenomena, in dynamic and interactive visual settings. They note that while line graphs provide good examples of iconic symbols, students still need to be taught psychologically distancing strategies such as "learn(ing) how to identify relevant components of the visual display, as well as how to relate spatially and temporally separated components to one another" so that they are able to "fully process interactive and dynamic visualizations" (Ploetzner et al., 2009, p. 64).

2 A Progressive Formal Account of Symbols: Focus on School Algebra

In this section, we explore in some detail visuoalphanumeric symbols in school mathematical practices. Such symbols are the "fruits" of progressive mathematization. For our purpose, we exemplify these symbols in the context of school algebra, which most teachers consider to be variable driven and visual free. Our initial discussion below takes us back to the history of algebra viewed from the perspective of Katz (2007).

Katz (2007) paints a very interesting history of algebra. He notes that algebra in its early phase of development was actually concerned with solutions of equations. Solving for unknowns in equations involved manipulations, using either symbols or words, that generalize arithmetic operations. Reflecting on the history of algebra from its beginnings with early Egyptians to the present time, Katz (2007) saw changes in our perceptions and practices concerning the nature of unknowns to include numbers, geometric quantities, and algebraic functions.

Katz proposes four stages in the evolution of algebraic practices relevant to solving equations. The first stage involves solving for unknowns using proportions. What was interesting about the *proportion stage* was the extent to which this pan-cultural method was also evident among the early Egyptians, the Chinese, and the Indians. The Egyptians used the "false position" method, while the Chinese and Indians employed "the rule of three." The algebra content in Europe at the beginning of the thirteenth century also emphasized the proportion method in solving equations. The early Babylonians began to concern themselves with other types of equations beyond what could be solved by the proportion method. Motivated by concerns about land division, their algebra was geometric in content as it has been documented in clay tablets. Thus, the *geometric stage*, which took place in early Babylonia and Greece, was about solving for unknowns in which the quantities were geometric in content. What distinguished the Greeks from the Babylonians was the introduction of axioms that enabled them to manipulate different geometric situations in algebraic terms.

When Islamic mathematicians appropriated algebra, Katz notes a shift occurred from the geometric to the *numerical equation solving stage*. The content of algebra books during this time focused on algorithms that were used to solve linear, quadratic, and systems of linear and/or quadratic equations whose solutions were numbers and not geometric quantities. Even when they established methods for simplifying and operating on polynomial expressions, including rules in dealing with positive and negative exponents, "the goal of these manipulations was to solve equations" (Katz, 2007, p. 8). Consequently, algebra in Europe in the twelfth and thirteenth centuries was also numeric.

There were changes, however, in the sixteenth century. While the focus remained on refining existing appropriated algorithms for solving equations, new ones emerged. Complex numbers came into existence and common mathematical words were assigned abbreviated symbols.

In the early seventeenth century, mathematicians became concerned with matters involving astronomy and physics. This meant that the types of numbers that mattered to them, in particular, to Galileo and Kepler, were curves such as those that involved the conic sections. This was, according to Katz, the *dynamic function stage* in which concerns in physics and motion drove the changes in the kinds of unknowns they wanted to solve. When Pierre Fermat employed algebra to represent curves and René Descartes used it to solve geometric problems, these methods furnished mathematicians, in particular, Newton, and scientists with techniques for representing motion.

Much of eighteenth-century mathematics focused on strengthening the link between algebraic and geometric methods (early part) and refining the tools (later part). But the primary focus was clear, that is, it was all about addressing dilemmas in physics and solving equations whose solutions were not merely numbers but curves and functions.

While Katz's historical account of the development of algebra in four stages is a powerful telling of ongoing shifts between numeric and geometric objects, it is also interesting to interpret such shifts within Freudenthal's (1981) interpretation of the history of mathematics as being about "a learning process of progressive schematizing" (p. 140), which made him suggest that our students.

> need not repeat the history of mankind but they should not be expected either to start at the very point where the preceding generation stopped. In a sense [they] should repeat history though not the one that actually took place but the one that would have taken place if our ancestors had known what we are fortunate enough to know.
>
> (Freudenthal, 1981, p. 140)

Visuoalphanumeric symbols are derived from the above perspectives of Katz and Freudenthal. Katz's historical account, on the one hand, provides a compelling story of how algebra progressed as a consequence of its numerical and geometrical foundations that mutually determined the evolution of its content and symbols. Freudenthal, on the other hand, saw progress in the history of mathematics and emphasized the need to move forward in our understanding of mathematical knowledge not by repeating "the one that took place" but by appropriating the one that would have taken place in light of what we are fortunate enough to know.

Visuoalphanumeric symbols incorporate both visual and alphanumeric components. While it seems easy either on the surface or at least initially to categorize such symbols as being either visual or alphanumeric, they are actually, like the three Peircean modes of signification, irreducible to one another with each component playing a significant role.

To illustrate, consider once again the situation in the introduction when my Algebra 1 students tried to make sense of what it meant to solve the equation $5x + 4 = 3x + 8$ or, more generally, $Ax \pm B = Cx \pm D$. Figure 4.13a shows how they visually solved the specific equation by utilizing algeblocks in the initial stage and the process of undoing in the final stage. Figure 4.13b shows how I recorded with the use of alphanumeric objects what I saw them do at every stage. Repeatedly doing this over several more examples that grew in complexity (Fig. 4.4) eventually led to a shared institutional mathematical practice – that is, a collateral knowledge in cognitive activity – that could not be simply characterized as an algebraic method involving alphanumeric symbols alone. Thus, while there was evidence of visual fading in the abstract phase, it does not actually fade but becomes embedded in the alphanumeric symbols. Mason (1987), like Wittgenstein's view of signs, considers symbols to be "artifacts resulting from recording perceptions on paper, mere vestiges of a complex inner experience" (p. 75). We extend the experience to include institutionally drawn experiences and visual symbols.

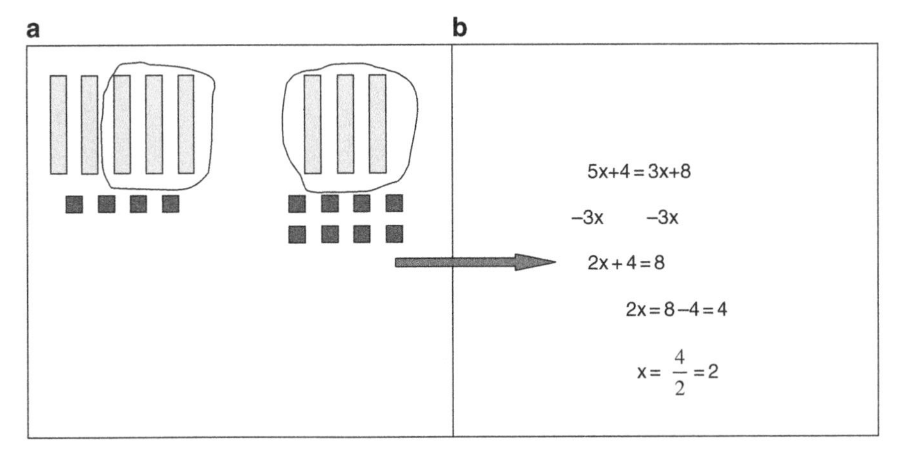

Fig. 4.13 **a** Visual process used in solving. **b** Alphanumeric process in solving $5x + 4 = 3x + 8$ 5x + 4 = 3x + 8

Another illustration that involves visuoalphanumeric symbol is shown in Fig. 4.14. When my Cohort 1 participated in an earlier teaching experiment on integers in sixth grade, we used binary chips to explore integer operations. In a written assessment, I asked them to obtain answers to a set of integer expressions and also to explain how they arrived at them. While on surface they appeared competent in obtaining the answers, their written explanation conveyed the use of visual thinking in relation to the given problems.

An interesting aside is worth discussing at this stage. Certainly, the argument in favor of visuoalphanumeric symbolic practices could simply be dismissed as something that teachers already know and do. Unfortunately, this is not the case.

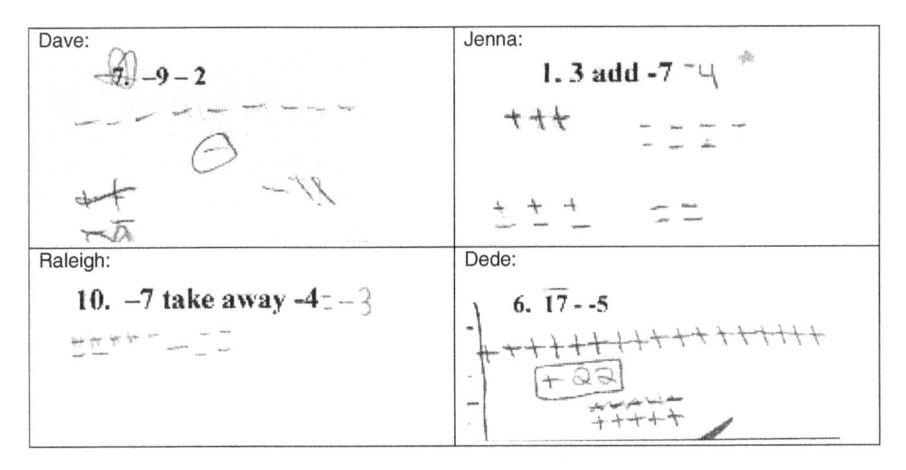

Fig. 4.14 Students' written work on integer problems

Fig. 4.15 Linear equation tasks in Vlassis's (2008) study

$4-x=5$

$-6x=24$

$-x=7$

$12-x=7$

$-32=-8y$

$5x=40$

To illustrate, Vlassis's (2008) account of the role of mathematical symbols in the development of number conceptualization began with a historical and psychological analysis of a particular case, that of the minus sign. While she discussed the importance of both concrete and sociocultural contexts in her analysis, none of the characterizations (or functions) of the minus sign (which she refers to as "negativity") that comprised her theoretical framework actually addressed either concrete or sociocultural meanings associated with the negative sign. Hence, when she asked her sample of 17 eighth graders (eight of low ability; five of medium ability; four of high ability) to solve for the unknown in each equation in Fig. 4.15, their responses reflected an overvaluing of alphanumeric control in solving for the unknowns. It seems to be the case that they never used visual symbols when they first learned about negative integers and formal algebra in seventh grade. Also, the interview protocol did not mention the use of relevant concrete objects. So, for example, in the case of the equation $4-x=5$, the main sources of difficulty that Vlassis identified were all numerically driven. In the interview below with a low-ability student whose answer was $x=0$, he or she resorted to trial and error.

Interviewer: Why have you put $x=0$*?*
Student: I just couldn't find the answer. I couldn't find four minus something to make five; you can't find it with five.
Interviewer: Ah I see, you found four minus something equals five strange?
Student: Yes.

In my Algebra 1 class, the above equation would have been a simple case of situation (1), which would involve undoing, that is, taking away 4 from 5 and then dividing the result by -1.

The reported case involving the equation $-x=7$ was even more troubling. While some students gave the correct answer of -7, they did not understand how it came to be that way and did not have other means of explaining why it had to be so. Consider the interview below with a medium-ability student:

Interviewer: You need to write out the equation again, replacing x with its value of -7*. So write out the whole equation, and when you get to x, put* -7 *instead.*
Student: I'll end up with $-7=7$*. My solution isn't right!*

Interviewer: But have you replaced just x itself with −7? What does your equation start with?
Student: x.
Interviewer: What's the first thing that you see in your equation?
Student: . . . −x.
Interviewer: No. You shouldn't put −7 instead of –x, but instead of x alone. What is there in front of x?
Student: Minus.
Interviewer: Write it down, and then put the value of x.
Student: 7.
Interviewer: Be careful: your value for x is −7!
Student: Yes, but the minus sign is already written.
Interviewer: Is it the same minus as the minus sign in front of 7? In the equation, the first thing you see is a minus sign. Write it down. Afterwards, you have x, and the value of x which you have found is −7. Write down this value.
Student: Then I'll have 7, because a minus times a minus makes a plus.
Interviewer: Ok, but I want to see the two minus signs.
Student: But then I'll have − −7!
Interviewer: And what do you need to do in algebra when you have two signs in a row?
Student: Use brackets.

What was actually valued in the discussion above, including the interpretive analysis that followed, was the significance of understanding the role of brackets in making sense of the correct answer. But the situation would have been easily resolved if the activity was accomplished with the action of undoing and/or with the aid of algeblocks: since $-x = 7$ meant $-1 \cdot x = 7$, divide 7 by -1, so $x = -7$. Also, $- -7$ meant pulling 7 green cubes in the negative region to the opposite region, hence $x = 7$ (Fig. 4.16).

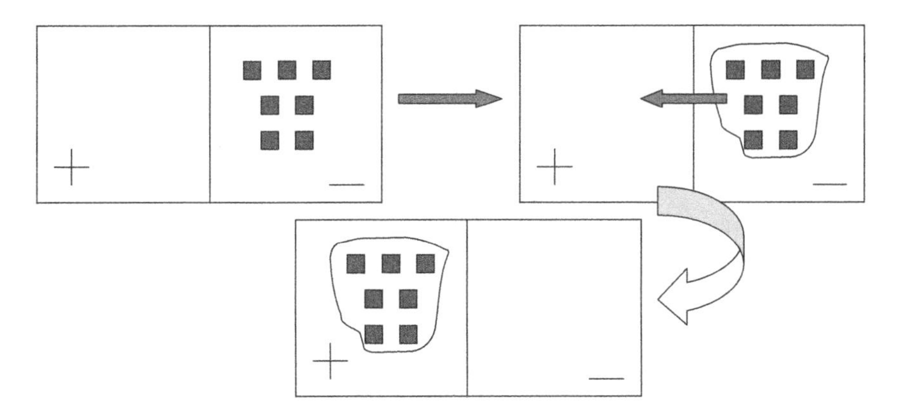

Fig. 4.16 Demonstrating $--7 = 7$ using algeblocks

3 Some Issues in Inter- and Intra-semiotic Transitions

All visuoalphanumeric symbols that are used in a formational context exist primarily to fulfill the function of reification. In fact, all definitions and theorems used in school mathematics are better understood when visuoalphanumeric objects are used to demonstrate them. However, the situation is different and sometimes more difficult in many transformational contexts. Using visuoalphanumeric symbols in problem-solving situations might serve a heuristic function, but there is always a foreshadowing of doubt that comes with the "fallibility of intuitions" that guides the construction of such symbols (Brown, 2005, p. 65).

In cases when the relevant domains are finite, as in many problem situations in K–12 discrete mathematics and data analysis and probability, or in cases when the theorems or propositions have a simple structure such as those found in school geometry, it might perhaps suffice to construct an isomorphic or a homomorphic visuoalphanumeric symbols (Barwise & Etchemendy, 1991) that could then be manipulated leading to a picture-based explanation.

However, in cases when propositions involve domains that are infinite such as the ones shown in Figs. 4.17 and 4.18, the visuoalphanumeric symbols are not really proofs in the rigorous sense (Cellucci, 2008; Jaffe & Quinn, 1993). The pebble arithmetic demonstration in Fig. 3.16 is an exclusively visual representation of a particular case and does not provide a rigorous proof involving the general case. Visuoalphanumeric or visual symbols in such contexts play an important role in developing structural awareness (Mason, Stephens, & Watson, 2009) and relevant pattern-imagery schemes (Presmeg, 1992) or in conveying an observation about an intuition. They alone, however, are not able to express the full range of their corresponding generalities. Hence, in such cases, visuoalphanumeric symbols are not "representation(s) in any strict sense, but rather something like a telescope that helps us to 'see' . . . a device for facilitating a mathematical intuition" (Brown, 2005, p. 66) and, thus, offer us with an "explaining in the sense of making a fact more intuitively compelling to a person than it was before" (Giaquinto, 2005, p. 79).

The nature of the mode(s) of signification of visuoalphanumeric symbols depends on their location on the visual–abstract spectrum in Fig. 2.8. Certainly, students make individual decisions in terms of how such symbols appeal to them,

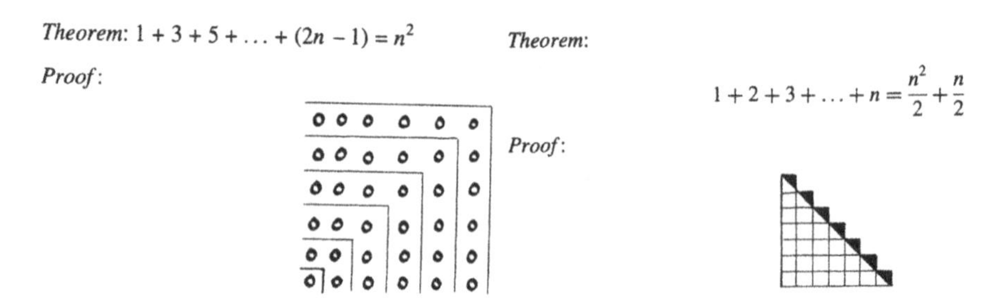

Fig. 4.17 Simple cases of picture proofs (Brown, 1997)

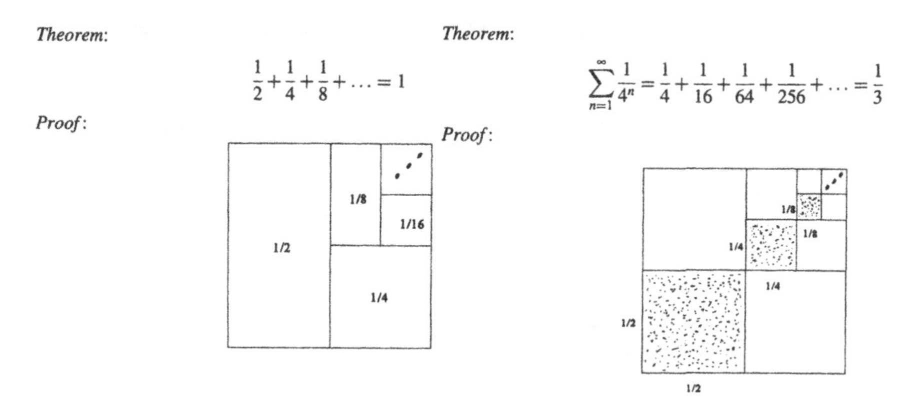

Fig. 4.18 Complex cases of picture proofs (Brown, 1997)

which also means to say that the modes themselves are not pre-determined in one way or another. Presmeg (2008), for example, captures this condition in relation to the quadratic formula as follows:

> The roots of the equation $ax^2 + bx + c = 0$ are given by the well known formula $x = (-b \pm \sqrt{b2 - 4ac})/2a$. Because symbols are used, the interpreted relationship of this inscription with its mathematical object may be characterized as symbolic, involving convention. However, depending on the way the inscription is interpreted, the sign could also be characterized as iconic or indexical. The formula involves spatial shape. In my original research study of visualization in high school mathematics, many of the students interviewed reported spontaneously that they remembered this formula by an image of its shape, an iconic property. However, the formula is also commonly interpreted as a pointer (cf. a direction sign on a road): it is a directive to perform the action of substituting values for the variables a, b, and c in order to solve the equation. In this sense the formula is indexical. Thus whether the inscription of the formula is classified as iconic, indexical or symbolic depends on the interpretation of the sign.
>
> (Presmeg, 2008, p. 4)

However, based on my own classroom research, it is more productive to think in terms of transitions in modes of signification, where visuoalphanumeric symbols evolve from being solely visual images to their "full expressions" (Peirce, 1957, p. 239) in abstract thought. *Pace* Peirce:

> Thought, however, is in itself essentially of the nature of a sign. But a sign is not a sign unless it translates itself into another sign in which it is more fully developed. Thought requires achievement for its own development, and without this development it is nothing. Thought must live and grow in incessant new and higher translations, or it proves itself not to be a genuine thought.
>
> (Peirce, 1957, p. 239)

For example, consider once again the quadratic formula. Figure 3.12b represents a particular case of a simple quadratic equation that could be solved visually. The steps in the alphanumeric formula structurally resemble the visual process, thus iconicity is evident. However, in more complicated cases of the equation $ax^2 +$

$bx + c = 0$ (e.g., in cases when $a \neq 1$), the visual process becomes tedious and unreasonable to implement. Hence, the process shifts in mode from being iconic to symbolic with the application of an alphanumeric formula. Here, the symbolic mode of the formula does not convey the absence of a visual model. What takes place, instead, is the incorporation of a particular visual experience in what is considered to be a more general situation that serves a symbolic function.

Some visuoalphanumeric symbols could start at the indexical phase before transitioning to symbolic. For example, it is not apparent among beginning learners that the function $y = mx + b$ represents a graph corresponding to a set of points that lie on a straight line, hence the relationship or the association may have to be pointed out. While the y-intercept (b) is a part of the graphical representation, the situation is not the same in the case of the slope (m) since it is a description of the inclination of a line in terms of ratio. There is, thus, a dyadic relationship between sign (i.e., graph of a line) and object (linear function and its components). But, in its symbolic mode, the students have to see the form $y = mx + b$ as an abstract representation of any nonvertical line.

The indexical phase in my Algebra 1 class took place in the context of the two activities shown in Fig. 4.19a, b that focused on slopes and y-intercepts of a line. In transitioning from the indexical to the symbolic phase, they used the transformation application activity in Fig. 4.20 to explore various situations of the slope and y-intercept of a line. In the symbolic phase, each student developed his or her own “lin-art” (Fig. 4.22) on a task (Fig. 4.21) that required him or her to bound his or her region by line segments and then to identify the linear system of equations that describe the boundary. (In a later task, the students revisited their lin-art and established a linear system of inequalities.) In accomplishing the task, the students manipulated their linear equations on the basis of an acquired symbolic understanding that went beyond their initial index-oriented view of linear functions. Figure 4.23 (embedded in a student’s written work) was a closure activity that had the students analyzing a linear model problem.

Some visuoalphanumeric symbols could also start at the iconic level before transitioning to indexical and finally to symbolic. For example, binary chips (Fig. 4.10a, b), paired words such as win-loss used in a metaphorical sense, and the green alge-block cubes all serve as iconic models for positive and negative integers. They provide “simple” or “monadic experiences” (Peirce, 1958b, p. 315) in cases involving simple integer operations such as $-7+5$, $+15+7$, $3+-4$, $8+-2$, $-7--2$, 5×3, $2-4$, $15 \div 3$, . and $-8 \div 4$ The number line and the binary chip models in more complicated cases of integer operations (e.g., $-2--7$, -3×5, -4×-7, $-4 \div -2$, and $4 \div -2$) seem to operate more at an indexical mode of signifying than iconic (Fig. 4.9a, b).

Hence, when students perform operations on integers, they could retain the iconic and indexical experiences or transition to symbolic understanding. To illustrate, when my Cohort 1 students were in sixth grade, they used the binary chips in making sense of $-12 - -8$, which meant taking away 8 black chips from 12 black chips resulting in 4 black chips (i.e., -4). When they started to generalize, the iconic character of the results was preserved in relatively reasonable cases that involved taking away a smaller number of negative chips from a larger number of negative chips. But they also had to deal with the issue that the iconic interpretation (i.e., -12 is "more than" -8) with the binary chips (or with the algeblocks for that matter) was not consistent with the mathematical interpretation (i.e., $-12 < -8$). Also, an additional conceptual dilemma was how to deal with the reverse case of subtraction, say, $-8 - -12$. In these situations, the students' iconic experiences had to transition into indexical and then much later to symbolic experiences.

Figure 4.24 shows how we addressed the particular situation –8 – –12 in class with the binary chips. First, the students gathered 8 black chips. Second, since they saw that it was not possible to take away 12 black chips from 8 black chips, they gathered 12 white chips and 12 black chips in two piles. Third, they took away the 12 black chips. Fourth, since there were 8 zero pairs that were all equal to 0, 4 white chips were left (i.e., –8 – –12 = 4). In the last stage, the students visually saw that –8 – –12 and –8 + 12 were equivalent expressions. Hence, when they dealt with

a

Consider the graph of the straight line below.

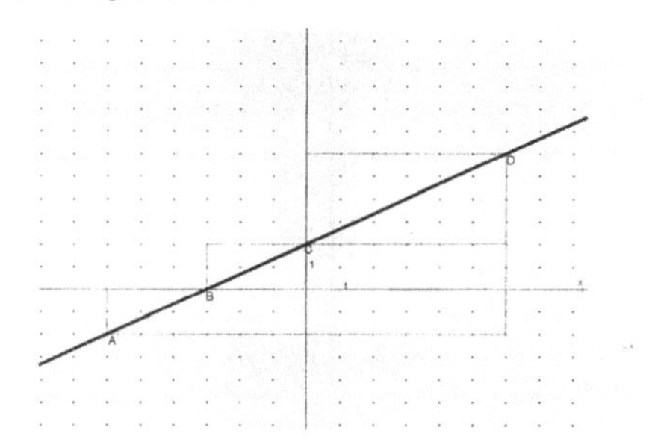

Fill in the missing values.

	Vertical direction or change (***Rise***)	Horizontal direction or change (***Run***)	m = slope = Rise / Run
From point A to point B	2	3	2/3
From point B to point C	2	3	2/3
From point C to point D	4	6	4/6 = 2/3
From point A to point D			
From point B to point D			
From point D to point C	–4	–6	–4/–6 = 2/3
From point C to point A			
From point D to point A			
From point B to point A			
From point D to point B			

A. What can you conclude about the slope of the given line for any pair of points on the line?

B. Imagine a horizontal line. What is its slope? Explain.

C. Imagine a vertical line. What is its slope? Explain.

Fig. 4.19 (continued)

b

Remember:

If you know the y-intercept and the slope of a line, you can easily find the equation of that line.

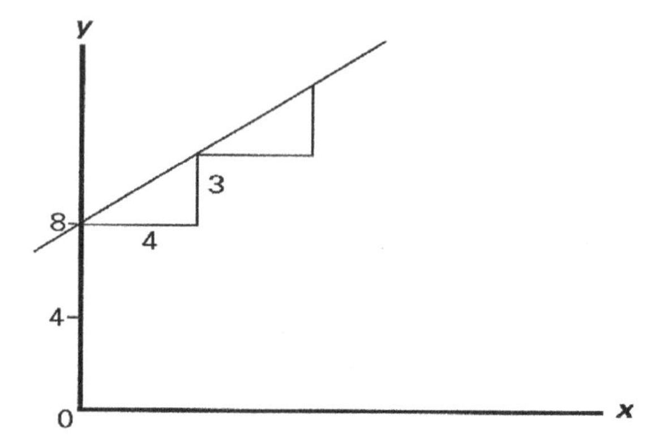

Example: y-intercept = 8, slope = $\frac{3}{4}$, equation: $y = 8 + \frac{3}{4}x$

- Complete the table.

slope	y-intercept	x-intercept	equation
	24	-8	
$\frac{3}{4}$	9		
$\frac{3}{4}$		−8	
	10	10	
			$y = 10 + 2x$
			$y = 2(x + 5)$
10		1	
			$y = \frac{5}{8}x$

Fig. 4.19 **a** Indexical phase involving the linear function. **b** Follow-up activity involving linear functions (MiC, 2006)

the general case of subtracting integers and, later, real numbers, it was the symbolic experience that enabled them to focus on the rules and conventions for combining any pairs of numbers that were extremely large (or extremely small), including real number. But the symbolic phase never meant the absence of a visual representation among the students. Instead, the visual was intertwined with the symbolic.

1. Compare the two graphs:

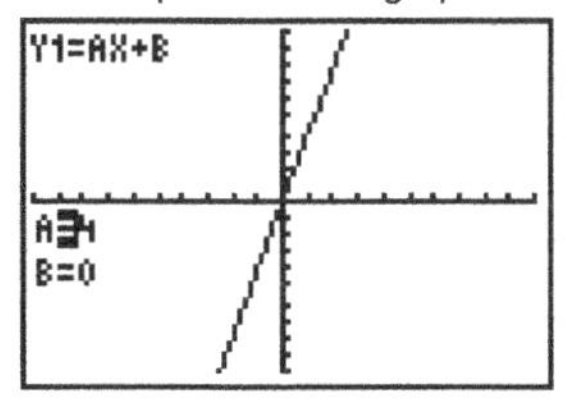

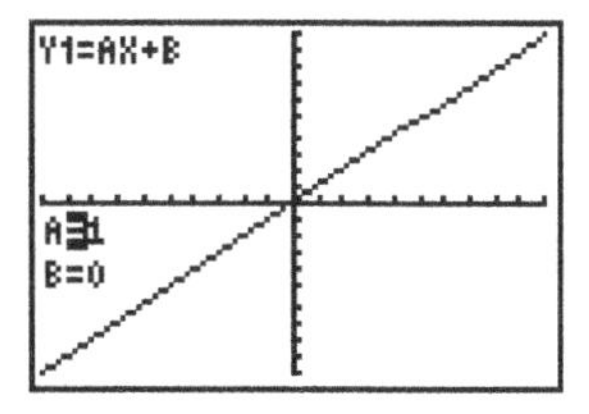

Which of these two graphs has a larger slope?

Explain how you determined your answer.

2. Compare the two graphs:

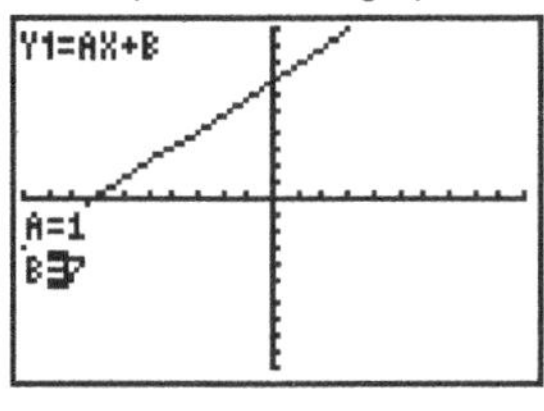

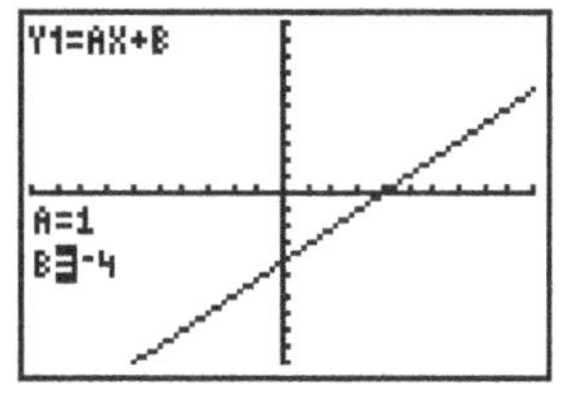

How are these two graphs similar?

What is the difference between these two graphs?

What caused the change to take place?

3. Examine the graph:

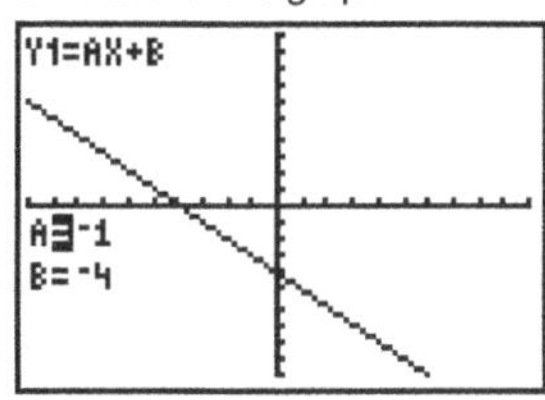

Describe two things you can determine about the graph just by looking at it and not doing any calculations.

Fig. 4.20 Transformation application activity involving linear functions (O'Brien, n.d.)

When Cohort 1 was in seventh grade, the students used the number line model in exploring once again integer operations; however, they started at the indexical phase. To illustrate, $-8 - -12$ meant starting at -8 on the number line and facing the positive axis, turning around, and then moving backward 12 units. (See also Fig. 4.9a, b.) The visual experience, repeated over several different pairs of numbers, allowed them to make sense of the sign ("direction") of the result in various situations of a and b when combined using either addition or subtraction and also to focus on finding the result ("magnitude") in a simpler way. So, for example, when the students dealt with the subtraction problem involving $-31 - -57$, they initially imagined the situation on a number line, predicted the sign to be positive ("start on negative axis, turn around, move backward several times, and you end up somewhere along the positive axis"), and then reinterpreted the difference

CLASS WORK

MATHEMATICAL OP-ART AND COLOR FIELD DESIGNS

Instructions:

1. Draw x-y axes on your centimeter grid paper.

2. Design an Op-Art or Color Field whose boundaries are diagonal lines. No horizontal lines and vertical lines allowed. Use a maximum of four colors to shade regions. See models on the board to help you develop your own design. BE CREATIVE. A minimum of ten lines is required in your design. Number each line.

3. For each numbered line, write the correct equation of the line. Show complete work below to receive full credit.

Fig. 4.21 Lin-art activity

as the distance between the absolute values of the two numbers but expressed in the context of the constructed equivalent expression $-31 + 57$ or $57 - 31$. Again, repeated action on similar situations allowed them to transition to a symbolic mode of signifying operations involving integers and, more generally, real numbers.

In Algebra 1, both Cohorts 1 and 2 used algeblocks in developing a shared sense – that is, the collective "seeing as" – of integer operations. Revisiting the mathematical processes involved actually enabled them to quickly understand the basic arithmetical operations that were used in dealing with polynomial expressions. Initially, at the indexical phase, integers were associated with small green unit cubes. A basic mat consisting of two regions, one positive and one negative, allowed the students to develop the visual image that negative (and positive) integers were indeed opposites of positive (and negative) integers by the concrete action of "pulling" the cubes on one region to the opposite region. Figure 4.25a, b shows how we visually tried to make sense of the rules for $-\mathrm{a} \times \mathrm{b}$ and $-\mathrm{a} \times -\mathrm{b}$, where a and b are positive integers, from an indexical standpoint. In Fig. 4.25a, -3×2 was associated with the action of "pulling 3 groups of 2 on the positive region to the negative region," which explains the negative result. In Fig. 4.25b, -3×-2 was interpreted as "pulling 3 groups of -2 on the negative region to the positive region," which explains the positive result. But the consequent dilemma involved explaining why $-a \times b = -(a \times b)$ and $-a \times -b = -(a \times -b)$, which the Cohort 1 students also encountered in seventh grade in the case of the number line model. Resolving this dilemma necessitated making a transition to the symbolic mode. The steps in Fig. 4.26c illustrate how we accomplished this in a particular case, which eventually routed the students' symbolic experiences toward a deductive proof. In sixth grade, Cohort 1 established the two rules symbolically in a much simple manner that was appropriate at their grade level. With the binary chips as their iconic model, no indexical phase occurred. At the symbolic level, we used the commutative property for multiplication in establishing the arithmetical fact that $a \times -b = -b \times a = -ab = -ba$. We then used the numerical patterning activity in Fig. 4.26d in demonstrating the relationship $-b \times -a = ba$.

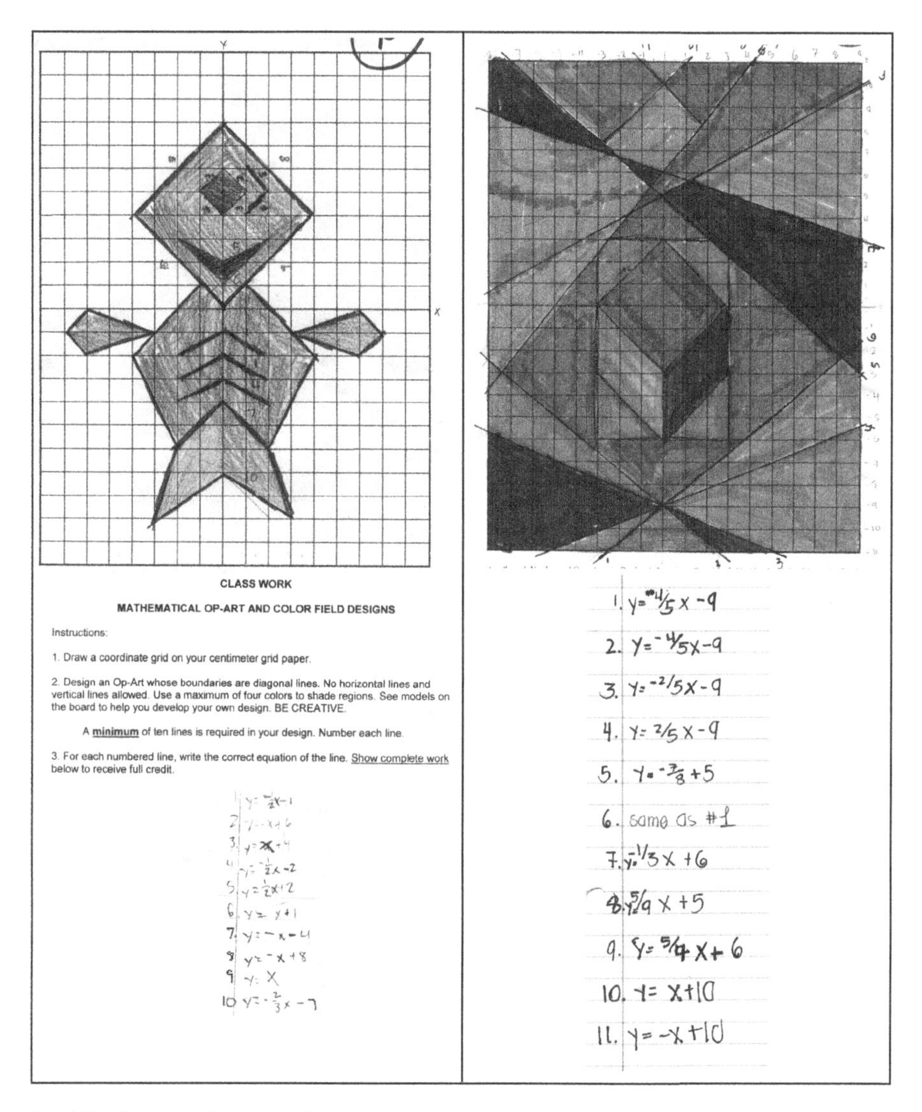

Fig. 4.22 Samples of students' lin-art

The foregoing discussion demonstrates the intricate relationship between iconic and indexical on one side and symbolic on the other. Students who operate mostly at the iconic and indexical modes without transitioning to the symbolic mode produce visuoalphanumeric knowledge that has the property of being constrained generalizations with very limited representational power. Students who operate mostly at the symbolic mode – that is, are able to use the rules and conventions – but are not provided with either transition opportunities or meaningful experiences at the

Fund Raising

The Middle Grades Math Club needs to raise money for its spring trip to Washington, D.C., and for its Valentine's Day dance. The club members decided to have a car wash to raise money for these projects. The club treasurer provided the following information:

- Charge per each car washed: The usual charge for washing one car is $4.00.
- Materials expense: The total cost of sponges, rags, soap, buckets, and other materials needed will be $117.50.

As a member of the club, your job is to produce a mathematical analysis. Do as follows.

1. Complete the table below. When done, enter the completed table in your graphing calculator and then graph.

Number of cars washed (c)	0	1	2	3	4	5	6	7
Amount of money raised (M)	-117.5	-113.5	-109.5	105.5	-101.5	-97.5	-93.5	-89.5

+4 +4 +4 +4 +4 +4 +4

2. Why is the amount of money raised with no cars washed a negative value? What happens to the club's finances? It is a negative value because they spent money on the materials. They will have to wash 30 cars to earn back the money they spent on materials before they earn money.

3. Explain how the value of -$113.50 was obtained for 1 car.
you added four dollars to -117.5 dollars. one car wash cost 4dollars.

4. What kind of pattern does your table above represent? How do you know for sure?
a positive pattern because the graph goes up. Every time they wash a car, they earn four dollars.
increasing pattern

5. Find a direct formula that enables you to obtain the amount of money raised (T) for any number of cars washed (c).
$T = -117.5 + 4c$

6. What stays the same and what changes in the table above? How does knowing what stays the same and what changes relate to your direct formula? For every car they wash, they earn $4. but the #'s of car they wash grows so the money they earn increases.

Fig. 4.23 Diana's work on a linear function problem

iconic or the indexical mode are likely to demonstrate the use of alphanumeric symbols with very little to no understanding of the relevant relationships involved in the development of such symbols.

It is perhaps hard to believe that such obtuse phenomena are possible but decades of classroom research demonstrate this to be the case (e.g., Nickson, 1992). Also, the two students above who were interviewed by Vlassis (2008) in relation to solving linear equations exemplify learners at the symbolic mode who do not have any visual means to make sense of equations; so they end up employing

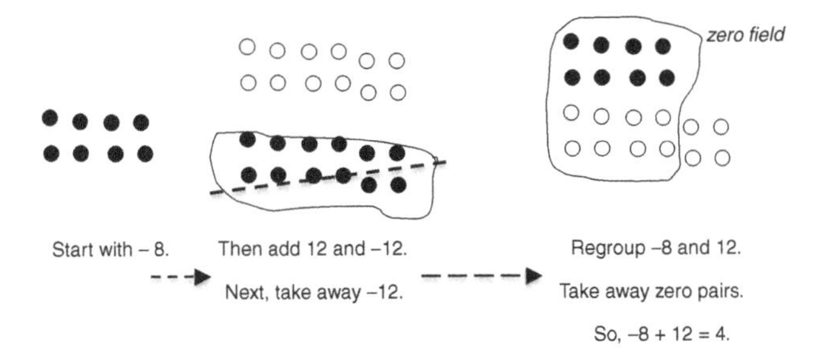

Fig. 4.24 Binary chip model for $-8 - -12 = 4$

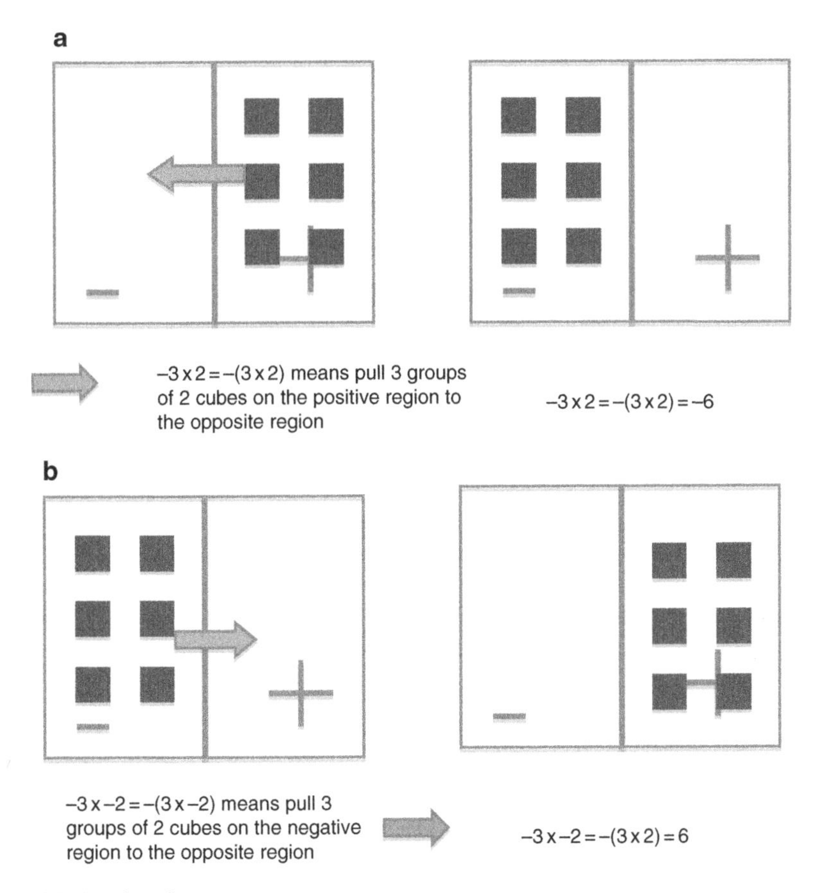

Fig. 4.25 (continued)

c

Let us illustrate why $-3 \times 2 = -(3 \times 2)$ in Figure 3.25a using properties of integers.

$-3 \times 2 = (-3 \times 2) + 0$	(Additive Identity Property)
$= (-3 \times 2) + [(3 \times 2) + -(3 \times 2)]$	(Additive Inverse Property)
$= [(-3 \times 2) + (3 \times 2)] + -(3 \times 2)$	(Associative Property)
$= [(-3 + 3) \times 2] + -(3 \times 2)$	(Distributive Property)
$= 0 + -(3 \times 2)$	(Additive Inverse Property)
$= -(3 \times 2)$	(Additive Identity Property)

d

Since $4 \times -5 = -5 \times 4 = -20$ (by the commutative property), then

$$-5 \times 3 = -15$$

$$-5 \times 2 = -10$$

$$-5 \times 1 = -5$$

$$-5 \times 0 = 0$$

$$-5 \times -1 = 5 = 5 \times 1$$

$$-5 \times -2 = 10 = 5 \times 2$$

$$-5 \times -3 = 15 = 5 \times 3, \text{ etc.}$$

The last three instances in the pattern demonstrate why $-b \times -a = ba$ for any integers a and b.

Fig. 4.25 **a** Visual actions relevant to -3×2 on an algeblock mat. **b** Visual actions relevant to -3×-2 on an algeblock mat. **c** Numerical demonstration showing why $-3 \times 2 = -(3 \times 2)$. **d** Numerical patterning activity illustrating why $-b \times -a = ba$

case-sensitive and unreliable numerical strategies such as trial and error and bracketing. A few of my own middle school students in my 3-year study never seemed to have overcome symbol realism and symbol attachment, which prevented them from successfully making the transition. Davydov (1990) reminds us that alphanumeric expressions remain senseless entities unless they are "placed under" a "real, object-oriented, sensorially given foundation" (p. 34). Even for Mason (1980), such symbolic expressions have to be rooted in the enactive, routed in the iconic, and "ultimately become enactive" (p. 11), especially if they have "to be built upon or become a component in a more complex idea" (1980).

But, beyond the obvious rhetoric, the valuing of the symbolic mode and the transition phases in modes of signification share what Thurston (1994), Fields medalist and pioneer of low-dimensional topology, has eloquently articulated about the perennial human interest in the power of *understanding* in proof and more generally mathematics in the following sentences:

> (C)omputers and people are very different. For instance, when Appel and Haken completed a proof of the 4-color map theorem using a massive automatic computation, it evoked much controversy. I interpret the controversy as having little to do with doubt people had as to the veracity of the theorem or the correctness of the proof. Rather, it reflected a continuing desire for *human understanding* of a proof, in addition to knowledge that the theorem is true. On a more everyday level, it is common for people first starting to grapple with computers to make large-scale computations of things they might have done on a smaller scale by hand. They might print out a table of the first 10,000 primes, only to find that their printout isn't something they really wanted after all. They discover this by this kind of experience that what they really want is usually not some collection of "answers" – what they want is *understanding*.
>
> (Thurston, 1994, p. 162)

Of course, there are many ways to characterize mathematical understanding, in particular, symbol sense (e.g., Arcavi, 1994; Fey, 1990; Mason, 1980). However, Thurston's point above captures the purpose and significance of how rules and conventions in the symbolic mode in visuoalphanumeric representations are perceived in this book – that is, they convey a type of understanding that, like Davydov's (1990) notion of concept, reflects "the generality or essence" (p. 249) of a collection of answers that has been drawn by repeatedly acting on their iconic or indexical representations. We throw in as a pedagogical aside what Thurston believes to be the encompassing matter of "what is it that mathematicians accomplish" not in terms of how they either prove theorems or make progress in mathematics but how they "find ways for people to understand and think about mathematics" (Thurston, 1994, p. 162).

4 Development of Visuoalphanumeric Symbols in Pattern Generalization Activity in a Middle School Classroom

In the preceding section, we talked about the significance of visuoalphanumeric thinking in activities relevant to the recognition and interpretation of institutional symbols. In this section, which provides the context for Chapter 5, we focus on the construction of visuoalphanumeric symbols relevant to pattern generalization among my Cohort 1 students who participated in a 3-year longitudinal study that involved various aspects of pattern generalization activity. In particular, we describe the quality, content, and types of symbols that the students used to convey their generalizations on mostly linear and a few quadratic patterns. In Chapter 5, we provide a more comprehensive account of the development of pattern generalization and relevant issues in visualization among my Algebra 1 and Grade 2 students.

Table 4.1 is a summary of Cohort 1 students' preinstructional modes of generalization by type on the five patterns shown in Fig. 4.26a. The responses have been drawn from the Year 1 clinical interviews that occurred at the beginning of the school year and prior to a 4-week teaching experiment on patterning and generalization. Overall, about 51% provided superficially iconic responses and about 24% suggested structurally iconic ones. To illustrate, the first three responses in Fig. 4.26b are *superficially iconic*, that is, they exhibit external generality that is

a

1. Consider the sequence of figures below.

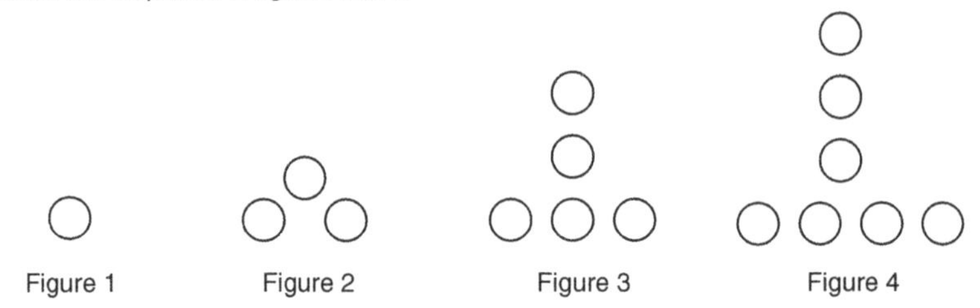

A. How many circles would figure number 10 have in total? Explain howyou obtained your answer in detail.

B. How many circles would figure number 100 have in total? Explain how you obtained your answer in detail.

C. You are now going to write a message to an imaginary Grade 6 student clearly explaining what s/he must do in order to find out how many circles there are in any given figure of the sequence. Message:

D. Find a formula to calculate the number of circles in the figure number "n".

2. Toothpicks are used to build shapes to form a pattern. The table below shows the number of toothpicks used to build a particular shape.

Shape number	1	2	3	4	5	6	7	8	20	60	n
Number of toothpicks	3	7	11	15	19	23					

A. Fill in the missing number of toothpicks in the case of shape number 7, 8, 20, and 60. Explain how you obtained each answer.

B. Find a formula to calculate the number of toothpicks in Shape "n." How did you obtain your formula?

3. Tiles are arranged to form pictures like the ones below.

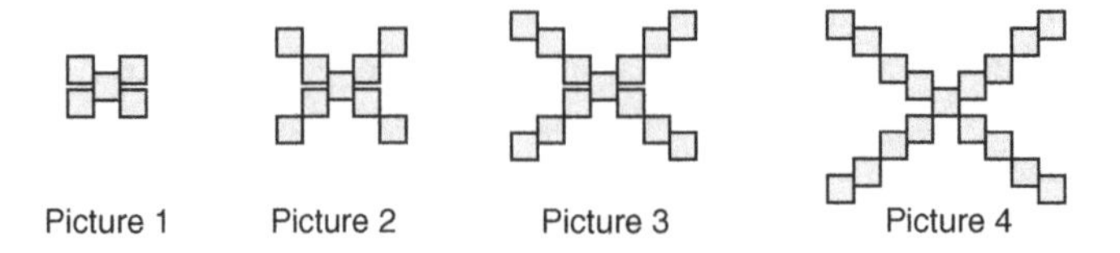

Find a formula to calculate the number of tiles in Picture "n." How did you obtain your formula?

Fig. 4.26 (continued)

4. In the diagram below, a 3 x 3 grid of squares is colored so that only outside squares are shaded. This leaves one square on the inside that is not shaded and 8 squares that are shaded.

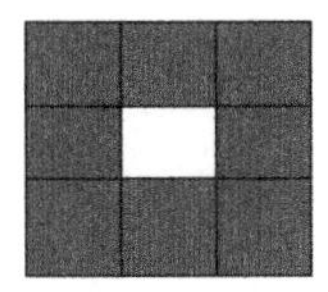

A. If you had a 25 x 25 grid of squares and only the outside edge of squares are shaded, how many squares would be shaded? How did you obtain your answer?

B. If n represents the number of squares on a side and you have all the outside squares of an n x n grid shaded, write an algebraic expression representing the total number of shaded squares in the figure. How did you obtain your expression?

5. Blocks are packed to form pictures that form a pattern as show below.

A. How many blocks are needed to form Picture 8? How did you obtain your answer?

B. How many blocks are needed to form Picture 35? How did you obtain your answer?

C. Find a formula to calculate the number of blocks in Picture "n." How did you obtain your formula?

b

Dianara (Superficially Iconic, Verbal) on Task 1

Message: you start at one and keep adding two until you get the right number of circles in all.

Anna (Superficially Iconic, Visual) on Task 1

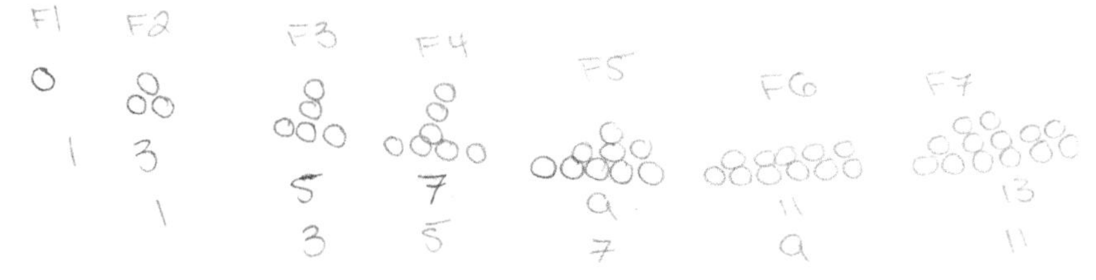

F8 15
F9 17
F10 19
× 2
38

Fig. 4.26 (continued)

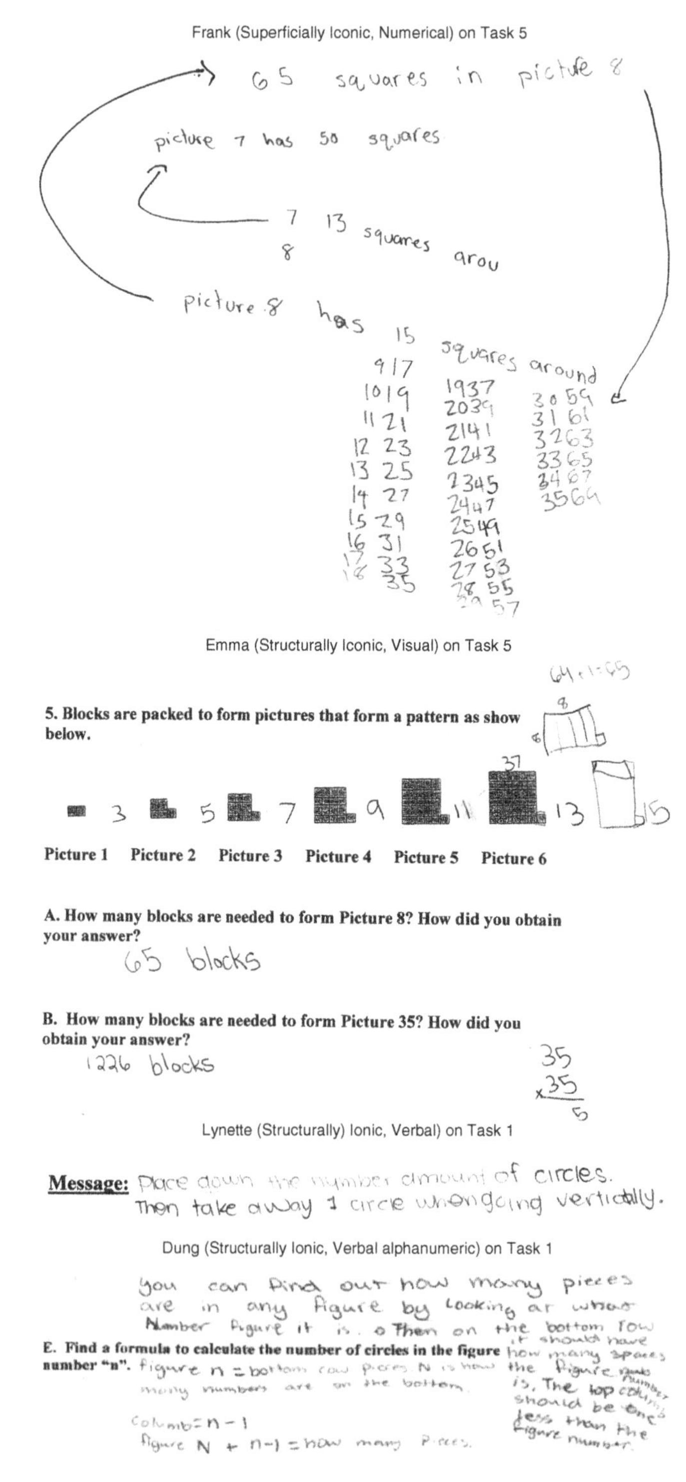

Fig. 4.26 **a** Year 1 preinstructional patterning tasks. **b** Student responses in relation to some of the pattern tasks in Fig. 4.26a

Table 4.1 Summary of sixth-grade students' preinstructional modes of generalization

Pattern task	Superficially iconic generalizations (%)	Structurally iconic generalizations (%)	Could not do (%)
Task 1	66	24	10
Task 2	97	3	0
Task 3	62	24	14
Task 4	3	7	90
Task 5	28	62	10

Mean age of 11; 10 males, 19 females; mostly Asian-Americans

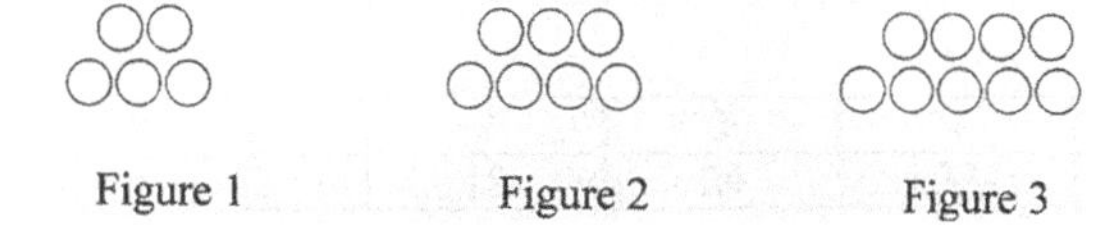

Fig. 4.27 Two-layer circle pattern

characterized by a surface understanding of a perceived general relationship across the pattern stages. The last three responses in Fig. 4.26b are *structurally iconic*, that is, they exhibit internal generality that focuses on a perceived structure within and across the pattern stages. In either case of iconicity, the students' descriptions of their generalizations focused on relationships within, between, and across pattern stages. Further, while they appeared generally unsystematic and minimally parsimonious, the generalizations conveyed abstractions of characteristics they perceived in the patterns.

Clinical interviews on tasks analogous to Fig. 4.26a were conducted immediately after the Year 1 teaching experiment. Not surprisingly, none of the generalizations were expressed in words. Instead, all the generalizations were *alphanumerically drawn symbolic* representations. The students were aware of the rules and conventions in constructing a direct formula (i.e., a closed formula in function form) that they initially developed in class and later emerged as a classroom mathematical practice. In the classroom episode below, the class was exploring the two-layer circle pattern in Fig. 4.27. Ana and her group suggested a numerical symbolic strategic that involved generalizing using differencing, which shares many of the features of the institutional mathematical practice called finite difference method.

Anna: We made up a formula. Like we got the figures until figure 5, and we tried it with other ones. We got $n \times 2 + 3$, where n is the figure number and timesed it by 2. So 5×2 equals 10, plus 3, that's 13. So for figure 25, it's 53.
FDR: I like that formula. So tell me more. So your formula is?
Anna: $n \times 2 + 3$.
FDR: So how did you figure this out?
Anna: First we were like making the numbers to 25. We kept adding 2 and for figure 25, it was 53.

FDR: Wait. So you kept adding all the way to 25?
Anna: Yeah.... Then we used our chart. Then finally we figured out that if we timesed by 2 the figures and plus 3, that would give us the answer.
FDR: Does that make sense? [Students nodded in agreement.] So what Anna was suggesting was that if you look at the chart here, Anna was suggesting that you multiply the figure number by 2, say, what's 1 × 2?
Tamara: 2.
FDR: 2. And then how did you [referring to Anna's group] figure out the 3 here?
Anna: Because we also timesed it with figure number 13.
FDR: What did you have for figure 13?
Anna: That was 29. And then 13 × 2 equals 26 plus 3.
FDR: Alright, does that work? So what they were actually doing is this. They noticed that if you look at the table, it's always adding by 2. You see this? [Students nodded.] They were suggesting that if you multiply this number here [referring to the common difference 2 by figure number, say figure number 1, what's 1 × 2?
Students: 2.
FDR: Now what do you need to get to 5? What more do you need to get to 5? [Some students said '3' while others said '4.'] Is it 4 or 3?
Students: 3.
FDR: It's 3 more. So what is 1 × 2?
Students: 2.
FDR: Plus 3?
Students: 5. [The class tested the formula when n = 2, 3, and 25.]

One unfortunate consequence of the above alphanumeric practice was the gradual loss of the students' ability to justify their direct formulas. While they could construct direct formulas using differencing, they became inconsistent in justifying them. As they internalized the alphanumeric process and neglected the visual aspect altogether, this particular symbolic appropriation, in fact, constrained them. In the words of d'Alembert (1995), "(m)athematical abstractions help us in gaining this knowledge [i.e., general relationships], but they are useful only insofar as we do not limit ourselves to them" (p. 21; quoted in Alexander, 2006, p. 722). To illustrate, Dung, prior to the Year 2 teaching experiment on pattern generalization (about 7 months after the Year 1 teaching experiment), dealt with the adjacent toothpick pattern task in Fig. 4.28 in the following manner: He initially set up a two-column table of values, listed down the pairs (1, 4), (2, 7), and (3, 11), and noticed that "the pattern is plus 3 [referring to the dependent terms]." He then concluded by saying, "the formula, it's pattern number × 3 plus 1 equals matchsticks." In writing $T = (n \times 3) + 1$, he noted that the coefficient referred to the common difference and the y-intercept as the adjustment value that was needed in order to match the dependent terms. When he was then asked to justify his formula, he provided the following faulty reasoning in which he projected his formula onto the figures in a rather inconsistent manner (see Fig. 4.29 for a visual illustration):

> *For 1 [square], you times it by 3, it's 1, 2, 3 [referring to three sides of the square] plus 1 [referring to the left vertical side of the square]. For pattern 2, you count the outside sticks and you plus 1 in the middle. For pattern 3, there's one set of 3 [referring to the last three sticks of the third adjacent square], two sets of 3 [referring to the next two adjacent squares] plus 1 [referring to the left vertical side of the first square].*

Consider the sequence of toothpick squares below.

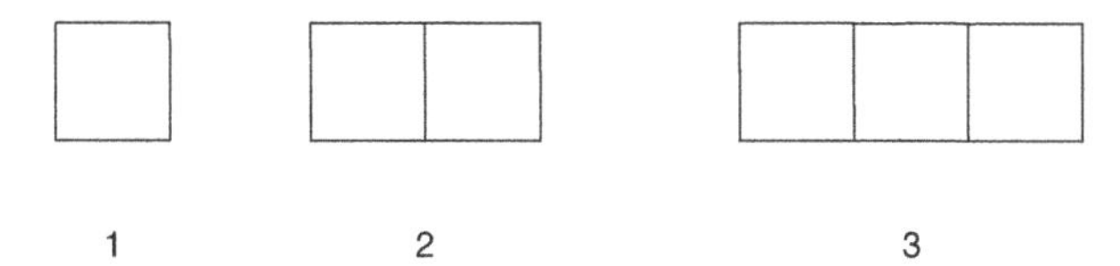

A. How many toothpicks will pattern 5 have? Draw and explain.

B. How many toothpicks will pattern 15 have? Explain.

C. Find a direct formula for the total number of toothpicks T in any pattern number n. Explain how you obtained your answer.

D. If you obtained your formula numerically, what might it mean if you think about it in terms of the above pattern?

E. If the pattern above is extended over several more cases, a certain pattern uses 76 toothpicks all in all. Which pattern number is this? Explain how you obtained your answer.

F. Diana's direct formula is as follows: $T = 4 \cdot n - (n - 1)$. Is her formula correct? Why or why not? If her formula is correct, how might she be thinking about it? Who has the more correct formula, Diana's formula or the formula you obtained in part C above? Explain.

Fig. 4.28 Adjacent toothpick pattern task

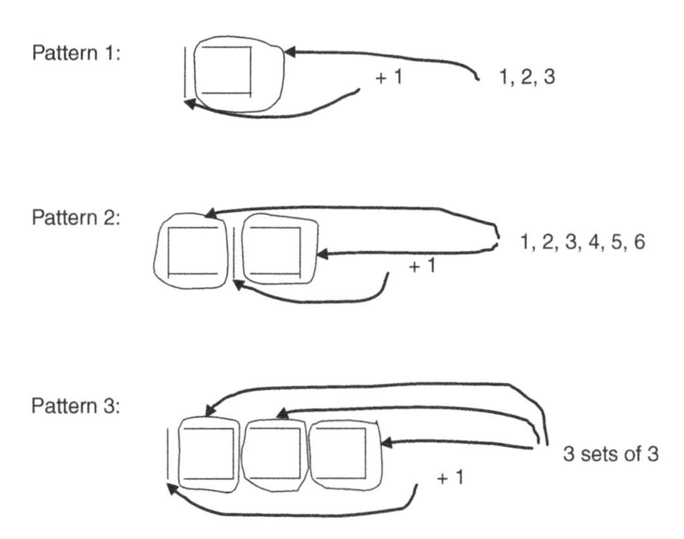

Fig. 4.29 Dung's justification of his direct formula in relation to the pattern in Fig. 4.28

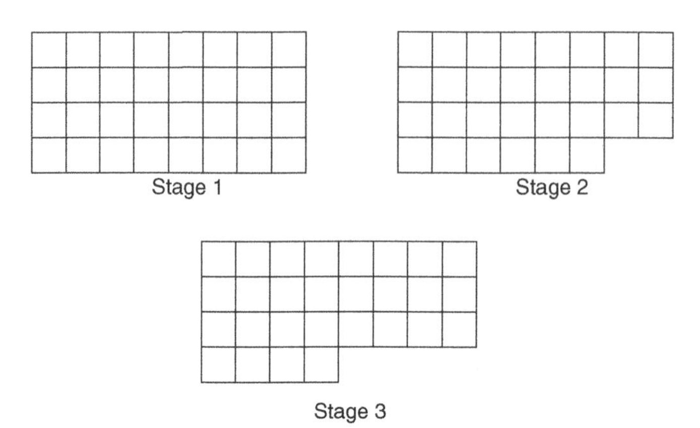

Fig. 4.30 Losing squares pattern

But the students also found the above alphanumeric method too difficult to use or to "transfer" in the Year 2 study when they had to deal with decreasing linear patterns that involved negative differencing. For example, in a clinical interview prior to a teaching experiment on decreasing patterns, Tamara was first asked to establish and justify generalizations for two increasing linear patterns, which she accomplished successfully. When she was then asked to obtain a generalization for the losing squares pattern in Fig. 4.30, she first saw that every stage after the first involves "minusing 2" squares. She then used multiplication to count the total number of squares at each stage. When she then proceeded to obtain a direct formula, she was perturbed by the negative value of the common difference and said, "I was trying to think of, just like the last time, I was trying to get a formula. . . . I was thinking of trying to do with the stage number but I don't get it." The presence of the negative difference, including the necessity of multiplying two differently signed numbers, partially and significantly hindered her from applying what she knew about constructing direct formulas in the case of increasing patterns. In fact, she had to first broaden her knowledge of multiplication to include two factors having opposite signs before she was finally able to state the form $S = -2 \times n + 34$. Further, while she could explain what the slope and y-intercept in her direct formula meant in the case of increasing patterns, she was unable to justify the forms she established for decreasing linear patterns.

In Year 3, the students' generalizations were also symbolic; however, they were *visuoalphanumeric symbolic* representations. That is, the content of their constructed direct formulas involving alphanumeric objects was rooted visually on what they interpreted to be the structure of a pattern across the known stages and assumed to hold for the unknown stages as well. For example, in the clinical interview after the teaching experiment on patterning and generalization, several students developed at least four different visuoalphanumeric symbolic representations on the T star pattern task shown in Fig. 4.31a. In Fig. 4.31b, each direct formula actually combined construction and justification with the alphanumeric expressions conveying a perceived structure of the pattern. Dung cognitively perceived a middle star that stayed the same from stage to stage and three groups of neighboring stars that grew by the stage number. Jenna saw overlapping groups of stars. Tamara offered two formulas as a result of interpreting two structures for the same pattern. Her first formula, $s = 3n + 1$, was the same as Dung's. However, her mode of grouping stars was different from Dung's since she saw n groups of three stars instead of three groups of n stars. In her second formula, she initially added n stars in order to justify the expression $4n$. She then took them away and then added the middle star that stayed the same from stage to stage.

While the Year 1 rules for constructing alphanumeric symbolic generalizations have been drawn from the numerical method of differencing that Anna and her group shared in class, the Year 3 rules for developing visuoalphanumeric symbolic generalizations have been drawn from the students' experiences in two classroom events. The first classroom event took place about 4 weeks before the teaching experiment on pattern generalization. In the classroom episode below, the students were asked to use their conceptual understanding of multiplication of two integers to obtain a mathematical expression for the two sets of figures shown in Fig. 4.32. The activity was meant to prepare them later when they had to perform multiplication of integers with the algeblocks.

> *FDR: So what mathematical expression corresponds to what you see here [referring to the first set]?*
> *Francis: 6 circles.*
> *FDR: Yes, there are 6 circles but I want a mathematical expression that shows how you got 6.*
> *David: Add them one by one.*

Consider the pattern shown below.

1. What stays the same and what changes?

2. Find a direct formula for the total number of stars T at any stage number *n*. Explain why your formula makes sense. If needed, use the duplicate copy below to help you explain.

3. Find a second direct formula that is equivalent to the one you obtained in item 2 above. Explain why it works. If needed, use the duplicate copy below to help you explain.

4. If your two direct formulas above do not involve overlaps, find a direct formula that takes into account overlaps. Explain why it works. If needed, use the duplicate copy below to help you explain.

Fig. 4.31 (continued)

Dung

2. Find a direct formula for the total number of stars T at any stage number *n*. Explain why your formula makes sense. If needed, use the duplicate copy below to help you explain.

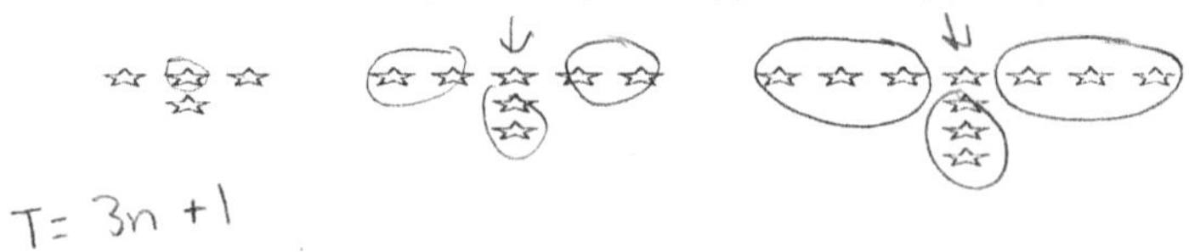

Jenna

3. Find a second direct formula that is equivalent to the one you obtained in item 2 above. Explain why it works. If needed, use the duplicate copy below to help you explain.

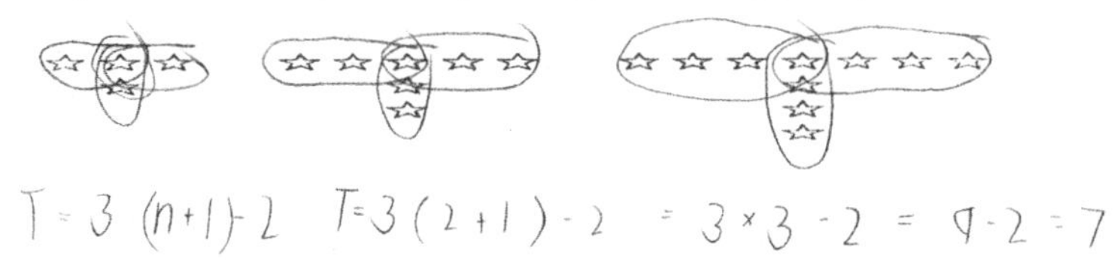

Tamara

2. Find a direct formula for the total number of stars T at any stage number *n*. Explain why your formula makes sense. If needed, use the duplicate copy below to help you explain.

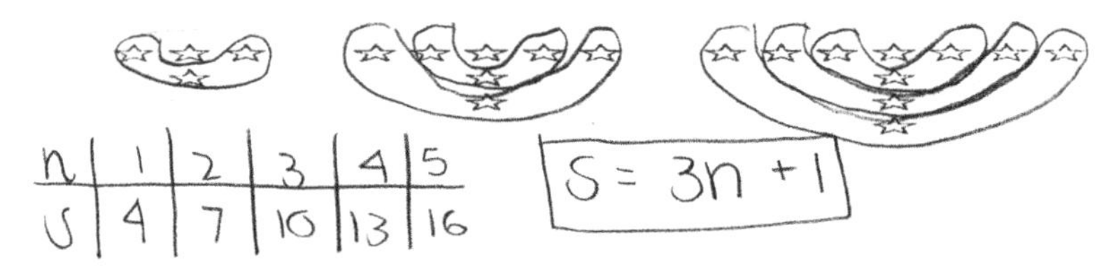

3. Find a second direct formula that is equivalent to the one you obtained in item 2 above. Explain why it works. If needed, use the duplicate copy below to help you explain.

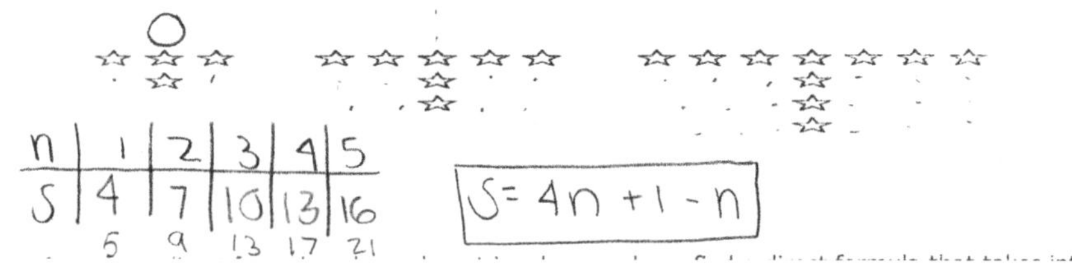

Fig. 4.31 **a** T star pattern task. **b** Four visuoalphanumeric direct formulas for the pattern in Fig. 4.31a

FDR: Yes, you can certainly add them one by one. But are there other ways of getting 6?

Eric: 2 times 3. [FDR writes the answer on the board.]

FDR: So what do you mean by 2 × 3, Eric?

Eric: It means 2 threes.

FDR: Uhum, 2 threes or we say 2 groups of threes. Does that make sense? [Students nod in agreement.] Okay, so now what expression works with the second item here?

Salina: Three groups of 6.

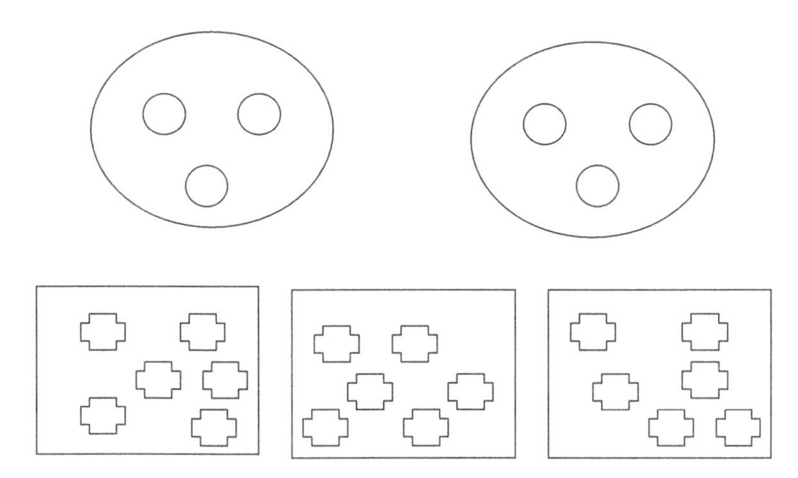

Fig. 4.32 Two multiplication tasks

FDR: Uhum, and how do we write that using multiplication?
Students: 3 times 6.
FDR: Times meaning what?
Salina: Groups of.

The second classroom event took place in relation to the tile patio pattern task shown in Fig. 4.33 that asked them to obtain several different equivalent direct formulas for the same pattern. Initially, the students worked in pairs to

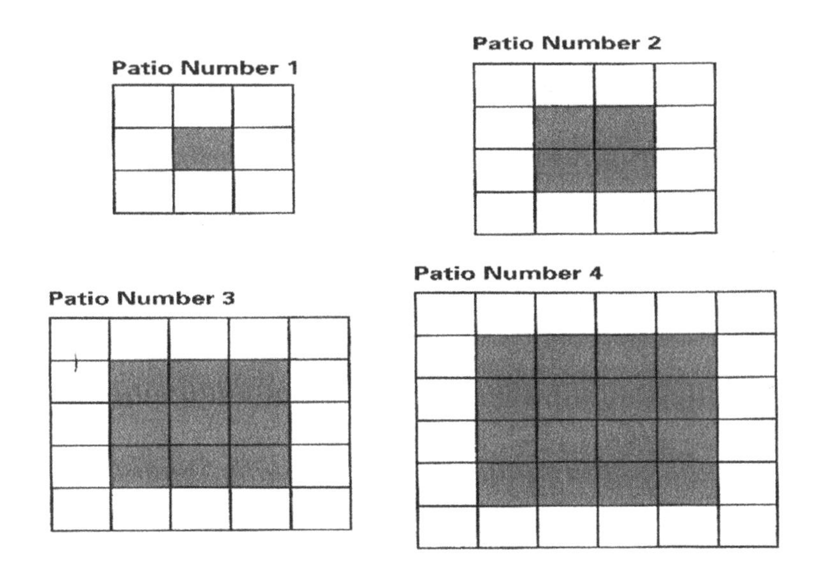

Fig. 4.33 Tile patio pattern

Che: "W = 4n + 4. So there's like 4 squares and then you add 4 to each one [corner]. ... And for patio #2, 2 x 4 is 8, so 1, 2, 3, ..., 8, then you add 4. [There are] 4 groups of n."

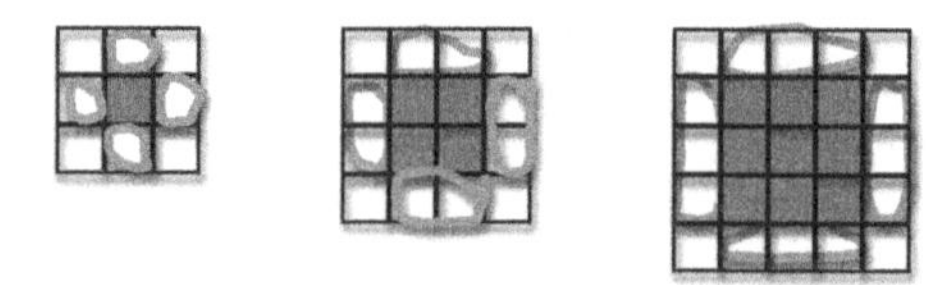

Dina: "4(n + 1) coz patio #1 you have two groups of 4, patio #2, 3 groups of 4, etc."

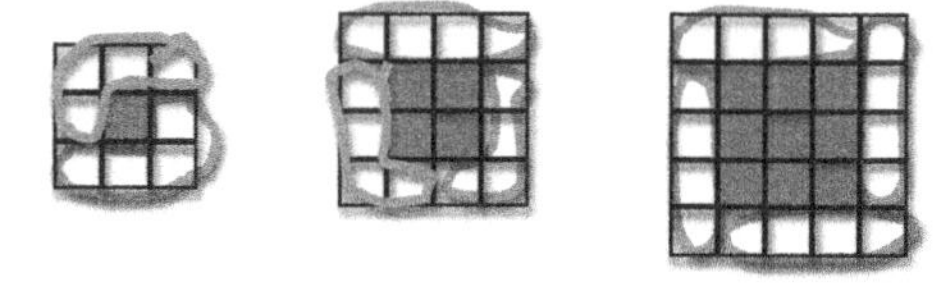

Dave: "T = 2(n+2) + 2n. The top part, 2 + 1 = 3. Then I multiplied by 2, the bottom, so that's 6. And the 2n, so here's 1 [row 2 column 1 square] and 1 here [row 2 column 3 square], etc."

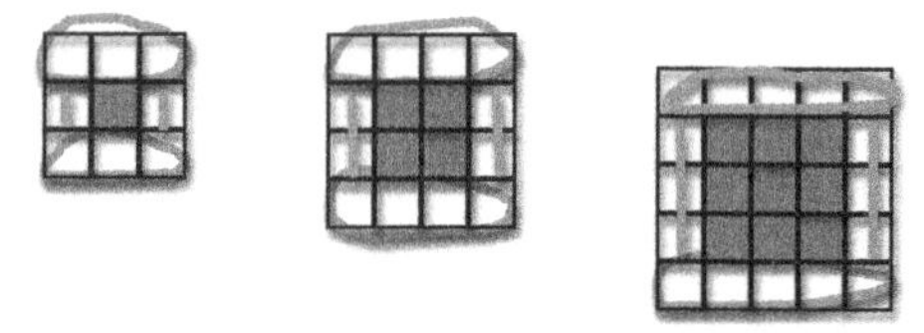

Fig. 4.34 Four direct formulas for the pattern in Fig. 4.33

establish a structural unit for the pattern. The structural unit addressed issues relevant to aspects in the pattern that for them stayed the same and changed. Each alphanumeric formula shown in Fig. 4.34 conveyed an embedded visual description of a structural unit – and thus, a visuoalphanumeric symbol – that is stage-induced and multiplicative in most cases.

We should note that the use of visuoalphanumeric symbols has been documented in many other studies done with several different age-level groups [see, e.g., the ZDM (2008) special issue on generalization]. But three points are worth noting. *First,* in many of these reported studies, the emphasis has been about extrapolating semi-cognitive processes relevant to "capturing" in mathematical form what is noticed, observed, or abduced in an emerging generalization. Consequently, such studies have not explicitly articulated the central role of rules and conventions that might have given some indication of, or insight into, the symbolic nature of direct formula construction and justification. *Second,* we underscore the significance of purposefully orchestrated joint activity and classroom interaction in enabling intra-semiotic transitions to occur in pattern generalization activity. Among Cohort 1, their 3-year story could be summarized as one in which their direct formula construction and justification as a symbol system progressively transitioned from the visual to the alphanumeric and finally to the visuoalphanumeric. This positive account illustrates the view that signs and their modes of signification are neither predetermined nor arbitrary but negotiated and conventional. *Third,* in this section we started to address the important role of structural unit and multiplicative thinking as rules and conventions that the students used in their simultaneous construction and justification of visuoalphanumeric-drawn direct formulas. We explore these ideas further in the next chapter.

5 Overview of Chapter 5

Chapter 5 provides details concerning the central role of visual thinking in the construction and justification of simple and complex algebraic generalizations relative to linear and nonlinear figural patterns. While numerical- or symbolic-driven processes help students fulfill aspects of school mathematical activity, the findings in this chapter indicate that the absence of any visual-based context in learning mathematics is likely going to prevent them from further developing useful insights and meaningful understanding. In Chapter 5, we explore in some detail the meaning and significance of abductive reasoning in induction and generalization processes. We dwell for the most part on figural patterns and address the primary dilemma of how to develop reasonable inferences and algebraic generalizations on the basis of limited information (i.e., the known stages in any pattern) that could then be applied to the unknown stages.

Also, in Chapter 5, we provide a visually drawn empirical account of progressive evolution of structural thinking involving patterns. Such an account demonstrates how accounts of progressive formalization, schematization, symbolization, and mathematization all involve accounts of progressive abductions. In two sections, we distinguish between entry-level abductions and mature abductions that produce visuoalphanumeric representations. We also discuss constraints and difficulties in making abductive transitions that have implications in the content and quality of structural thinking. Finally we consider ways in which visuoalphanumeric representations in pattern generalization activity at the middle school level and structured visual thinking in patterning activity at the elementary level mediate in students' understanding of functions.

Chapter 5
Visuoalphanumeric Representations in Pattern Generalization Activity

How general is general?
(Bastable & Schifter, 2008, p. 166).

The most important operation of the mind is that of generalization
(Peirce, 1960, p. 34).

(A)lgebra ... not as symbol manipulation, not as arithmetic with letters, not even as the language of equations, but as a succinct and manipulable language in which to express generality and constraints on that generality
(Mason, 2008, p. 77).

In Chapter 4, we drew on a few examples from my own classroom work in discussing a progressive account of symbol formation in school mathematics. A visually grounded approach provides an alternative and effective route that could assist students in understanding mathematics better. Otte (2007) notes how mathematical knowledge seems to be already "everything [that] just is and thus means itself" (p. 243). The more pressing issue appears to be "not that of rigor but the problem of the development of meaning" (Rene Thom quoted in Otte, 2007, pp. 244–245).

School mathematics content has been, and continues to be, depicted and taught in solely alphanumeric terms. One effect of this valued perception toward alphanumeric-driven rules and conventions involves the widely accepted binary practice concerning the separation of the visual and referential, on one side, and the abstract and operative, on the other (Fig. 2.8). Moyer's (2001) study with 10 middle school teachers as well as my own in the context of a state-funded professional development grant for Grades 5 through 9 teachers provide sufficient indications of this continued practice, where visual learning is associated with the view that "math is fun" and alphanumeric learning assigned the serious task of "doing real math."

The empirical evidence presented in several sections in Chapter 4, however, offers a counterstance, a way out of dichotomized thinking in favor of more embedded and distributed mathematical practices in which mathematical knowledge emerges from transitions in symbolizing from the iconic and/to indexical to symbolic and, consequently, from visual images to more structured visual representations we call visuoalphanumeric symbols. I have also underscored the significance

F.D. Rivera, *Toward a Visually-Oriented School Mathematics Curriculum*, Mathematics Education Library 49, DOI 10.1007/978-94-007-0014-7_5,
© Springer Science+Business Media B.V. 2011

and influence of appropriate teacher orchestration and psychological distancing in helping students overcome cut points (Filloy & Rojano, 1989) and make transitions at both intra- (visual to visuoalphanumeric) and inter-semiotic (symbol, image, and language) levels. In this chapter, we provide a visually drawn empirical account of progressive evolution of structural thinking in one particular content strand in school mathematics, that is, patterns in both figural and numerical forms. Central in this account is Peirce's notion of *abduction*.

This chapter is divided into six sections. Section 1 involves clarifying the meaning of patterns. Section 2 introduces readers to abductive reasoning or, simply, abduction. This is an important section in light of its powerful role in the development of inductive reasoning and, more generally, in mathematical reasoning. Further, we establish a stronger claim in which accounts of progressive formalization, schematization, symbolization, and mathematization all involve abductive actions. Section 3 addresses issues surrounding entry-level abductions. Section 4 deals with visual templates as exemplifying instances of mature abductions that produce visuoalphanumeric representations. In Section 5, we discuss constraints and difficulties in making abductive transitions, which have implications in the content and quality of structural thinking. We close this chapter with Section 6 that deals with the relationship between visuoalphanumeric generalization and linear modeling in Algebra 1 and between structured visual thinking about patterns and functional thinking in Grade 2.

1 Assumptions About Patterns

Meaningful patterns convey structures. *Pace* Resnik (1997), "the primary subject-matter [of mathematics] is not the individual mathematical objects but rather the structures in which they are arranged" (p. 201). In the context of patterning activity, we assume that the known stages or objects in a pattern convey positions in some interpreted structural relationship, which also means to say that they do not have an identity or distinguishing features outside of that relationship. By *structure*, we mean an arrangement of objects expressed in either figural or numeric form.

The goal in every pattern is to obtain an appropriate generalization. By *generalization*, we follow Peirce's definition, as follows:

> *Generalization* in its strict sense, means the discovery, by reflection upon a number of cases, of a general description applicable to them all. This is the kind of thought movement which I have elsewhere called formal hypothesis, or reasoning from definition to definitum. So understood, *it is not an increase in breadth but an increase in depth.*
>
> (Peirce, 1960, p. 256; italics mine)

Further, when a generalization has been drawn from a larger (versus small) number of known cases, it signals "an increase of definiteness of the conceptions [we] apply to [the] known things" (Peirce, 1960, p. 256).

At least in this chapter, patterns in a generalizing or structuring task have the following properties:

1. There is a closed or a direct formula that can be derived from the given stages.
2. The known stages, together with an interpreted formal hypothesis or generalization, can assist in pattern extension, that is, generate a range of terms.
3. The stages resemble each other in some way.

For example, the sequence $\{2, 4, 6, 8, \ldots\}$ is a numerical pattern. The direct expression $2n$ is one way of describing the structure of the numbers in the sequence that allows us to say that the numbers 10, 12, 14, and so on are reasonable extensions. Also, it is perceptually apparent that evenness is one characteristic that is common to each number in the sequence.

The Fibonacci sequence $\{1, 1, 2, 3, 5, 8, 13, \ldots\}$ is a famous example of a recursive pattern; it can be generally described by the rule $a_n = a_{n-1} + a_n$ (where $a_1 = a_2 = 1$ and n is an integer greater than 1). Assuming that the pattern obeys the recursive rule, its closed formula is $\frac{1}{\sqrt{5}}\left[\left(\frac{1+\sqrt{5}}{2}\right)^n - \left(\frac{1-\sqrt{5}}{2}\right)^n\right]$. The direct expression $5n - 2$ (where $n \geq 1$) is a generalization associated with the arithmetic sequence $\{3, 8, 13, 18, \ldots\}$ with the added condition that the pattern is an increasing sequence.

Resemblance encompasses implicit (deep) and explicit (surface) properties that the stages in a pattern have in common. The properties are not inherently *a priori* but interpreted – that is, depending on an individual learner's knowledge and experiences, he or she assumes a justified property relative to the pattern that he or she then projects onto both known and unknown stages. Projecting involves employing abductive processes that generate hypothesized properties that are or are not directly knowable due to the incompleteness of the pattern stages. Abduction, for the time being, pertains to viable inferences routed through generalizations that individual learners "plausibly" claim in a pattern despite the fact that the information presented to them is incomplete (i.e., only a few initial or given stages in the pattern are known). "Plausibility," Peirce (1958a) writes, "is the degree to which a theory ought to recommend itself to our belief independently of any kind of evidence other than our instinct urging us to regard it favorably" (p. 173).

When students perform a *pattern generalization,* it basically involves mutually coordinating their perceptual and symbolic inferential abilities so that they are able to *construct* and *justify* a plausible and algebraically useful structure that could be conveyed in the form of a direct formula (Lee, 1996; Radford, 2008; Rivera, 2010a). Especially in cases involving visual patterns, among the most important perception types that matter involves visual perception. Visual perception involves the act of coming to see; it is further characterized to be of two types, namely sensory perception and cognitive perception (Dretske, 1990). Sensory (or object) perception is when individuals see an object as being a mere object in itself. Cognitive perception goes beyond the sensory when individuals see or recognize a fact or a property in relation to the object. For example, in my Grade 2 class, 12 of the 20 students who were individually interviewed and assessed for preinstructional competence on pattern generalization exhibited sensory perception in the case of the triangular pattern shown in Fig. 5.1a. Taking the stages as mere sets of objects, the 12 students produced a variety of extensions in the shape of lines, rectangles, and triangles

Consider the following steps in a pattern

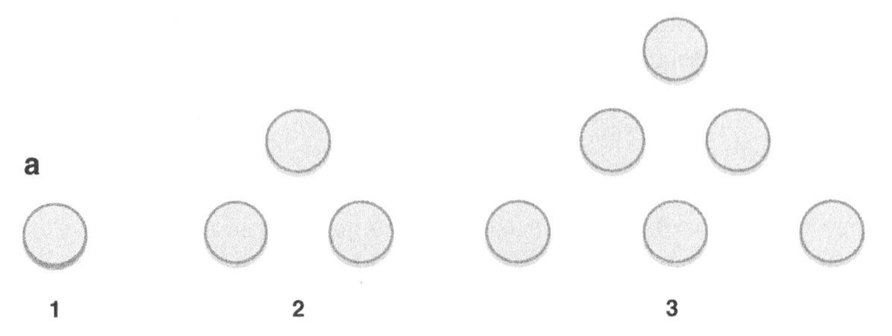

1. Use the circle chips to draw the next two figures.
2. Without using the chips, draw or describe stage 10 of the pattern.

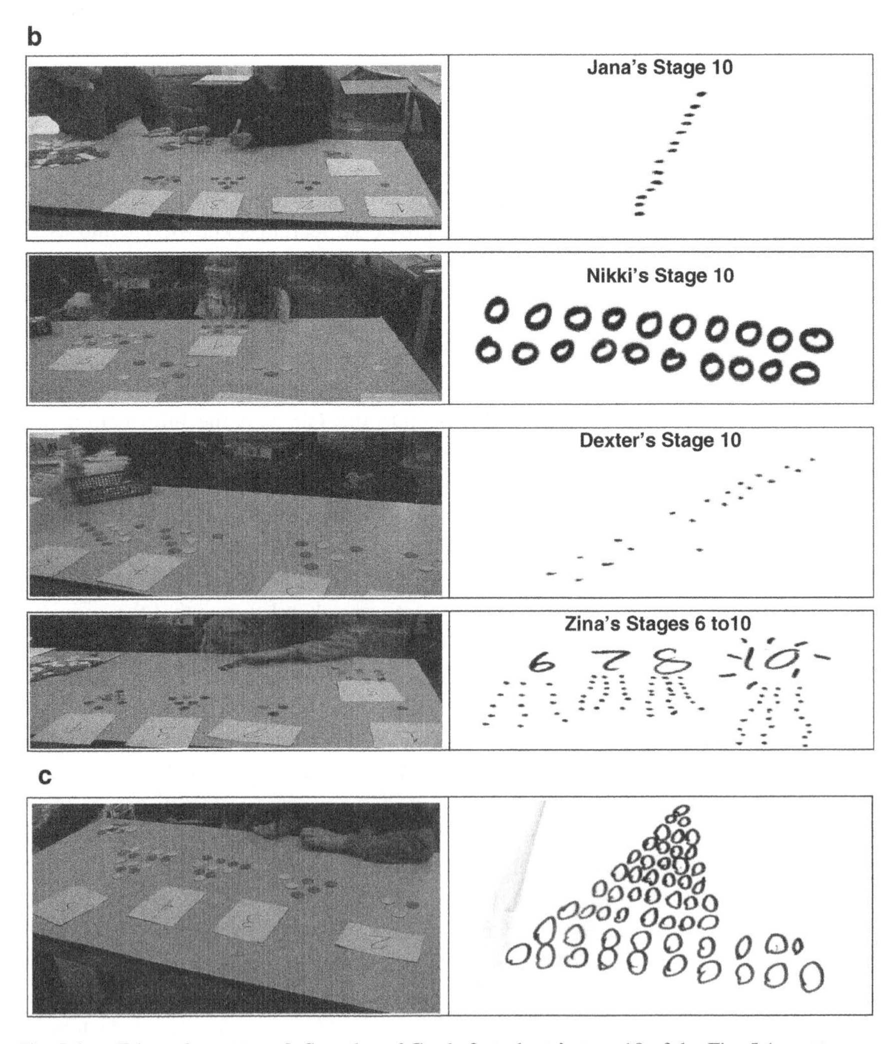

Fig. 5.1 **a** Triangular pattern. **b** Samples of Grade 2 students' stage 10 of the Fig. 5.1a pattern. **c** Joshua's (Grade 2) stage 10 of the Fig. 5.1a pattern

(Fig. 5.1b). Seven other students exhibited cognitive perception in varying stages of structural awareness. Among the seven, only Joshua produced a consistent cognitive perception on the basis of his written work in the case of stage 10 (Fig. 5.1c). Cognitive perception necessitates the use of conceptual and other cognitive-related processes that enable individual learners to articulate what they choose to recognize as being a fact or a property of a target object. It is mediated in some way through other types of visual knowledge that bear on the objects, and such types could be either cognitive or sensory in nature.

There are two kinds of patterns that we analyze in this chapter, which are as follows.

Numerical patterns involve numbers as the primary objects of generalization. Four points are worth noting with respect to all the numerical patterns that I used in all my studies. *First*, when the students were presented with tables of numbers such as the ones shown in Fig. 5.2, they had sufficient knowledge of the contexts, assumptions, and conditions that were needed in order to make sense of them.[1] *Second*, since most linear function modeling problems could be effectively and efficiently analyzed as numerical patterns, numerical modes and strategies of generalizing were encouraged, in particular, in the constructive aspect of pattern generalization. *Third*, numerical tables of values were either presented or student generated. *Fourth*, numerical tables of values can be presented in at least two ways (cf. Rivera, 2009). Figure 5.3a are two examples of a table that fosters an *inductive-structuring strategy*, while Fig. 5.3b exemplifies a table that employs a *finite differencing strategy*. The tables were generated in relation to the circle fan pattern task shown in Fig. 4.1.

Figural patterns involve shapes as the primary objects of generalization. The shapes need to be assessed for meaningfulness prior to generalization, which means they have to be analyzed in terms of parts or subconfigurations that operate or make sense within an interpreted structure. We use the term *figural patterns* to convey what we assume to be the "simultaneously conceptual and figural" (Fischbein, 1993, p. 160) nature of mathematical patterns. The term "geometric patterns" is not appropriate due to a potential confusion with geometric sequences (as instances of exponential functions in discrete mathematics). The term "pictorial patterns" is also not appropriate due to the Peircean fact that figural patterns are not mere pictures

[1] Parker and Baldridge (2004) emphasize the need for tables to have real (or experientially real) and predictable contexts with a sufficient number of particular instances in order to generate a reasonable algebraic expression (or formula). For example, the table below with the context about rainfall could not be assessed correctly, that is, even if the expression appears to take the algebraic form $(1/2)h$, "there is no reason why the rainfall will continue to be given by that expression, or *any* expression. This question cannot be answered" (p. 90). Other examples include stock market prices and gas prices, where tables could be generated but oftentimes do not lead to correct and justifiable algebraic expressions.

It started to rain. Every hour Sarah checked her rain gauge. She recorded the total rainfall in a table. How much rain would have fallen after h hours?

Hours	Rainfall
1	0.5in.
2	1 in.
3	1.5 in.

4. There are 4 apples in each packet.

(a) How many apples are there in n packets?

Number of packets	Total number of apples
1	$4 \times 1 = 4$
2	$4 \times 2 = 8$
3	$4 \times 3 = 12$
4	$4 \times 4 = 16$
5	$4 \times 5 = 20$
n	$4n$

We write $4 \times n$ as $4n$.

(b) If $n = 8$, how many apples are there altogether?
(c) If $n = 11$, how many apples are there altogether?

Fig. 5.2 Sample of a context-based numerical table (Curriculum Planning and Development Division, 2008, p. 8)

of objects but exhibit characteristics associated with visually drawn schematic representations.

While all patterns for generalization purposes need to be well defined, ambiguous ones allow learners to value this necessary condition. The term "ambiguous" shares Neisser's (1976) general notion of ambiguous pictures as conveying the "possibility of alternative perceptions" (p. 50). The term "well-defined pattern" means there is an unambiguous and unique structure that could be associated with the pattern and described accordingly in an algebraically useful manner (i.e., a direct formula in function form). We note, of course, that an individual learner's level of cognitive perception is likely to determine how well defined a pattern is, which explains why we need to have both constructive and justificatory phases in pattern generalization.

The above point raises the important issue of ease in structure discernment relative to (figural) pattern generalizing. We discuss briefly for now the well-established *Law of Good Gestalt* (Metzger, 2006, pp. 19–27). Later, in Sections 4 and 5, we use the terms *pattern goodness* and *Gestalt effect* interchangeably and they both refer to the same Gestalt law. This law pertains to the order in which we perceive, discern, and organize a figure, a picture, or an image – that is, we are predisposed to organizing on the basis of what naturally belongs or fits together, including that which is simple and recognizable enough that enables us to associate or specify a geometric shape or an algebraic formula. Hence, a figural pattern that is *high in Gestalt goodness* tends to have an interpreted structure that reflects the orderly, balanced, and harmonious form of the pattern, which allows learners to easily specify an algebraically useful formula. A figural pattern that is *low in Gestalt goodness* is interpreted as being disorganized with a complex (unbalanced) structure that either

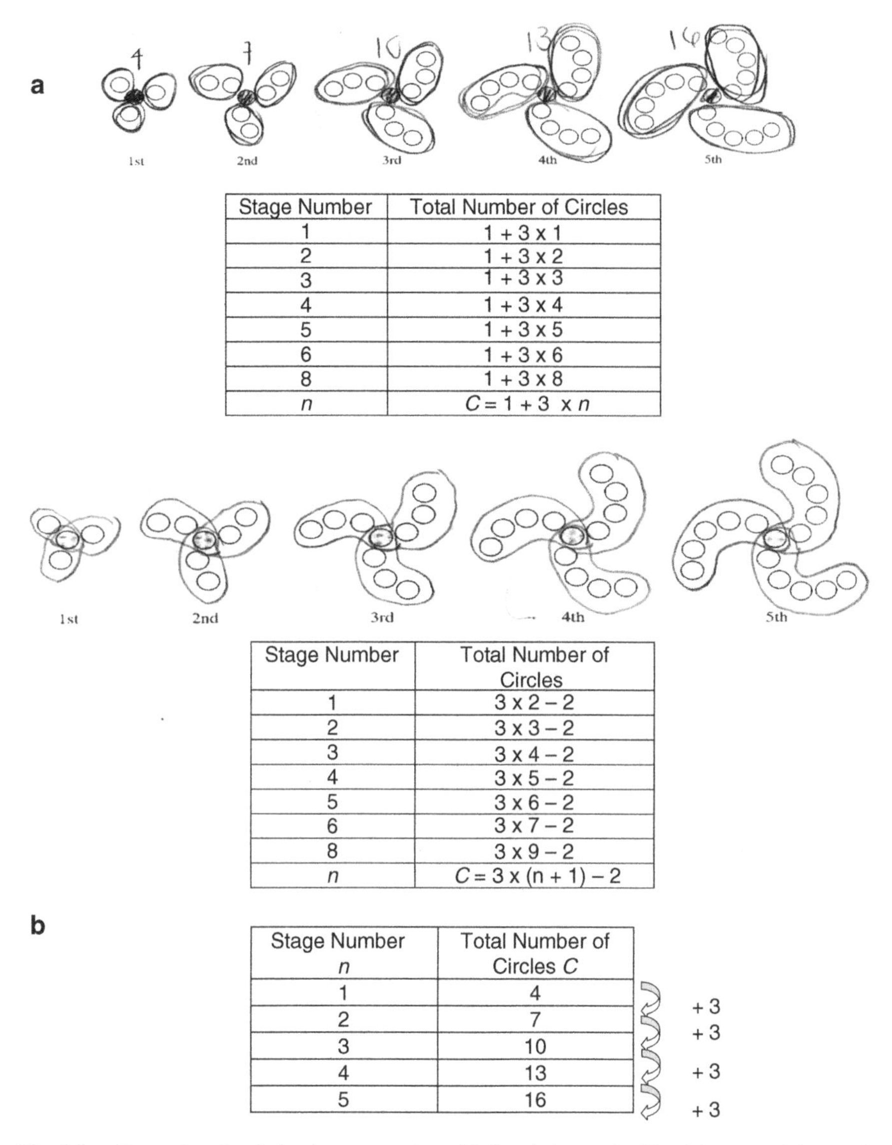

Stage Number	Total Number of Circles
1	1 + 3 x 1
2	1 + 3 x 2
3	1 + 3 x 3
4	1 + 3 x 4
5	1 + 3 x 5
6	1 + 3 x 6
8	1 + 3 x 8
n	$C = 1 + 3 \times n$

Stage Number	Total Number of Circles
1	3 x 2 – 2
2	3 x 3 – 2
3	3 x 4 – 2
4	3 x 5 – 2
5	3 x 6 – 2
6	3 x 7 – 2
8	3 x 9 – 2
n	$C = 3 \times (n + 1) - 2$

Stage Number n	Total Number of Circles C
1	4
2	7
3	10
4	13
5	16

Fig. 5.3 **a** Examples of an inductive-structuring table in relation to the Fig. 4.1 pattern. **b** Example of a finite difference table in relation to the Fig. 4.1 pattern

has no easily discernible parts or consists of parts that have no "natural divisions," which makes the task of constructing an algebraically useful formula difficult to accomplish.

Thus, in both cases of pattern goodness, visual actions that are conveyed by their individual cognitive perceptions play a significant role in the construction of an algebraic generalization. Certainly, grade-appropriate tasks matter as well; however, it is still a matter for individual learners to decide how they want to perceive, construct, and interpret (stages in) a figural pattern. For example, prior to formal instruction

Consider the following steps in a pattern

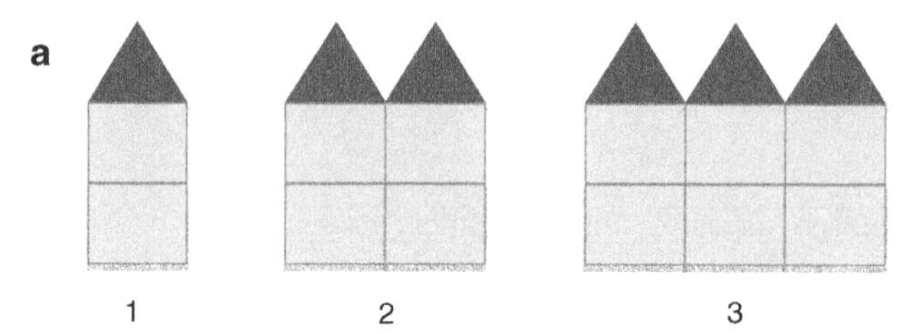

1. Use the patten blocks to draw the next two figures.
2. Without using the blocks , draw or desceibe stage 10 of the patten.

Consider the following steps in a pattern

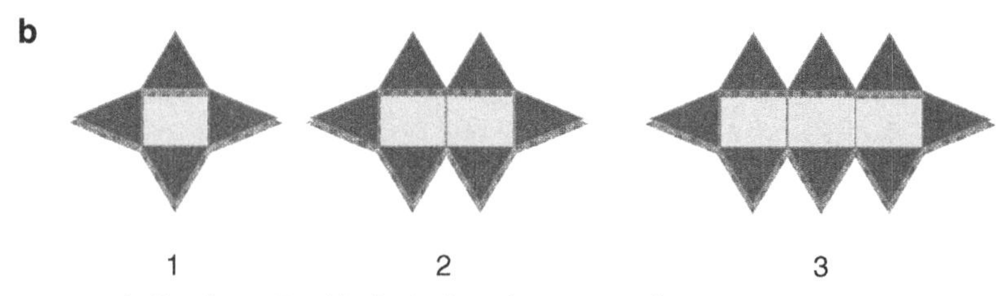

1. Use the patten blocks to draw the next two figures.
2. Without using the blocks , draw or desceibe stage 10 of the patten.

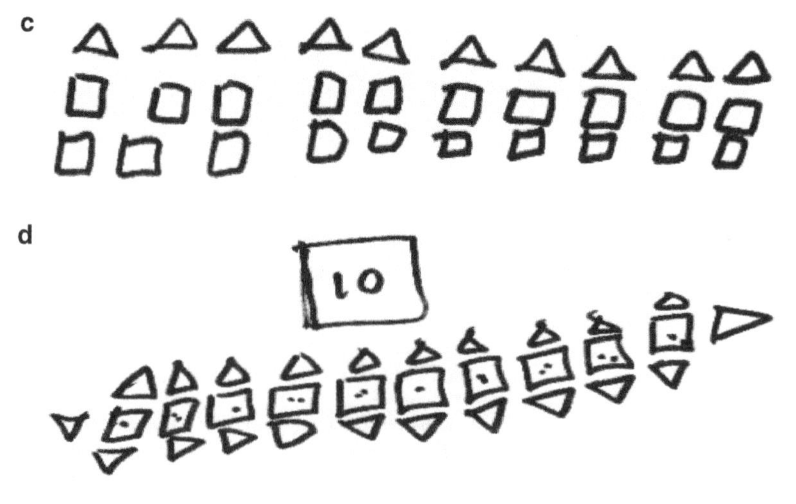

Fig. 5.4 **a** House pattern task used with Grade 2 students (pattern taken from Greenes et al., 2001, p. 79). **b** Star pattern task used with Grade 2 students (pattern taken from Greenes et al., 2001, p. 79). **c** Sylvia's (Grade 2) stage 10 of Fig. 5.4a. **d** Nikki's (Grade 2) stage 10 of Fig. 5.4b

on pattern generalization, second grader Joshua in Fig. 5.1c interpreted the triangular pattern in Fig. 5.1a as having a well-defined structure that reflects what more experienced and older students are likely to produce (i.e., the sequence of triangular numbers). Unlike Joshua, however, the remaining 19 students found Fig. 5.1a to be low in Gestalt goodness. In the same class, Joshua and 11 others considered the house pattern and the star pattern in Fig. 5.4a, b, respectively, as being high in Gestalt goodness and, thus, well defined on the basis of how they quickly saw a correct structure for stage 10 in each pattern. They initially inspected the given stages 1 through 3, then extended the pattern to stages 4 and 5 with pattern blocks, and finally drew on paper what they inferred to be the structure of stage 10. Figure 5.4c, d provides examples of their drawn work on the two patterns. The remaining eight students sensed the two patterns as having a low Gestalt effect.

The idea behind ambiguous patterns such as the one shown in Fig. 3.9 has been drawn from Dörfler's (2008) insights on the current state of patterns research. He notes that patterns with well-defined stages impress on learners the view that "there is an expected direction of generalizing," which could "intimate one and only way

Fig. 5.5 Dexter's (Grade 2) extensions of the Fig. 3.9 ambiguous pattern task

[of continuing] a figural sequence" (p. 153). Consequently, such patterns might induce "a strong regulating or even restrictive impact" on their thinking (p. 153). He then recommends the use of "free generalization tasks" that ask them to think about (figural) patterns in a different way, as follows:

> How otherwise can one ask for, say, the number of matchsticks ... in an "arbitrary" item of the sequence? The situation would presumably be much more open if one asked simply "How can you continue?" or "What can you change and vary in the given figures?" ... I rather want to hint to possible further directions for research ... a plea for "free" generalization tasks not restricted by pre-given purposes.
>
> (Dörfler, 2008, p. 153)

Figure 3.10 shows samples of well-defined figural patterns that four of my Algebra 1 students developed relative to the Fig. 3.9 task. Figure 5.5 shows second grade Dexter (7 years old) who, in a clinical interview session, produced the same growing L-shaped pattern in Fig. 3.10 prior to formal instruction at the beginning of the school year. While he was unable to establish an algebraic generalization (i.e., using variables to convey a direct formula), his incipient generalization on stage 10 of his constructed pattern provides a glimpse of his emergent structure, that is, "[in stage 10,] there are 10 across and 10 up."

2 Abductive Reasoning in Pattern Generalization Activity

When Cohort 1 and 14 others were in sixth grade, results of the clinical interviews prior to a teaching experiment on pattern generalization indicate all but one extended the incomplete pattern in Fig. 2.6a by adding two circles, one on the vertical side, and the other on the horizontal side, from stage to stage. Figure 5.6a shows Shawna's extensions relative to the Fig. 2.6a pattern. While she initially stated that the pattern "adds by 2," she then focused on stage 4 and saw that "it adds one on here [top of the column] and one on here [right end of the row]." Having that in mind, she constructed stages 5, 6, and 7 and eventually stated her generalization in words, shown in Fig. 5.6b. We classify her incipient generalization as conveying *structural iconicity*, that is, she noticed and interpreted an internal generality that applies within and across the given stages. In light of her generalization, she then constructed stage 100 and said, "the bottom would be 100 and the vertical would be 99."

Like Shawna, Dina in Fig. 2.6b and Anna in Fig. 4.26b also perceived the same structure relative to the Fig. 2.6*a* pattern. Their thinking about the pattern differed in their constructed stages. Anna and Dina saw the constant addition of two circles and nothing more, which should explain why their constructed stages were as such, that is, sets of circles that grew by two circles per set. Their incipient generalization conveys *superficial iconicity*, that is, they noticed and interpreted an external generality across the given stages.

Figure 5.6c shows Jenna's extensions relative to the Fig. 2.6*a* pattern. Unlike Dina, Anna, and Shawna, Jenna initially saw the lengths of the row circles as somehow related to the stage number and the lengths of the column circles oscillated over a cycle of 0, 1, 2, and 3 circles. Figure 5.6d shows her incipient generalization, which conveys *structural iconicity*. Unfortunately, Jenna found it difficult to extend

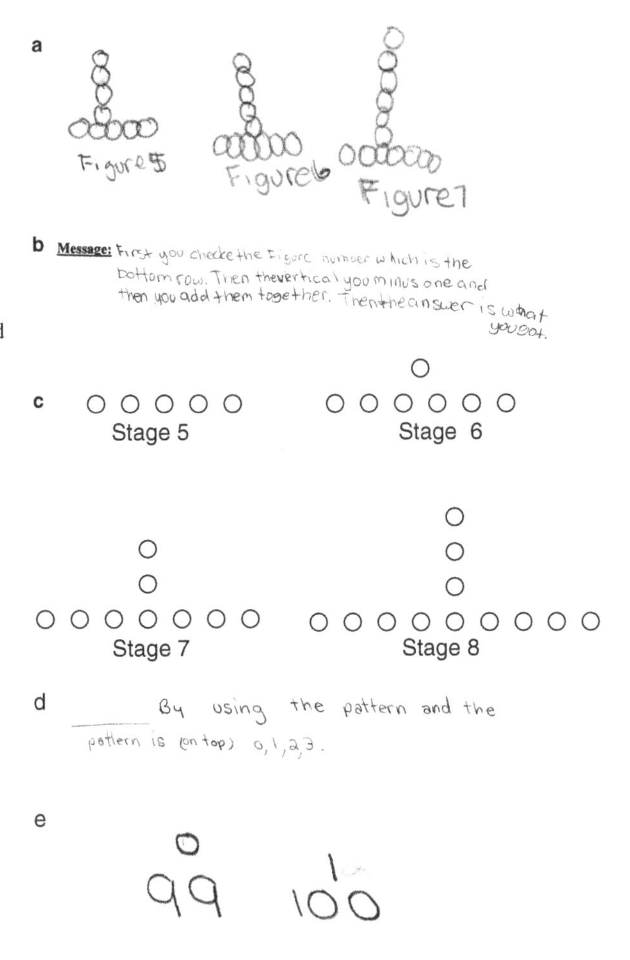

Fig. 5.6 **a** Shawna's extension in relation to Fig. 2.6a. **b** Shawna's verbal generalization of her pattern in Fig. 5.6a. **c** Jenna's extension of the pattern in Fig. 2.6a. **d** Jenna's verbal generalization of her pattern in relation to Figs. 2.6a and 5.6c. **e** Jenna's predicted stages 99 and 100 relative to her generalization in Fig. 5.6d

her stages correctly. In Fig. 5.6e, she guessed that stage 99 would have a row of 99 circles and no circle on the column, while stage 100 would have 100 circles on its row and 1 circle on its column, etc.

Figure 5.7 is a diagram that shows phases in pattern generalization, which I have drawn on the basis of the clinical interviews that were conducted with my sixth-grade class relative to the five tasks shown in Fig. 4.26a. In the diagram, abduction is situated at the kernel of the generalizing process. Individual learners initially explore a plausible rule (i.e., a tentative generalization) that might explain the known stages and then use it to construct the unknown stages in any given pattern. A signpost for abductive reasoning is when they offer an explanatory hypothesis or rule for a given pattern on the basis of the available stages. They then use the abductive claim to extend the pattern ("near generalization tasks" such as determining the fourth or the fifth stage) and repeatedly test – that is the inductive phase – that ultimately enables them to either confirm the rule (preferably a direct formula) or see the necessity of developing a further abduction. When the rule is confirmed, a generalization emerges that allows them to deal with far generalization tasks such as obtaining the

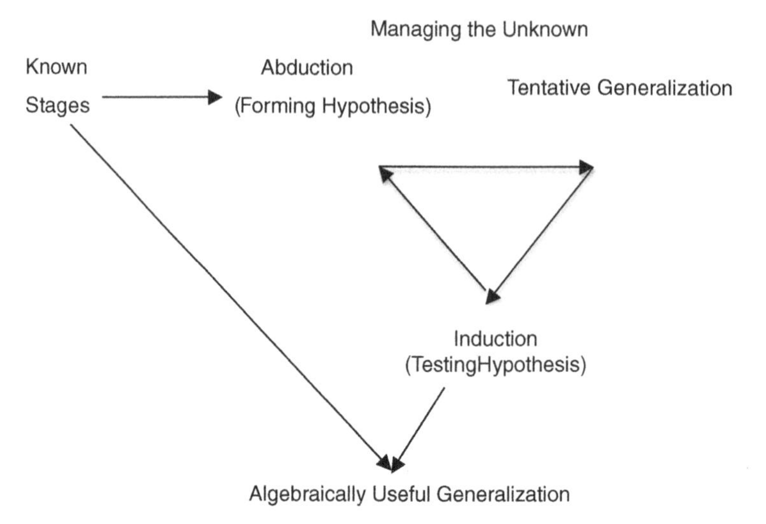

Fig. 5.7 Diagram of phases in generalization

77th stage of the pattern without laboriously constructing the preceding 76 stages. A further abduction is warranted when the rule makes it almost difficult for them to successfully deal with any far generalization task. In the above situation, for example, Shawna first established two generalizations on the basis of the known stages (adding by 2 s in the case of stages 1–4 and then extending the generalization to stages 5–7; seeing a relationship between the number of circles on each row and circle in stage 7 and then generalizing), which she then used to extend her pattern to stage 100. Jenna, Dina, and Anna, however, needed to refine their abductions in an algebraically useful manner.

Figure 5.7 should help understand why students need to accomplish both near and far generalization tasks beyond the busy work of calculating dependent outcomes. Near generalization tasks help students to begin the verification process (inductive phase) following an explanatory hypothesis (abduction), a tentative generalization. With repetitive action over a few simple, manageable instances comes the secure sense of confirming. Consequently it means slowly producing and projecting a plausible or an appropriate shape or image relevant to a pattern and an interpreted structure. Being confronted with a far generalization task represents a moment of perturbation. At this stage, they test the projective power and validity of, including convenience in using, the hypothesis or tentative generalization in dealing with an extremely large stage. Repetitive success in far generalization items signals the moment of encapsulation and (empirical) justification in which a final generalization emerges and is seen as an object and a process, where the object (i.e., the direct formula) conveys the structure of the pattern and the process (i.e., the individual coefficients, terms, and indicated operations) refers to the arithmetical power of the established generalization in determining an exact outcome.

Thus, when students perform a mathematical generalization, they go through stages of abduction and induction in which abduction is the creative domain of rule inference and formation and tentative generalization and induction the confirmatory phase of testing and extending but only in relation to a stated abduction (cf. Rivera & Becker, 2007a). In other words, it is simply not possible to induce without stating an abductive claim. The original context in which abduction has been conceptualized situates it as always being prior to induction (Peirce, 1957). However, recently, abduction has been reinterpreted in a more dynamic manner that works simultaneously with induction. This is done to ensure the production of the best rule or explanatory hypothesis, which in many cases occurs after a series of abduction and induction (Arzarello, Micheletti, Olivero, & Robutti, 1998; Rivera, 2008, 2010a; Rivera & Becker, 2008, 2010).

Hence, in thinking about generalizing patterns on the basis of a few known stages such as the ones shown in Fig. 4.26a, abduction should help settle the issue of whether such patterns are well defined. The property of a pattern being well defined is not something that is inherent in any pattern. It becomes well defined only when students establish an abductive claim, which they then use to impose and project on both the known and unknown stages of the pattern. Sources of abduction are drawn in several places and could be developed depending on the kind and type of learning experience and relevant contexts that matter in patterning activity. Certainly, in figural patterns such as the ones shown in Fig. 4.26a, students' abductive claims about their structure, feature, property, or attribute within and across stages are likely going to depend on their prior and current mathematical knowledge and, especially, the Gestalt effect that such patterns have on them. Hence, progressive transitions in abductive reasoning are possible.

An important aside on the general nature of abduction is necessary and useful at this stage in order for readers to acquire a full sense of its value and, in fact, its central role in mathematical reasoning. An exemplary model of mathematical reasoning involves deduction, which involves inferring a conclusion from completely known premises. The monotonic canonical form in deduction, "if p, then q," conveys how, say, having a known rule and an observed case always logically infers a result. Consider, for example, the statements below taken from the pebble diagram shown in Fig. 3.16.

Rule or Law: The sum of any number of even integers is even.
Case: 8, 4, and 6 are even integers.
Result: $8 + 4 + 6 = 18$, an even sum.

Peirce (1960) in the late nineteenth century observed that we are by nature drawn to "perpetually making deductions" (p. 449). He writes:

> We conceive that Cases arise under these laws; these cases consist in the predication, or occurrence, of *causes*, which are the middle terms of the syllogisms. And, finally, we conceive that the occurrence of these causes, by virtue of the laws of Nature, results in effects which are the conclusions of the syllogisms.
>
> (Peirce, 1960, p. 449)

But Peirce also notes that both abduction and induction are reasonable and natural inferential structures except that the order in which we set up the statements depends on the context we are analyzing. While deductive reasoning proceeds from the rule and the case (i.e., the premises) to the result (i.e., the conclusion), inductive reasoning begins with the result and the case leading to the rule and abductive reasoning takes as given both the rule and the result and then concludes with the case. Thus, for Peirce (1934), "(d)eduction proves that something *must* be; induction shows that something *actually* is operative; abduction merely suggests that something *may be*" (p. 106).

Alternatively, it makes sense to view a deductive inference as primarily *predicting in a methodical way* an inevitable and valid result or conclusion because all the required premises (the rule and the case) are completely known. Further, Peirce notes how,

> (d)eduction is the only necessary reasoning. It is the reasoning of mathematics. It starts from a hypothesis, the truth or falsity of which has nothing to do with the reasoning; and of course its conclusions are equally ideal. The ordinary use of the doctrine of chances is necessary reasoning, although it is reasoning concerning probabilities. Induction is the experimental testing of a theory. The justification of it is that, although the conclusion at any stage of the investigation may be more or less erroneous, yet the further application of the same method must correct the error. The only thing that induction accomplishes is to determine the value of a quantity. It sets out with a theory and it measures the degree of concordance of that theory with fact. It never can originate any idea whatever. No more can deduction. All the ideas of science come to it by way of abduction. *Abduction consists in studying facts and devising a theory to explain them. Its only justification is that if we are ever to understand things at all, it just be in that way.*
>
> (Peirce, 1934, p. 90; italics mine)

We note at this stage two important clarificatory points on abduction raised by Thagard and Eco. For Thagard (1978), an abductive process involves developing and entertaining inferences toward a rule (i.e., an explanatory theory), which is expected to undergo testing via induction leading to an inference about a case. However, Eco (1983) makes a stronger claim, that is, "it is important to stress that the real problem is not whether to find first the Case or the Rule, but rather how to figure out both the Rule and the Case *at the same time*, since they are inversely related, tied together by a sort of chiasmus" (p. 203).

In pattern generalization activity, students always begin with the result (i.e., the given stages in a pattern) and then proceed to find the appropriate rule (i.e., the direct formula of the pattern). Consequently, they use these premises (i.e., result and rule together) to explain why the conclusion (i.e., the case) is the way it is relative to the pattern. The nature of such explanation is, of course, a matter of knowing that it is but a "formulat[ion of] a general prediction" and carries with it "no warranty of a successful outcome" (Sebeok, 1983, p. 9). For an abductive reasoning process is "always-already" constrained by an incomplete knowledge base (the "result") that is used to determine an appropriate rule (at least for the time being). This is, unfortunately, its weakness: it has the quality of being fallible because the source of the case is not completely available. The inference that is constructed is, therefore, ampliative and deductively invalid (Peirce, 1960, pp. 446–447). Taken together, the rule and the result in abduction have a rather perfidious nature or, in Cifarelli and Saenz-Ludlow's (1996) words, are "plausible hypotheses on probation" (p. 161).

Inductive reasoning, which Peirce (1960) also classifies as a type of ampliative inference, proceeds in the reverse order of deduction. The hypotheses involve the result and the case. Further, through repetition of several more cases, a rule can be inferred. Where Peirce sees a difference between abduction and induction is on matters involving strength of inference. The conclusion in an induction has already been drawn from repetition, which could be evaluated as being false, true, or true to some degree. The conclusion in an abduction, however, is "truth producing" (Josephson & Josephson, 1994) in the sense that it fuels discovery on the basis of a few known instances (i.e., stages in a pattern). Hence, the essence of induction lies in verifying an abductive inference through repeated testing.

To sum up, while abduction involves discovering a new hypothesis, induction establishes the strength of the hypothesis through an experimental confirmatory process that produces tendencies in support of the hypothesis. There is closure in induction with a statement of a probable obvious generalization unlike abduction that never seems to do so since it works primarily in discovering or suggesting new events (Abe, 2003) or novel actions (Cifarelli, 1999). For Flach (1996), an abductive inference is always confronted through inductive testing, and increased inductive success means increased confidence in the abduced claim. Hoffmann's (1999) point astutely captures the original Peircean sense in which abduction forecloses any illusion of a perspective-free induction: "Induction is not what can be generalized from a sample of data, but only a quantitative determination of what is already given by abduction" (p. 272).

Hence, in pattern generalization, there is a mutual, complicit relationship between abduction and induction. Josephson articulates it clearly in the following manner:

> [I]nductive generalizations derive their epistemic warrants from their natures as abductions. . . . [The] mechanisms for inductive generalization must be abductively well-constructed, or abductively well-controlled, if they are to be smart and effective.
>
> (Johnson, 2000, p. 44)

We end this section with three important points concerning abductive reasoning in cognitive mathematical activity. These points are actually raised in various sections in this book, especially in the next chapter that deals with visual diagrams in mathematical reasoning. *First,* we adhere to the notion of *progressive abduction.*[2] Consistent with our proposed traveling theory in this book, progressive abductions are necessary in any account of progressive formalization. In the case of pattern generalization activity, in particular, the construction and the justification of a direct formula relative to patterns explored in the school mathematics curriculum symbolize mature, complete, and structural abductions. The use of a recursive rule, like in the case of Anna and Dina above, is indicative of entry-level, superficial abductions that could still evolve into structural abductions as was the case with Shawna. More generally, Josephson and Josephson (1994) have proposed the following neo-Peircean version of abduction below that I associate with mature,

---

[2]This notion conceptually shares Thagard's (1978) interpretation of Peirce's abduction as "cover[ing] both the act of arriving at plausible new hypotheses and the act of entertaining them for the sake of further investigation" (p. 166). The end result (i.e., at least in provisional terms) of progressive abduction is an *inference to the best explanation* ("*IBE*;" Hartman, 1965).

structural abductive claims. The variable H stands for "*the inference that yields the best explanation*."

Case: D is a collection of data (facts, observations, givens).
Rule: H explains D (would, if true, explain D).
Strong Claim: No other hypothesis can explain D as well as H does.
Result: H is probably true.

To illustrate, while Anna and Dina produced an abductive claim, Shawna developed a better abduction. Jenna also produced a reasonable abductive hypothesis; however, her structural abduction did not progress into an algebraically useful abduction. Hence, as shown in Fig. 5.7, pattern generalization involves the development of an algebraically useful structural abduction that could withstand inductive testing prior to being classified as an algebraic generalization for the pattern. Additionally, the strength of H could be evaluated by considering

- how good H is by itself, independently of considering the alternatives;
- how decisively H surpasses the alternatives, and;
- how thorough the search was for alternative explanations

(Josephson, 1996, p. 3).

Second, abductive reasoning offers explanations that do not prove, that is, are not deductively sufficient (Josephson, 2000). Certainly, some deductive proofs do not necessarily explain as well, so we need to be clear about the nature of explanations in abductive reasoning. For Josephson (2000), such explanations primarily assign causal responsibility; hence, reasoning proceeds from effect to cause.

(E)xplanations give causes. Explaining something, whether that something is particular or general, gives something else upon which the first thing depends for its existence, or for being the way that it is. .... It is common in science for an empirical generalization, an observed generality, to be explained by reference to underlying structure and mechanisms.

(Josephson, 2000, pp. 38–39)

Since there is a strong relationship between abduction and induction, Josephson (2000) notes that "smart inductive generalizations are abductions." An inductive generalization is

an inference that goes from the characteristics of some observed sample of individuals to a conclusion about the distribution of those characteristics in some larger population.

(Josephson, 2000, p. 40)

Examples include categorical statements ("All A's are B's"), statistical generalizations ("75% of A's are B's"), and example-driven concept learning in artificial intelligence. The generalizations as they are routed in the conclusion basically explain the characteristics of the available sample and do not necessarily explain why the instances must be so. Josephson writes:

"All A's are B's" cannot explain why "This A is a B" because it does not say anything at all about how its being an A is about why this one is, except that it suggests that if we want to know why this one is, we would do well to figure out why they all are. Instead, all A's are B's helps to explain why, when a sample was taken, it turned out that all of the A's

> in the sample were B's. A generalization helps to explain some characteristics of the set of observations of the instances, but it does not explain the instances themselves. That the cloudless, daytime sky is blue helps explain why, when I look up, I see the sky to be blue, but it doesn't explain why the sky is blue. Seen this way, an inductive generalization does indeed have the form of an inference whose conclusion explains its premises.
>
> (Josephson, 2000, p. 41)

Third, since the content focus in this chapter is pattern generalization, where justification means the same way that Josephson (1996, 2000) viewed explanation in the preceding paragraphs, we skip situations relevant to deductive closure in abductive and inductive reasoning activity. An example of deductive closure in my Algebra 1 occurred when they tried to make sense of the rule $-(a \times b) = -a \times b$ and $-(a \times -b) = -a \times -b$, where a and b are integers. Figure 4.25a, b enabled the students to initially abduce and induce the appropriate rules using algeblocks. Then we shifted to a deductive mode in Fig. 4.26c when we tried to understand the equivalent statements using a particular example that modeled the deductive steps involved in the actual proof with variables. Deductive closures drawn from processes of abduction and induction are exemplified in recent work on exemplary uses of dynamic technologies in school geometry (e.g., Arzarello et al., 1998; Boero, Garuti, & Mariotti, 1996; Pedemonte, 2007) that demonstrate the cognitive unity thesis between and among the three types of reasoning. Suffice it to say, an ideal full progression in the school geometry curriculum involves transitions in thinking and reasoning from the "truth-producing" (Josephson & Josephson, 1994) mechanisms of abduction and induction to the "truth-preserving" (Josephson & Josephson, 1994) methods of deduction. In the current school algebra content curriculum, in particular, pattern generalization, we appear to be more concerned with situations that produce algebraically useful explanatory hypotheses that could be explained in ways that are appropriate to grade-level expectations (e.g., concrete structural justifications; cf. Pylyshyn, 2006; Rivera, 2007b, 2010b, 2010d). Deductive closure in the form of, say, the proof of mathematical induction is not usually pursued.

3 Entry-Level Abductions

Table 4.1 (p. 135) is a summary of the preinstructional modes of generalization that my sixth-grade class exhibited during the clinical interviews relevant to the five patterning tasks shown in Fig. 4.26a. There were more students who produced superficially iconic generalizations on the first three tasks (about 75%) than those who produced structurally iconic ones (about 17%). However, in the case of the fifth task, there were more structurally iconic responses (about 62%) than superficially iconic ones (about 28%). In the case of the fourth task, none of the students were able to establish a generalization. How might one explain such seemingly divergent findings? Certainly, it is reasonable to expect that superficially iconic generalizers would consistently produce the same type of response on all tasks. One plausible response we explore in some detail involves the students' visual predisposition toward seeing in either additive or multiplicative terms. We first clarify what we

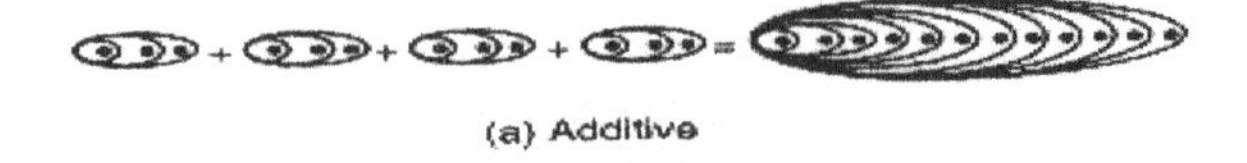

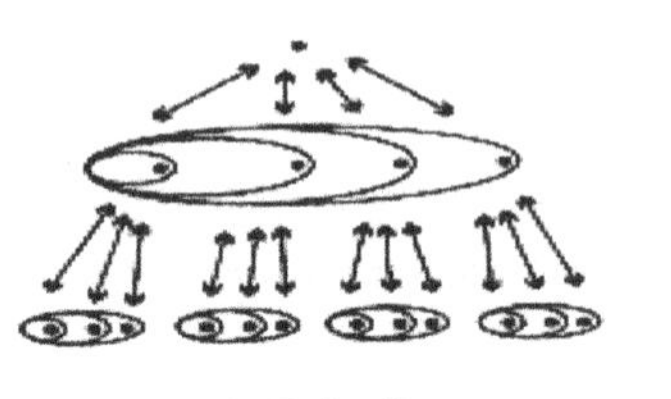

Fig. 5.8 Additive versus multiplicative schemes for counting whole numbers (Clark & Kamii, 1996)

mean by additive and multiplicative thinking in the context of students' experiences with whole numbers.

Figure 5.8 illustrates the fundamental difference between additive and multiplicative thinking involving whole-number objects in terms of level of abstraction and the complexity of inclusion relations involved. Drawing on Piaget's (1987) conceptualization, Clark and Kamii (1996) share the same view that multiplication is an "operation that requires higher-order multiplicative thinking that children construct out of their ability to think additively" (p. 42). They note that while addition "is inherent in the construction of number, which is accomplished by the repeated addition of ones," multiplication "is a more complex operation that is constructed out of addition at a higher level of abstraction" (p. 42). As shown in Fig. 5.8, additive thinking necessitates only one level of abstraction and one inclusion relationship (e.g., each unit of three consists of ones; one gets included in two, two in three, three in four, etc.). In the case of multiplicative thinking, there is a need to establish *simultaneously* (as indicated by the arrows) the following levels of abstraction and inclusion relationships: a many-to-one mapping (e.g., between three units of one and one unit of three) and at least two levels of relationships (e.g., horizontally by ones: one in two, two in three, etc.; horizontally by threes: 1 three in two threes, 2 threes in 3 threes, etc.; vertically by threes: 1 threes, 2 threes, 3 threes, etc.).

In Fig. 5.8, the notion of a *unit* in the context of counting sets of objects plays a central role in additive and multiplicative thinking. Sophian (2007) also shares this view, that is, counting involves "the choice of a unit .... In principle, provided we are consistent about what we take as a unit, we can count any sort of discriminable element in any kind of array" (p. 64). For Sophian,

> mathematical thinking requires a concept of *unit* that goes beyond our everyday notions of objects and groups of objects. Specifically, it requires flexibility in the choice of units, together with a concern for equivalences among units. ... Notably, in counting, *it is the identity of the unit rather than its size that matters*: Some of the shoes may be baby shoes and others large men's shoes, but we treat each of them as one shoe (or one half of a pair of shoes).
>
> (Sophian, 2007, p. 66; italics added for emphasis)

When we engage in additive thinking, which involves a single-level abstraction, we employ a unit that may or may not be independent of the quantities involved in addition (and subtraction). For example, in the comparison situation, "Maria is two inches taller than her sister Jana," the unit – inches – pertains to height that conveys a linear measure. We could, of course, use other linear units such as centimeters. In the combination situation, "There are 29 students in a room with 12 boys and 17 girls," the unit – number of people – is independent of either quantity being added. When we engage in multiplicative thinking, which involves multi-level abstractions, we employ a unit that is actually drawn from one of the quantities involved in multiplication (and division). For example, in the comparison situation, "Maria is twice as tall as her sister Jana," the unit is Jana's height which we then use to compare Maria's height with. In summary, both types of thinking necessitate the use of a *common unit* that further enables us to understand the nature of the inclusion relationships shown in Fig. 5.8. In Fig. 5.8a, the common unit may or may not be independent of the quantities involved. In Fig. 5.8b, the many-to-one map initially relies on the choice of a common unit that is then used to systematically count the quantities involved.

Going back to the results in Table 4.1, at the preinstructional phase, superficially iconic generalizers could only think *additively*, that is, they see relations of up to one level of abstraction. Visually, it means they see parts in a stage number as disconnected and unrelated from each other, which explains why their attention, say, in the case of increasing linear patterns was focused on the stable action of adding the same number of objects from stage to stage. For example, the incipient generalizations of Dina in Fig. 2.6b and Dianara and Anna in Fig. 4.26b relative to Task 1 in Fig. 4.26a all convey a singular focus on the constant addition of two circles from one stage to the next. For Frank, whose written work is shown in Fig. 4.26b relative to Task 5 in Fig. 4.26a, he saw the constant addition of two squares after he obtained the initial differences between two consecutive stages in the pattern and nothing else beyond that observation.

Structurally iconic generalizers, on the other hand, think in either structurally additive or structurally multiplicative manner depending on pattern goodness. A *structurally additive generalization of a pattern* is an additive arrangement of at least two *un/related* parts in a figural stage. For example, Diana initially obtained the total number of square tiles in stage 10 of Task 3 in Fig. 4.26a by adding, which is as follows: "10 + 10 + 10 + 10 + 1 = 41." Because the interviewer wanted to determine whether she could see anything else, Diana was then asked to find the total in the case of stage 100 in which case she replied, "100 + 100 + 100 + 100 + 1 = 401" tiles. From her last response, Diana gave no indication that she was seeing a relationship among the parts in a figural stage. The incipient generalizations of Shawna and Dung in the case of Task 1 in Fig. 4.26a could be classified as being structurally additive; however, they both saw parts that were related in some way. Dung's reasoning is as follows:

> *Dung: Like these all have ahm the bottom has the same number of figures and the top has the same number too [counting the overlapping circle on the bottom row. He then builds 10 horizontal circles and then adds 9 circles vertically.]*

JRB[3]*: So that's it? How many [circles] are there altogether?*
Dung: 19. [Dung then deals with stage 100.]
JRB: So we probably will not make figure 100, are we? We probably don't have enough [circle chips on the table] for that. But can you picture?
Dung: You have like 100 to be on the bottom and 99 on top. So there'd be 199.

A *structurally multiplicative generalization of a pattern* is a multiplicative arrangement among parts in a figural stage. The arrangement, which is not needed in additive structures, involves establishing a many-to-one correspondence between congruent parts. It requires abstraction on two levels, that is, seeing each part or copy as consisting of equal subparts and then seeing all the relevant congruent copies together as a single copy. For example, Jennifer's visual reasoning below in the case of Task 3 in Fig. 4.26a progressed from being superficially additive to structurally multiplicative.

JRB: Do you see any patterns with these tiles? *Jennifer: You add four more squares on each corner.* *JRB: Okay, does that keep going on each picture number?* *Jennifer: Uhum.*	Superficially iconic abduction
JRB: So how many tiles altogether are in picture 10? [Jennifer starts to build picture 6 from the tiles.] Could you tell me how many tiles are there in picture 6 altogether? *Jennifer: 24.* *JRB: 24? Are you sure there's 24? You did that real fast. How did you count it?*	Structurally iconic abduction
Jennifer: There's six right here [referring to one arm] and there's four lines, 25, counting the middle [square]. [In seeing multiples of the same arm, Jennifer then determines the number of tiles for pictures 9 and 10.] 37 [referring to the number of tiles for picture 9]. *JRB: How did you get that so fast?* *Jennifer: Coz there's 9 here, 9 there. So there's 9 on every arm. Then I multiplied by 4 and then I counted this [referring to the middle square].* *JRB: Good. So how [many] would you get for picture 10?* *Jennifer: Add another one [on each arm of picture 9]. There's 41. There's 10 here, 10 there, 41.*	Induction Phase

[3]JRB stands for Joanne Rossi Becker, interviewer.

So, central to Jennifer's structurally multiplicative generalization was her cognitive perception of seeing two levels of abstraction, that is, an arm that consists of n square tiles and all four arms that taken together are multiple copies of the same arm each having length n.

In the case of Task 5 in Fig. 4.26a, we note that there were more structurally iconic generalizations than superficially iconic ones. Liana's reasoning below exemplifies a structurally multiplicative generalization.

Liana: Picture 1 has 2. Picture 3 has 3 going across and 2 squares going up. Picture 3 has 3 going up and 4 going across. Picture 4 has 4 going up and 5 going across. Picture 5 has 5 going up and 6 going across. So for picture 8 [there'd be] 8 boxes going up and 9 boxes going across. .... Altogether 8 times 8. Is it 8 × 9? 8 × 8.
JRB: Why don't you check on for picture 6?
Liana: Six times 6 ... and that'd be equal to 36.
JRB: And what about [referring to the adjacent square block]?
Liana: 37. ... Picture 8 would be 8 times 8 ... [uses a calculator] 65. ... Picture 35, 35 times 35 plus 1.

Emma's written work in Fig. 4.26b and Dave's reasoning below exemplifies a shift in generalization from superficial to structural iconicity.

Dave: I just have to figure out the values first. ... I was thinking if I just add it by 2? ... Or added by 3.
JRB: What were you seeing in picture 1?
Dave: 2. Picture 2 has 5.
JRB: What did you see?
Dave: See coz it adds 1 up here [referring to picture 3] every single time, add 1 row. So one right here and add 1 row up there. But that's how high each goes. So there is a pattern but it changes.
JRB: It's not the same amount each time you mean?
Dave: Yes, it's not the same amount coz this [referring to picture 2] adds 3, 5, [for picture 4].
JRB: Uhum, and then how much to add the next time?
Dave: So this [referring to picture 6] is 6 plus 5 equals 11. 5, 4, 4, 3, 3, 2, 2, 1.
JRB: So what is it that you are seeing?
Dave: All of these together [referring to pictures 1 through 5].
JRB: Sort of build up? Is there a way that can help you find picture 8?
Dave: Yeah, it keeps on adding. In picture 6 add 7 more. [Dave uses the cubes on the table to build picture 7.] Nah. Too many. I'll draw it. [Dave uses a graph paper and starts to build picture 6].
JRB: Are you starting on where on picture 6, on the bottom or the top?
Dave: Ahm, on the side.
JRB: So how many on the side?

Dave: Six. [Dave draws 6 vertical, 6 horizontal lines forming a square. He then shades the square. Then he adds one column and one row forming picture 7. He next adds an adjacent square and adds 1 column and 1 row once more to form picture 8.] So 8 plus 7 is 15. 7 plus 6 is 13. Wait. [Dave gets another sheet.]

JRB: So what you're doing is you're adding the vertical and the horizontal on each of these? Is that what you're gonna do?

Dave: Yeah.

JRB: Is there an easier way?

Dave: I think there is.

JRB: How could that be?

Dave: Side by side. But then there'd be an extra one [square]. I shall just add that one.

JRB: So without the extra what do you have? [Dave writes: 8 × 8.] So what kind of figure is it though?

Dave: [Dave did not respond.] 8 times 9 minus 8 . . . plus 1. That'd be 65.

JRB: So how about picture 35. So what do you think?

Dave: I could write an expression.

JRB: That's a nice idea. What do you think?

Dave: Okay. So 35 times 35 plus 1.

JRB: So think about any picture of any size. So how many blocks would you have? Picture 8. Picture 35. Picture 100. Picture 1020.

Dave: So do you want a specific formula or the distance overall?

JRB: I'm not sure what you mean by specific formula. Not a specific number. We want a general one.

Dave: Overall?

JRB: It'd work for any number I would put in.

Dave: So n times n plus 1.

Compared with Tasks 1, 3, and 4 in Fig. 4.26a, the stages in Task 5 consist of squares whose basic structure has cohesiveness (i.e., in the sense of belonging together) and clearly redundant parts (i.e., repetition of the same part). Task 5 appealed to some students as being high in pattern goodness. Their interpreted structure enabled them to "specify a geometric or algebraic formula that stands out through its simplicity" (Metzger, 2006, p. 24), unlike the rest of the patterns which they saw as having parts that necessitated further conceptualization. For example, Dung's reasoning above in the case of Task 1 in Fig. 4.26a demonstrates two shifts in his cognitive perception of the parts in each figural cue. In thinking about stage 19, he initially perceived two overlapping sets of circles each having a cardinality of n and then taking away one circle corresponding to the overlap. He then changed his mind when he perceived a nonoverlapping relationship between the row and the column of circles. In stage 100, for example, he saw a row of 100 circles and a column of 99 circles. This conceptual kind of internal generality is difficult among students who merely perceive external generality in superficially iconic terms. Frank's written work relative to Task 5 in Fig. 4.26b demonstrates this perceived difficulty. Because his attention was fixed on the second-order difference of two square tiles, he merely kept adding. Certainly, it did not help that Frank (as well as a few others) harbored a view of mathematics that was all about manipulating numerical expressions and relationships as a consequence of his prior experiences. It constrained the manner in which he approached all the five tasks in Fig. 4.26a.

Among Tasks 1, 3, and 4 in Fig. 4.26a, the students found Task 4 to be the most difficult. Only one student produced a superficially iconic response that involved constantly adding four squares from stage to stage ("in a 4 × 4 grid, there are 4 + 3 + 3 + 2 = 12 shaded squares; in a 5 × 5 grid, there are 16 shaded squares, etc."). Two students offered the following structurally iconic response: "In a 3 × 3 grid, 3 × 4 – 4 = 8, so in a 25 × 25 grid, 25 × 4 – 4 = 96 shaded squares." The remaining 21 students did not see any relationship between a 3 × 3 and a 25 × 25 grid. Despite this fact, 13 of them correctly obtained the total count for the 25 × 25 grid by either unit counting ("1, 2, 3, . . ., 95, 96") or group counting ("23 + 23 + 25 + 25 = 96" or "25 + 24 + 23 + 24 = 96").

Pattern goodness could explain why they found it difficult to establish a generalization for the above pattern. The figural stages appealed to them as having a low Gestalt effect in the sense of not perceiving "harmony" among the parts in each stage. Pattern goodness was not immediate and conceptualizing the parts in, and relationships among, the stages could be accomplished in several different ways that necessitated more "acquired knowledge" unlike the situation in Task 5 in Fig. 4.26a. Metzger (2006) writes: "Knowledge becomes even more decisive a factor the less simple the structure of the image that requires completion" (p. 136). Thus, in some cases of visual patterns, the Gestalt effect is immediate. However, in some other cases, "the effects of knowledge and experience, and also of behavior" (Metzger, 2006, p. 137) play an equally significant role.

The idea of Gestalt effect plays a significant role especially in visual patterning tasks that are relatively unfamiliar to students. In Year 3 of my longitudinal study, I asked my Algebra 1 class to establish a generalization for the nonlinear task shown in Fig. 5.9a 3 months after a teaching experiment on pattern generalization that focused on linear patterns. The embedded rectangular pattern task in Fig. 5.9a was presented as a bonus item in an assessment that focused on factoring quadratic (polynomial) expressions. The success rate was 65%. The following week, I asked them once again to obtain a generalization for the same task in Fig. 5.9a as a bonus item. However, I presented it in a different way shown in Fig. 5.9b. This time, the success rate was 82%. Jackie's written work on both tasks is shown in Fig. 5.10a, b. Her written responses exemplify the thinking of those students who were unsuccessful in the Fig. 5.9a task but were successful in the Fig. 5.9b task. In Frank's case, he provided no response to Fig. 5.9a since the visual array did not make any sense with him. However, his written work in Fig. 5.10c showed a correct generalization. Further, the same generalization offered by Frank and Jackie in the case of the Fig. 5.9b task was the most frequent structurally iconic response (57%) followed by the formula $D = n\ (n + 1)$ (25%; see Fig. 5.10d for a sample of students' work). It is certainly interesting to note how 19 out of 34 students interpreted a $n \times (n + 1)$ rectangular cue as structurally consisting of a $n \times n$ square and a $1 \times n$ rectangle.

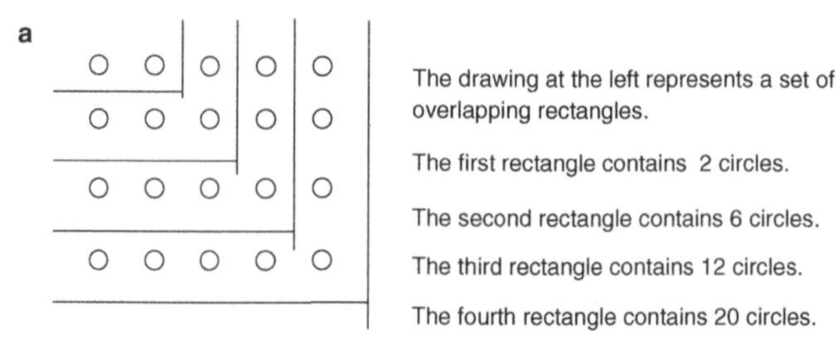

A. How many circles are there in the fifth rectangle? How do you know for sure?

B. Find a direct formula for the number of circles in the nth rectangle? Explain how you obtained it.

C. How many circles are there in the 100th rectangle? How do you know for sure?

b Below are the first four rectangular numbers.

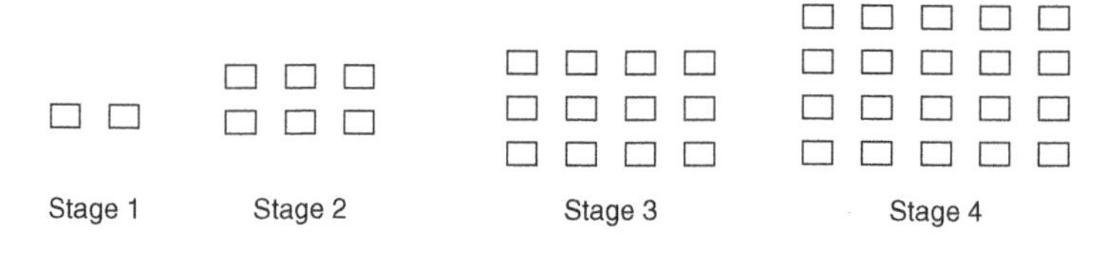

A. Describe stage 10 in a way that makes sense to you (for e.g., draw, etc.)

B. Find a direct formula that gives the total number of dots D at any stage n. Explain why your formula is true.

C. Is 9,900 a rectangular number? Explain your answer.

Fig. 5.9 **a** Embedded rectangular pattern task. **b** Separated rectangular pattern task

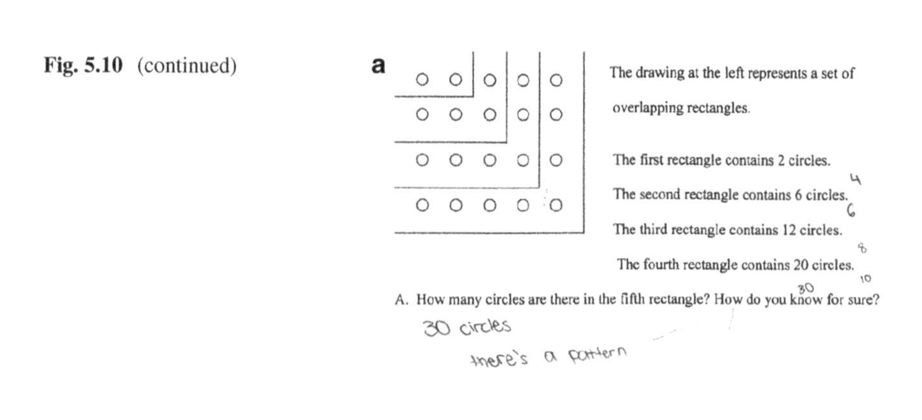
a

The drawing at the left represents a set of overlapping rectangles.

The first rectangle contains 2 circles.

The second rectangle contains 6 circles. 4

The third rectangle contains 12 circles. 6

The fourth rectangle contains 20 circles. 8

10

A. How many circles are there in the fifth rectangle? How do you know for sure? 30

30 circles

there's a pattern

Fig. 5.10 (continued)

B. Find a direct formula for the number of circles in the nth rectangle? Explain your formula.

can't find one

Fig. 5.10 (continued)

b

2. Below are the first four rectangular numbers.

2 6 12 20

Stage 1 Stage 2 Stage 3 Stage 4

A. Describe stage 10 in a way that makes sense to you (for e.g., draw, etc.)

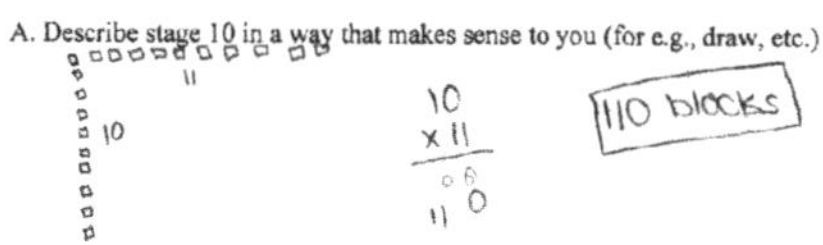

B. Find a direct formula that gives the total number of dots D at any stage n. Explain why your formula is true.

$D = n^2 + n$

Stage 2

$2^2 = 4$

$2^2 + 2 = 6$

C. Is 9,900 a rectangular number? Explain your answer.

c

2. Below are the first four rectangular numbers.

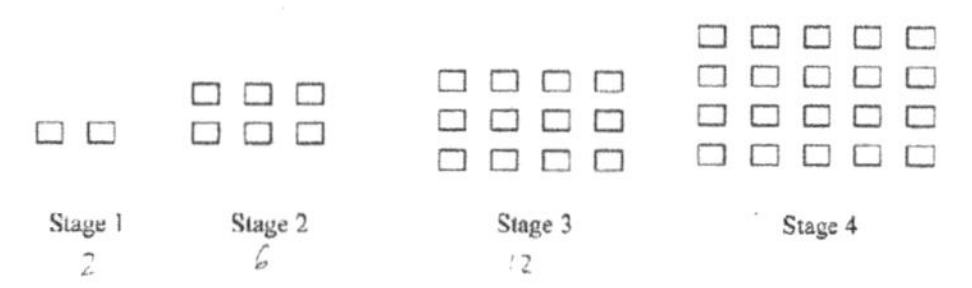

Stage 1 Stage 2 Stage 3 Stage 4

2 6 12

A. Describe stage 10 in a way that makes sense to you (for e.g., draw, etc.)

So you start off with ten squares. Then you would ad 10×10 squares to get the total

B. Find a direct formula that gives the total number of dots D at any stage n. Explain why your formula is true.

$D = n + n^2$

C. Is 9,900 a rectangular number? Explain your answer.

$\sqrt{9{,}900}$ 3,300

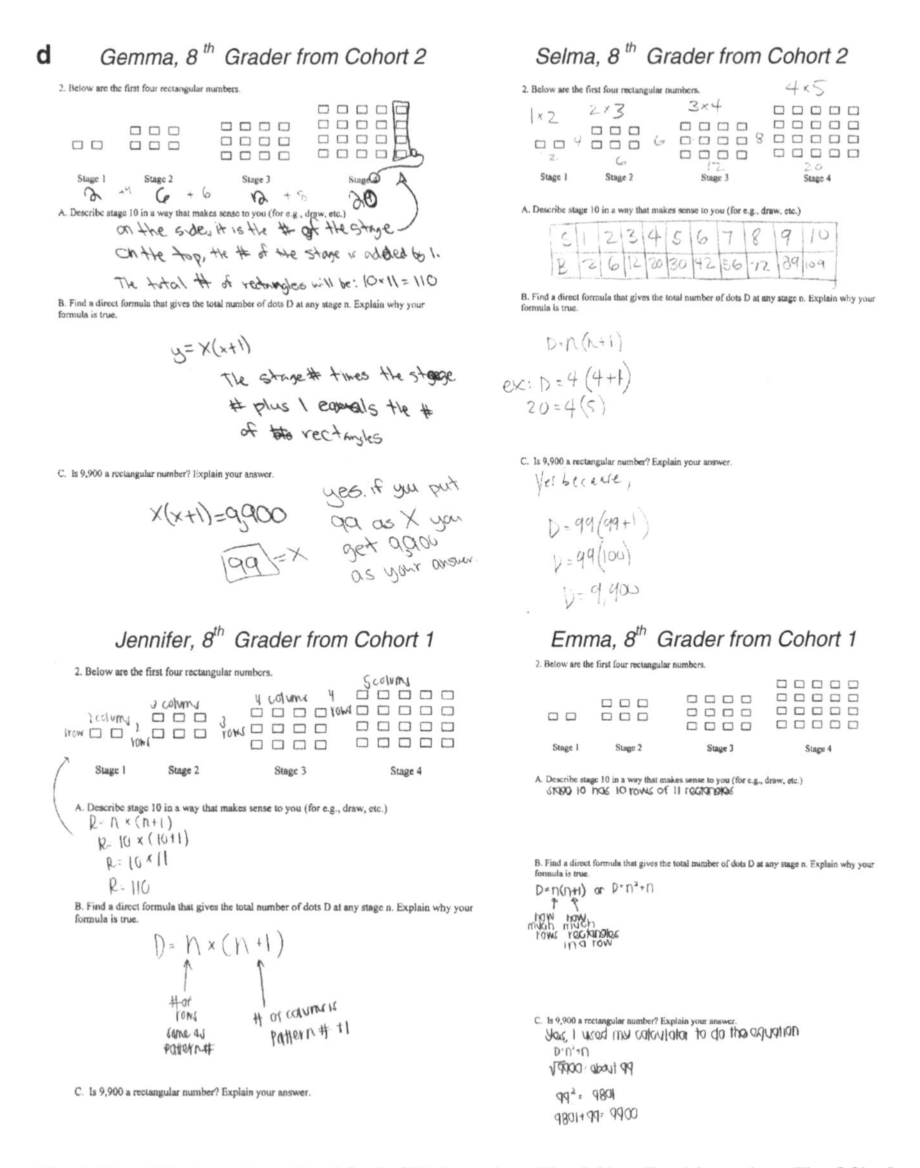

Fig. 5.10 **a** Ollie's work on Fig. 5.9a. **b** Ollie's work on Fig. 5.9b. **c** Frank's work on Fig. 5.9b. **d** Grade 8 students' work on Fig. 5.9b that follows the form $D = n(n + 1)$

Thus, it seems to be the case that the students' ability to pattern generalize was significantly influenced by the strength (or weakness) of the Gestalt effect in relation to figural patterns. In fact, the effect of knowledge was magnified in cases when visual pattern configurations (and not necessarily shapes) became more difficult. For example, when some of the students in the Year 3 study were asked to extend the semi-free patterning task in Fig. 3.9, 3 of the 11 students offered the triangular pattern shown in Fig. 5.11, which then necessitated additional knowledge in terms of how to obtain a pattern generalization for such a nonlinear pattern that appealed

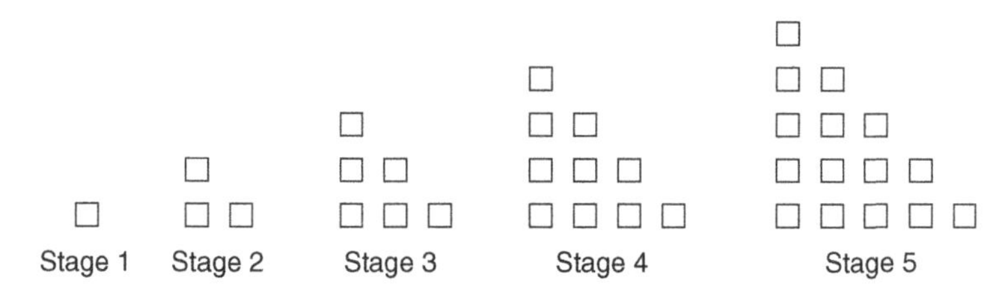

Fig. 5.11 Triangular pattern resulting from the semi-free patterning task in Fig. 3.9

to them as having a more complicated structure compared with the ones shown in Fig. 5.9a, b.

Within and beyond the Gestalt effect, we mention findings from vision studies that illustrate the human visual capacity to encode in small amounts and recall very little information when we perform a single glance on a figure (cf. Pylyshyn, 2006, p. 445). In the case of copying a figure, for instance, we tend to return our gaze several times on an aspect in the figure as we are constructing it. Hence, Gestalt effect in relation to patterning activity could also be influenced by the students' visual capacity to encode figural cues in memory. For example, when my Algebra 1 students encoded the structure of the figural cues shown in Fig. 5.9b, they actually pursued two different routes with one group seeing a rectangle (Fig. 5.10d) and another group seeing the union of a square and a rectangle (Fig. 5.10b), which then influenced the form of, as well as the time and ease it took them to develop, their algebraic generalization. Also, going back to Table 4.1, which shows the sixth-grade students' overall performance on the five tasks shown in Fig. 4.26a, the larger percentage of success in the case of Task 5 in comparison with the rest of the tasks could also be explained in qualitative terms by the relatively small amount of distractors that were present in each figural cue. In other words, the stages in Task 5 have a shared property that could be referred to as a "visual pop-out" in the sense that it could be detected rapidly without much effort unlike the figural cues in the case of the other tasks that require more effortful scrutiny (cf. Leslie, Xu, Tremoulet, & Scholl, 1998, p. 11).

4 Mature Abductions: Visual Templates

Entry-level abductions involving figural patterns among my sixth-grade students show the preponderance of superficially and structurally iconic generalizations. Mature abductions are characterized as being structurally symbolic generalizations. A direct, variable-based formula is used to express a generalization, which is expected to take the form of a function. Table 5.1 is a list of seven types of algebraic generalizations that my middle school students produced relative to figural pattern generalization at the end of a 3-year study, and they all exemplify visuoalphanumeric representations that have resulted from an evolved mode of abductive reasoning. The *constructive* dimension in both constructive standard and nonstandard generalizations refers to the cognitive perception that an interpreted structure relative to some pattern is seen as consisting of nonoverlapping parts that when added together form the perceived shape that applies across the stages in the pattern. The terms *standard* and *nonstandard* refer to the algebraic terms in a direct expression, that is, standard means the terms are already in simplified form, while nonstandard contains terms that can still be further simplified. *Constructive generalizations* reflect the use of either an additive or a multiplicative scheme. *Deconstructive generalizations* refer to the cognitive perception of seeing the known figural stages in a pattern as consisting of *overlapping parts* that can be decomposed quite conveniently. *Auxiliary-driven constructive or deconstructive generalizations* pertain to seeing each known figural stage in a pattern in the context of a larger configuration that has a well-known and/or simpler structure. Introducing an auxiliary set of objects strategically enables one to see a larger configuration as a way of obtaining an appropriate generalization for the pattern rather quickly and easily. *Transformation-based generalizations* are derived from initially performing actions of moving, reorganizing, and transforming parts in a figural stage of a pattern into some recognizable figure having a more familiar structure.

The existence of *visual templates* surfaced toward the end of my Year 3 study. I have mentioned in the Introduction that Cohort 1 participated in all 3 years of my study on pattern generalization with a Cohort 2 consisting of 19 seventh and eighth graders asked to join in the third year of the study. The teaching experiments on pattern generalization took place in the fall semester of each year with an average time span of about 4 weeks. Each teaching experiment in Years 1 and 2 occurred immediately after an 8-week teaching experiment that focused on integers and operations, while the Year 3 teaching experiment took place after an 8-week teaching experiment on integers, exponents, polynomials, and operations involving polynomials.

Section 4 in Chapter 4 provides details of the collective development of mathematical symbols relevant to pattern generalization activity in my 3-year design-driven longitudinal study. The progressive account surfaced the following three shifts in symbolic competence in relation to pattern generalization:

The symbols they used in the initial phase were visual, (structurally and superficially) iconic, and verbal prior to any teaching experiment in Year 1.

In the second phase, they favored numerical generalizing that lingered throughout the teaching experiments in Years 1 and 2. The symbols they used were alphanumeric that followed the differencing strategy suggested by Anna's group. Numerical tables of values were also favored due to the convenience in using the finite difference method in dealing with linear pattern problems (e.g., Fig. 5.3b).

In the third and final phase, they careened toward the visuoalphanumeric, which involves, in the words of Rotman (1995), "folding into each other [i.e., the visual and the alphanumeric] and are inseparable not only in an obvious practical sense, but also theoretically, in relation to the cognitive possibilities that are mathematically available" (p. 397). This phase took place in Year 3 when they began to think about pattern generalization activity in multiplicative terms.

Year 3 was particularly interesting since there were two cohorts of students with marked differences in their ability to generalize patterns. Hence, instead of reinventing the wheel, so to speak, I heeded Freudenthal's (1981) advice about repeating at

Table 5.1 Types of algebraic generalizations

Types of algebraic generalizations	Figural characteristics	Algebraic formula	Examples
Additive constructive standard	Seeing figural patterns as consisting of *nonoverlapping* parts	The terms in the formula are in simplified form This is the simplest case that applies to all linear figural patterns of the type $y = x + b$ (i.e., the constant addition of one object from stage to stage)	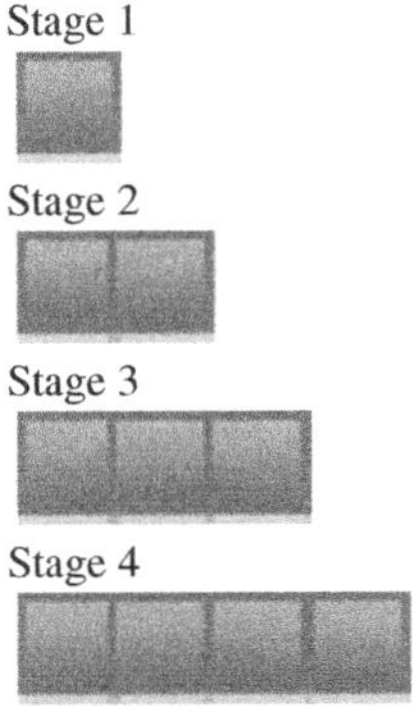 Stage 1 Stage 2 Stage 3 Stage 4
Additive constructive nonstandard	Seeing figural patterns as consisting of *nonoverlapping* parts	The terms in the formula are in expanded, nonsimplified form	Karen's formula in Fig. 4.16
Multiplicative constructive standard	Seeing figural patterns as consisting of *nonoverlapping* parts	The terms in the formula are in simplified form	Dung's formula in Fig. 3.31b
Multiplicative constructive nonstandard	Seeing figural patterns as consisting of *nonoverlapping* parts	The terms in the formula are in expanded, nonsimplified form	Dave's formula in Fig. 3.34

Table 5.1 (continued)

Types of algebraic generalizations	Figural characteristics	Algebraic formula	Examples
Deconstructive	Seeing figural patterns as consisting of *overlapping* parts. Hence, counting involves taking into account multiple counts that can be determined when the parts are appropriately decomposed	The terms in the formula could be standard or nonstandard in form The terms could convey the use of either an additive or a multiplicative scheme	Jenna's formula in Fig. 3.31b
Auxiliary-driven constructive or deconstructive	Seeing figural patterns as parts of a larger configuration that has a well-known and/or simpler structure. Introducing an auxiliary set enables learners to see the larger configuration more easily	The terms in the formula could be standard or nonstandard in form The terms could convey the use of either an additive or a multiplicative scheme	Diana's formula in Fig. 4.18
Transformation-based constructive or deconstructive	Seeing figural patterns in a different way by initially moving, reorganizing, and transforming a figural stage into some recognizable figure that has a structure that is familiar to the learner	The terms in the formula could be standard or nonstandard in form The terms could convey the use of either an additive or a multiplicative scheme	The formula in Fig. 4.19b

the point where I thought Cohort 1 benefited the most. In the Year 1 study, Cohort 1 spent a significant amount of time developing their understanding of a direct formula for a pattern, which actually emerged as a progressive formalization of notations from the use of arrow strings, numbers, and operations in a sequence and words (e.g., Fig. 4.3) to the use of the equal sign and variables in a function context (e.g., Fig. 2.7). In other words, direct formula construction in Cohort 1 emerged as a cultural and historical knowledge. So, in Year 3, I used this knowledge as my starting point in talking about pattern generalization. Also, both cohorts were already familiar with multiplicative thinking in relation to earlier activities where they used algeblocks in making sense of integers, exponents, and polynomials.

In the Year 3 classroom episode below, both cohorts were asked to obtain and justify an algebraic generalization for the tile patio pattern shown in Fig. 4.33. Five members from Cohort 1 initially clarified for the entire class the meaning of the term *direct formula* before they began working in pairs. Consequently, Cohort 2 was provided with the mathematical knowledge they needed to construct an acceptable algebraic generalization. The classroom event also assisted them in avoiding less efficient ways of expressing a generalization. The shared understanding that resulted from the social feedback provided by Cohort 1 in fact became the basis of communicative exchanges among the students in both cohorts. In the interchange below, the following aspects were discussed: the meaning of a direct formula; the elements that comprise a direct formula (equation form, two different variables); the difference between a direct and a recursive formula; and the shape of a direct formula.

FDR: What does direct formula mean?
Dave: A formula to find the [white tiles] in this problem [given] the patio stage number.
If you're given the patio number, you can find the number of tiles.
FDR: Okay. Meaning to say when you say direct formula, you need to find a formula that expresses the number of white tiles for any given patio number. So it should be like what? So how does it look then? . . . How do you know direct formula when you see one or when you're confronted with it?
Tere: Like you have to prove it. Like you have to do it.
FDR: Like how do you know one when you see one?
Tere: Coz it has variables in it.
FDR: So how? Describe those variables for me. How many variables do you see at least?
Tere: Two.
FDR: And where do you see those variables?
Tere: In the equation and other one at the end.
FDR: So there's an equation for one. . . . There should be one variable on the left side of the equation and there should be another one on the other side of the equation. [FDR makes a gesture by extending both his arms to indicate both sides in an equation.] Does that make sense? . . . If I have a recursive formula like add 2, is that a direct formula?

Emy: No.
FDR: Why not?
Emy: Because it doesn't have variables.
FDR: Well, for one it doesn't have variables, so that's one way to think about it. That's what you call a recursive formula. But I want you to come up with a direct formula. Alright. Give me an example of a direct formula just so we know.
Ford: n = 3n [FDR starts writing the partial formula.]
FDR: Same variable?
Ford: n = 3w plus 8. [FDR writes n = 3w + 8.]
FDR: Alright. This is an example of a direct formula. So there's a variable here on this side [referring to left] and a variable on the other side. What's another direct formula?
Dave: w = n times 4 plus 4. [FDR writes w = n × 4 + 4.]
FDR: Dave, is there another way to express n times 4?
Dave: n four.
FDR: So what's another way to say n four plus 4? [FDR writes w = n4 + 4.] If you want to be hip about it, what's another way to write n4?
Che: 4n.
FDR: 4n. Normally in algebra we start with the coefficient and then followed that up with a variable.

Figure 5.12 displays the written visuoalphanumeric generalizations of four Cohort 2 students in relation to the pattern shown in Fig. 4.33 that reflect the same responses offered by some of the Cohort 1 students as shown in Fig. 4.34. Their direct formulas conveyed how they cognitively perceived and abduced an interpreted structure that made sense with the pattern.

Figure 4.31b also shows the written work of three students from Cohort 1 who participated in a clinical interview immediately after the Year 3 teaching experiment on pattern generalization. Again, the consistency of the visuoalphanumeric process was evident in their work. Their direct expressions ranged in form from simple (Dung and Tamara) to complex (Jenna and Tamara). Further, while Dung and Tamara produced the same direct expression, $3n + 1$, the consistent manner in which they circled the objects empirically demonstrates differences in how they abduced and constructed a structure for the T stars pattern shown in Fig. 4.31a that influenced the form of their algebraic generalization.

Four and a half months after the Year 3 teaching experiment on pattern generalization, 8 eighth graders from Cohort 1 and 3 seventh graders from Cohort 2 (four males and seven females) on three nonroutine pattern tasks shown in Figs. 3.9 and 5.13a, b. The basic intent of the last clinical interview was to assess and describe the content and form of their pattern generalizations after a prolonged period of time in which no patterning activity was pursued on purpose.

As soon as I completed all the interviews, visual templates became particularly evident. To illustrate, when Karen (Cohort 2, Grade 7) extended the pattern

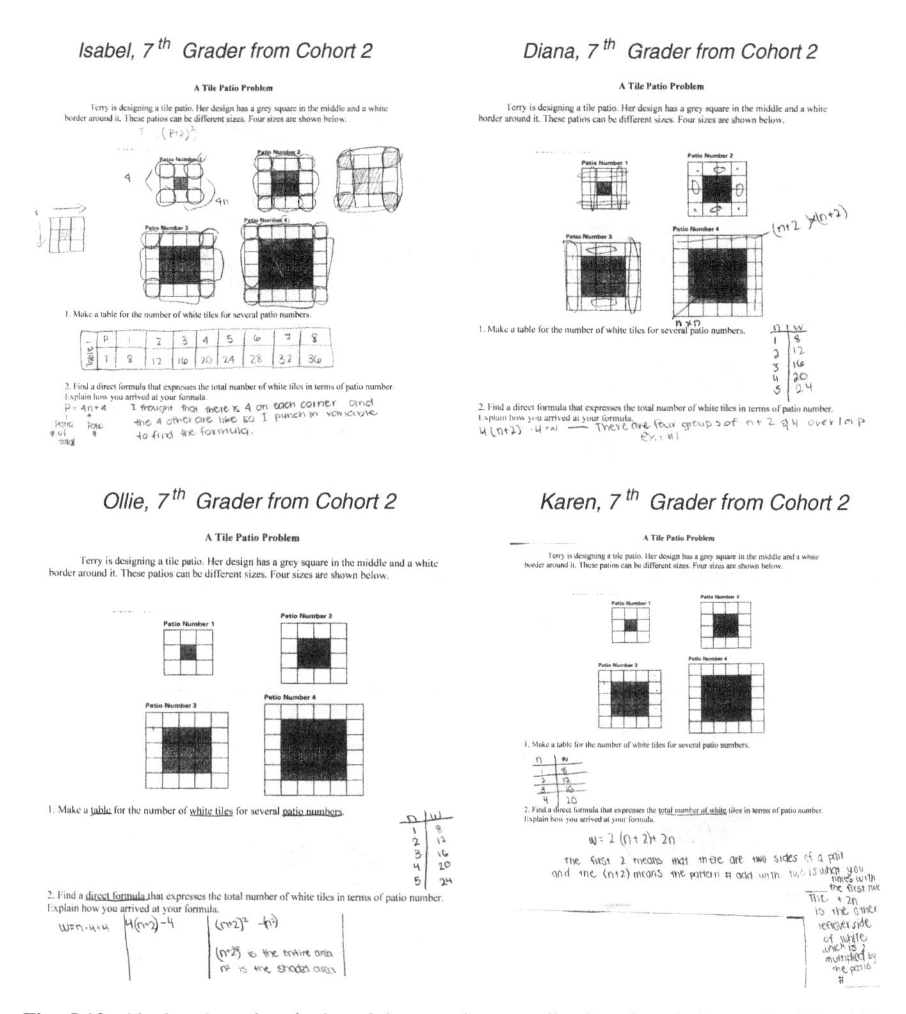

Fig. 5.12 Algebra 1 students' visuoalphanumeric generalizations in relation to the Fig. 4.33 pattern

in Fig. 3.9 to five stages and then established and justified a deconstructive generalization, she initially added a square on each row and column per stage number resulting in the growing L-shaped pattern shown in Fig. 5.14a. In justifying her direct formula, $s = 2n - 1$, we obtain a glimpse of her visual template below (see Fig. 5.14b for a graphical illustration).

> I visualized it in groups. So like for the 2n – 1, you take the stage number which are these two [i.e., the two circled groups in every stage] and then you subtract 1 because there's an overlap.

a Consider the pattern below.

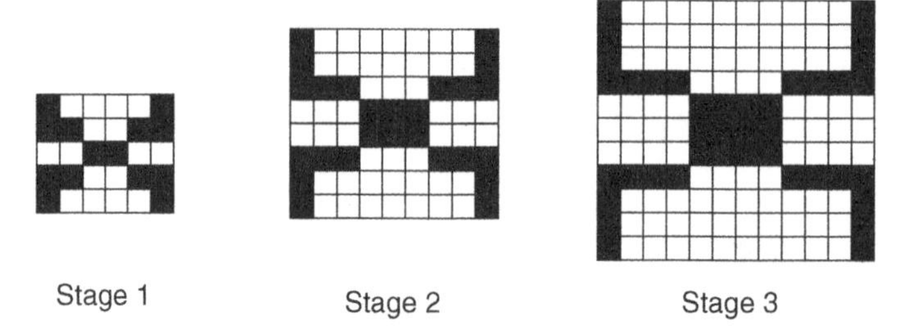

1. How does Stage 4 look like? Either describe it or draw it on a graphing paper.
2. Find a direct formula for the total number of gray square tiles at any stage. Explain your formula
3. How many gray square tiles are there in stage 11? How do you know?
4. Which stage number contains a total of 56 gray square tiles Explain.

b Consider the following array of sticks below

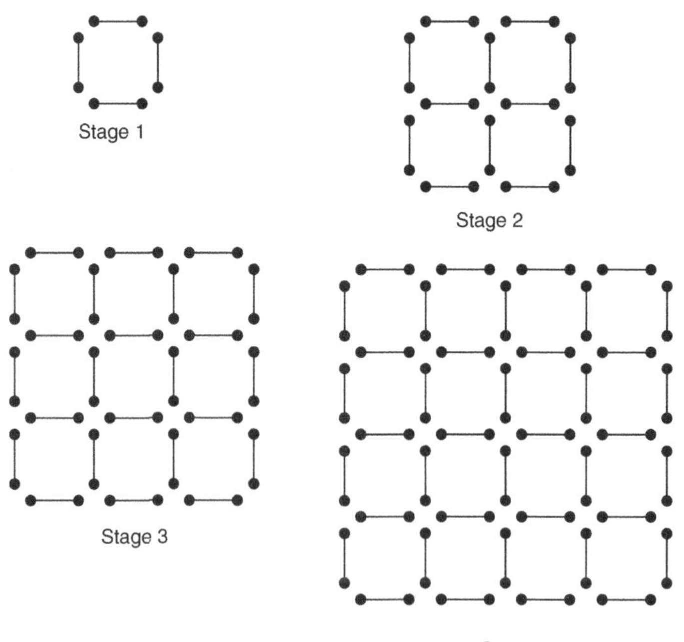

A. Find a direct formula for the total number of sticks at any stage in the patten. Justify your formula.

B. Find a direct formula for the total number of points at any stage in the pattern. Justify your formula.

Fig. 5.13 **a** Growing frog pattern. **b** Growing squares pattern

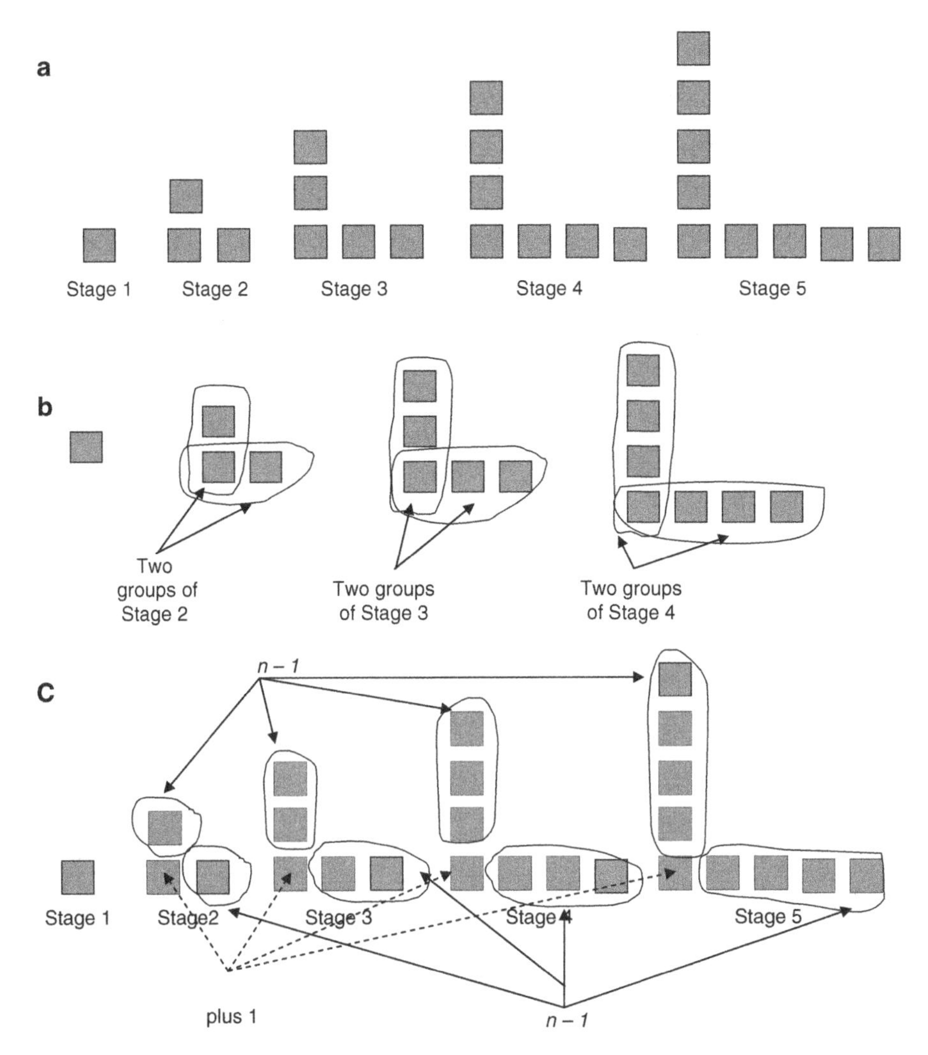

Fig. 5.14 **a** Karen's extended pattern in relation to the Fig. 3.9 task. **b** Karen's justification of her formula $s = 2n - 1$ in relation to Fig. 5.14a. **c** Karen's justification of her second formula $s = (n - 1) + (n - 1) + 1$ in relation to Fig. 5.14a

So, what are visual templates? Giaquinto (2007) initially talked about it in the context of structure discernment. He notes that when we investigate elements in a structured set, we either recognize the set as a particular instantiation of an established theory or at least share an isomorphic structure with the theory. Doing this helps us deal with the elements in the set in the same way we would manipulate the elements in the established theory. Giaquinto developed his notion of visual template from Resnik (1997), who describes a template as a blueprint, a "concrete device for representing how things are shaped, structured, or designed" (p. 227).

Neisser's (1976) notion of template matching is more general. At least in the context of pattern recognition, individuals tend to use prototypes or canonical forms as a standard or basic model that help them either learn a characteristic of a new object or compare the new with an existing object.

The visual aspect in visual templates emphasizes the role of the "'eye' as a legitimate organ of discovery and inference," thus "restor[ing the] balance between mathematical methodologies" (Davis, 1993, p. 342). Davis (1993) astutely points out how "discovery is not usually made in the deductive way" (p. 342), that is, not in the sole context of logical inference and verbal deduction but through our visual capacities of observing and seeing. Also, following Arcavi (2003), visual templates enable students to see the unseen of an abstract world that is dominated by relationships and conceptual structures that are not always directly evident. Thus, visual templates employ "visuospatial relations in *making inferences* about corresponding conceptual relations" (italics added; Gattis & Holyoak, 1996, p. 231).

In Karen's case, as soon as she constructed her pattern of five stages, she began to think multiplicatively in terms of groups of the stage number. Figure 5.14c conveys her second direct formula in additive form that we classify as a constructive nonstandard generalization. Here pattern generalization in Fig. 5.14a, b, c demonstrates how she also employed stage-induced grouping in making sense of her Fig. 5.14a pattern.

Figure 5.15 is a three-dimensional cognitive model of the visual template that I have drawn from the clinical interviews with the 11 students. It is a refinement of the model shown in Fig. 5.7. A more efficient and coordinated visual template in pattern generalization activity has the following five properties that operate in a distributed and dynamically embedded manner: structural unit, analogy, stage-driven grouping, Gestalt effect, and knowledge/action effects. *Gestalt effect* addresses issues relevant to visual attention and structural familiarity with pattern goodness as a criterion for algebraic usefulness. *Knowledge effect* is related to the degree of goodness of a Gestalt effect. The dashed segment that connects both effects indicates a spectrum of favored modality. For example, when the Gestalt effect is high in goodness, then there is less a need for knowledge mediation. When the Gestalt effect is low, then additional knowledge is needed in terms of how to make an interpreted structure of a pattern become more familiar to the perceiver. *Structural unit* pertains to a perceived interpreted general identity of a pattern within and across stages. It gives meaning to constructed units and addresses issues of invariance and change. Thus, it deals with the issue of stability of form or shape, property, attribute, or relationships in a pattern. *Analogy* addresses redundancy, consistency, and coherence among parts in a pattern. *Stage-driven grouping* shapes the content of an emerging template and, consequently, addresses operations that are used reflective of the (sequence of) grouping actions that are performed within and across stages. Central to the triad are the abductive–inductive and symbolic actions that require effective coordination as well.

The operation/s, which could be either additive or multiplicative, actually depends on how the pattern stages appeal to the student at the moment of cognitive perception. In Karen's case, her pattern has a linear structure that to her was a good Gestalt, which enabled her to construct two equivalent direct formulas. Her basic structural unit in Fig. 5.14b is the L-shaped figure that consists of two overlapping

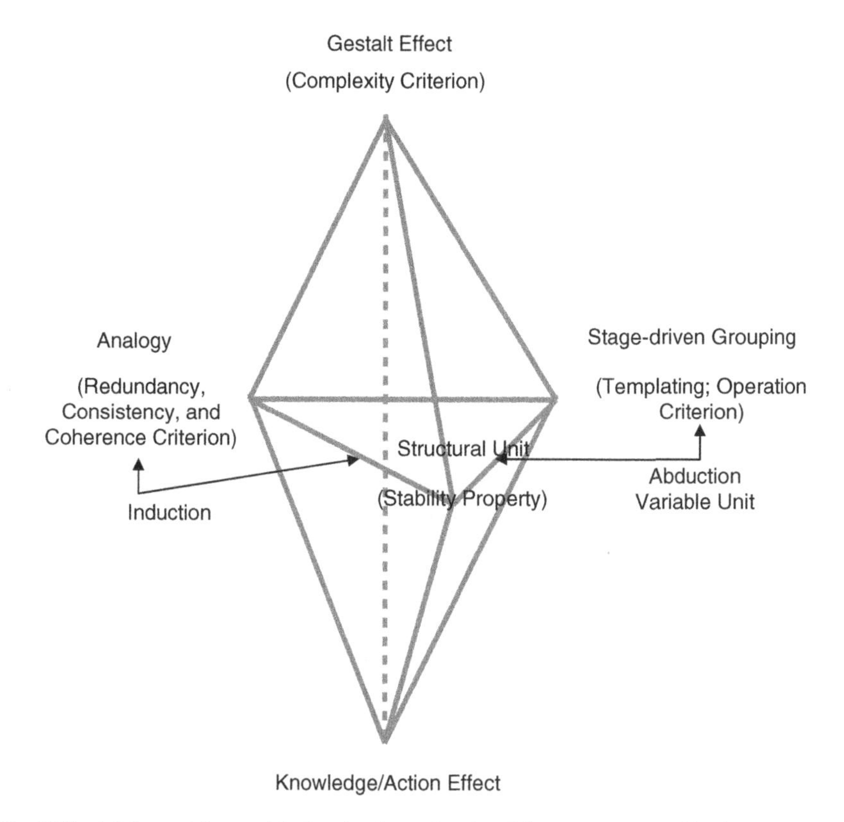

Fig. 5.15 A 3D cognitive model of a visual template involving pattern generalization

configurations or components. At the analogical phase, she repeatedly verified her structural unit over three stages. Consequently, her stage-induced grouping in Fig. 5.14b was multiplicative; she saw two copies of the stage number from one stage to the next. In Fig. 5.14c, her basic structural unit consists of a corner square that stayed the same from stage to stage and two nonoverlapping components each having a cardinality of $n - 1$. She again verified the same structural unit over three stages and then expressed her direct formula additively, $s = (n - 1) + (n - 1) + 1$, which is another example of a constructive nonstandard generalization.

Emma (Cohort 1, Grade 8) also produced the same pattern in Fig. 5.14a relative to the task shown in Fig. 3.9. Her pattern generalization was constructive nonstandard and additive, that is, $s = n + (n - 1)$, which she explained in the following manner below (see Fig. 5.16 for a visual illustration).

FDR: What helped you in transitioning from these visual squares to a direct formula?

Emma: Grouping it, I guess. This is stage 1 [referring to the one square]. This is stage 2 [the column of two squares]. This is stage 3 [the column of three squares] and this is stage 4 [the column of four squares]. And then so ahm

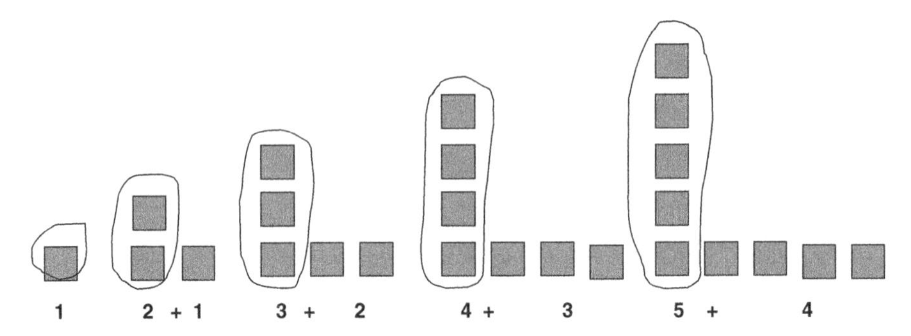

Fig. 5.16 Emma's justification of her formula $s = n + (n - 1)$ in relation to the pattern in Fig. 5.14a

> *when I figured that, I try to see what's left. So if it's 1 [the remaining square on the row of stage 2], if you subtract the stage number from 1, you get 1. If you subtract 1 from the stage number [stage 3] you get 2 [the two remaining squares on the row of stage 3]. If you subtract 1 from this stage number [stage 4], you get 3 [the three remaining squares on the row of stage 4].*

When Emma developed a second pattern for the task shown in Fig. 3.9, she produced the nonlinear sequence of triangular numbers shown in Fig. 5.11. In light of her existing knowledge and experience, the pattern would be low on Gestalt goodness, which meant conceptualizing about the parts and plausible relationships among them would likely require more effortful visual attentional mechanisms and acquired knowledge (in Metzger's sense). On her own, she established her initial structural unit upon seeing a growing vertical sequence of squares starting from the right, which represent consecutive numbers from 1 to n (the stage number) that she checked over several stages (e.g.: stage 2 has 1, 2 squares; stage 3 has 1, 2, 3 squares). Using stage-induced grouping, she suggested the following additive formula:

> *Emma: I know what to write but I don't know how to put it into a formula. Like so if there's n and then you add all the numbers before it. Like if it's 5, you put n – 1, n – 2, n – 3, and n – 4. But I don't know how to put that in a formula.*
> *FDR: So if it's stage 6, if you use your formula, it is?*
> *Emma: It's n plus n – 5 plus n – 4 all the way to 1.*

Because Emma was aware that she needed to construct a direct formula, she suggested a second structural unit that had her thinking about each pattern stage as the union of two components, that is, (1) an L-shaped figure at the base that she described by the rule $2n - 1$ and (2) an appended interior shape corresponding to "two stages before n." Figure 5.17 shows her written generalization in additive form.

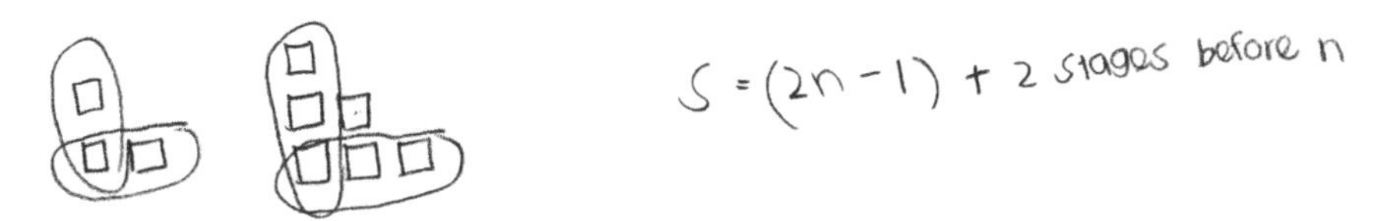

Fig. 5.17 Emma's written generalization in relation to her pattern in Fig 5.11

Thus, in the two situations above with Emma, her visual attentional mechanisms had her establishing two different structural units that resulted in two formulas either of which could not be expressed in closed form. Feeling frustrated, we both worked on her pattern together. Initially, I asked her to establish a relationship between a particular pattern stage number and the complete rectangle formed by joining two copies of the same stage number using two colored sets of unit squares. For example, in Fig. 5.18, the figure on the left is stage 2, while the figure on the right are two copies of stage 2 that together formed a rectangle with dimensions 2 $\times$ 3. She then employed the same visual strategy in dealing with stage 3. As soon as she noticed that the total number of squares in each pattern stage was one half the total number of squares in the complete rectangle, she concluded that stage n would have $s=(n(n+1))/2$ squares, an auxiliary-driven constructive generalization. She then used stages 4 and 5 to verify that her formula was correct.

Diana (Cohort 2, Grade 7) implemented a multiplicative-driven visual attentional mechanism on all three tasks in Figs. 3.9 and 5.13a, b that enabled her to establish an algebraic generalization effectively and efficiently. Instead of looking for parts that corresponded to a stage number, Diana sought out groups of parts that had the same count and then connected the count with the appropriate stage number. In the case of Fig. 3.9, Diana produced two patterns (Figs. 5.11 and 5.14a). The direct formula she constructed in the case of the Fig. 5.11 pattern also took place in joint activity with me. In the case of the Fig. 5.14a pattern, she produced two

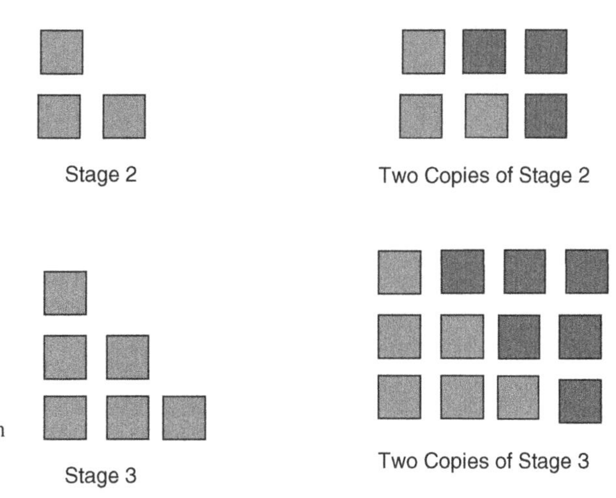

Fig. 5.18 Auxiliary stages in relation to the triangular pattern in Fig. 5.11

direct formulas, one deconstructive and the other auxiliary-driven constructive. In the interview transcript below, she described how she reasoned abductively about her two formulas.

FDR: Is there a formula here [referring to the pattern in Fig. 5.14a]?
Diana: Uhm, $n = 2x - 1$. This is step, stage 3. There's like 2 threes and this one [corner square] overlaps. [Refer to Fig. 5.14b for a visual illustration.]
FDR: So for stage 4? [Diana gestures two groups of four squares from the figure and the corner square.] Is there another formula for that?
Diana: Uhm, I'm not sure this is the simplest form. $n = x^2 - (x - 1)^2$. Like if you see a square and a square here [refers to stage 4 in Fig. 5.14a and gestures to indicate that she is taking away $(x - 1)^2$ squares].

In the interview segment below, Diana also used a multiplicative template in constructing and justifying her direct formula relative to the Fig. 5.13a pattern. Her first direct formula was $n = x(x + 1) + 4(2x + 1)$, a constructive nonstandard generalization, which she later simplified to $n = x^2 + 9x + 4$ and justified as a constructive standard generalization.

Diana: Well, basically you always, like, to this number here, to this part here [referring to stage 2], you added 1 and on this side you add 1 to make it longer [referring to the growing legs on every corner.] You always add 1 to everything to make the legs longer. Instead of like 2 × 2, you make it 3 × 3. And for this one, too [the middle rectangle], instead of 1 by 2, you make it 2 × 3. [She then finds a direct formula and obtains $x(x + 1) + (2x + 1) = n$].
FDR: Okay, so tell me what's happening there? Where did this come from, $x(x + 1)$?
Diana: This, the little square, x times x + 1.
FDR: So where's the x times x + 1 here [referring to stage 3]?
Diana: Like 3 × 3, or 3 × 4.
FDR: So where's the 2x + 1 coming from?
Diana: This. I mean I can look at it like 2 times (x + 1) minus 1 but I just made it, like, 3, 3, and 1, so 2x + 1. [She initially saw that each leg had two overlapping sides that shared a common square.]
FDR: But this 2x + 1 is just for this side [referring to one leg], right?
Diana: For all of the legs, oh, [then adds a coefficient of 4 to her formula: $x(x + 1) + 4(2x + 1) = n$].
FDR: Okay, so are you happy with your formula?
Diana: I think I could simplify it. I'd like to see what happens if I simplify it. [She then simplifies her formula to $x^2 + 9x + 4 = n$.]. 4 would be these [the corner middle squares] I'm pretty sure. 9x would be, 8, oh, yes, I see it. I see how it works. There's an x squared here [referring to the rectangle which she saw as the union of an n by n square and a column side of length n] if you see one square here and the 9x would be these legs [referring to

the (n + 1)th column of the rectangle of length n and the eight row and column legs minus the corner middle squares]. Plus 4 would be the center of each leg.

The consistent manner in which Diana used a multiplicative template became especially useful when she obtained her pattern generalization in relation to the Fig. 5.13b task. Initially, she obtained a formula for the number of sticks on the perimeter of each square, $4x$. Next, she counted the interior sticks as follows: (1) "[*in stage 2*] 2 minus 1 would be 1 so there would be 1 row going down and another row so that would be rows of 2 sticks"; (2) "in stage 3, there's 2 rows of 3 sticks"; (3) "[*in stage 4 there's*] three, okay, it's two sets of three rows of 4." She then wrote $4x + 2(x - 1)x = s$ which she simplified to $2x + 2x^2 = s$.

Table 5.2 provides a summary of the clinical interview results of the 3-year study on pattern generalization in relation to linear patterning activity. The row percentages under numerical and visual approaches by the year represent shifts from the visual to the alphanumeric and then finally to visuoalphanumeric generalizing. The row percentages under constructive standard generalizations were consistently high except in the case of decreasing linear patterns that a few students have found to be relatively difficult to justify. The row percentages under more complicated forms of generalizations such as deconstructive and constructive nonstandard generalizations significantly improved only in the Year 3 study when the focus of patterning activity shifted away from the use of difference-driven tables that engendered numerical generalizing to a purposeful reorientation to the properties described in Fig. 5.15. In Year 2, while none of the students interviewed could obtain a pattern generalization involving deconstructive generalization, they could, however, numerically verify their correctness on particular cases when the formulas were presented to them primarily to justify. In Year 3, at least 86% of their responses involved constructing and justifying deconstructive generalizations.

5 Situational Discords in Visual Templates

The 3D visual template in Fig. 5.15 consists of five properties that work dynamically together. A discord takes place when there is no harmony among them. We point out at least three situational discords, which are as follows: (1) having a pattern that has a low Gestalt effect but could be overcome on the basis of having visually accessible strategy; (2) having a pattern that has a high Gestalt effect but the knowledge that is needed to obtain a pattern generalization is not accessible using a visual strategy; and (3) having no structural unit with a significantly high knowledge effect.

Emma's initial pattern generalization of the triangular sequence she constructed in Fig. 5.11 above exemplifies situational discord (1). While her pattern was low in pattern goodness, she was, however, able to overcome her difficulty by having a visually accessible knowledge of the relationship between the known stages and the associated rectangles (Fig. 5.18) that led to a direct formula.

Table 5.2 Summary of Pattern Generalization

Year 1 Results	Before Teaching Experiment (n = 29)	After Teaching Experiment (n=11)
Overall Visual	63%	0%
Overall Numeric	37%	100%
Constructive Standard Generalizations	0%	100%
Constructive Nonstandard Generalizations	0%	0%
Deconstiuctive Generalizations	0%	0%
Year 2 Results*	**Before Teaching Experiment (n=8)**	**After Teaching Experiment (n=8)**
Overall Visual	12%	25%
Overall Numerical	88%	75%
Constructive Standard Increasing Patterns	100%	100%
Constructive Standard Decreasing Patterns	38%	75%
Constructive Nonstandard Generalizations	0%	0%
Deconstructive Generalizations	50%	100%
Year 3*	**Before Teaching Experiment (n= 18; 5 new**)**	**After Teaching Experiment (n = 14; 3 new**)**
Overall Visual	67%***	71%
Overall Numeric	33%***	29%
Constructive Standard Generalizations	100%	100%
Constructive Nonstandard Generalizations	6%	36%
Deconstructive Generalizations	11%	86%

(*Some tasks had multiple questions;
**Did not participate in earlier two-year interviews;
***More visual tasks than numerical).

It can also happen that some constructed patterns are high on Gestalt goodness but require some knowledge that is not visually accessible, that is, situational discord (2). This happened with Tamara (Cohort 1, Grade 8). Her extended pattern relative to the Fig. 3.9 task is shown in Fig. 5.19a. Comparing stages 1 and 2, she saw that stage 2 has "one group of two squares" added to stage 1. She then abduced that stage 3 would have "two groups of two squares" added to stage 2, stage 4 "three groups of two squares" added to stage 3, etc. In thinking about a direct formula, she saw the following sequence:

$$(2 \cdot 1 - 1, 2 \cdot 2 - 1, 2 \cdot 4 - 1, 2 \cdot 7 - 1, 2 \cdot 11 - 1, 2 \cdot 16 - 1).$$

In obtaining a direct expression, Tamara then focused on the length of each row. Unfortunately, the knowledge or the strategy she needed to obtain a formula for the sequence (1, 2, 4, 7, 11, 16, ...) was not visually accessible. Consequently, she changed her pattern to Fig. 5.14a when she realized that she needed to "find a way to relate grouping and stage number in some way."

Considering the above situations with Emma and Tamara, students need to know when, how, and why it is necessary to introduce auxiliary sets or, more generally, employ *figural transformations* or what Duval (2006) refers to as the action of *figural change* in apprehending figures. Visual actions of moving, reorganizing, and transforming, including the use of auxiliary sets and figural parsing, are powerful mathematical ways of inferring a plausible structure on a pattern with figural stages. Figure 5.19b shows one possible figural transformation of Tamara's pattern in Fig. 5.19a that produces a transformation-based algebraic generalization.

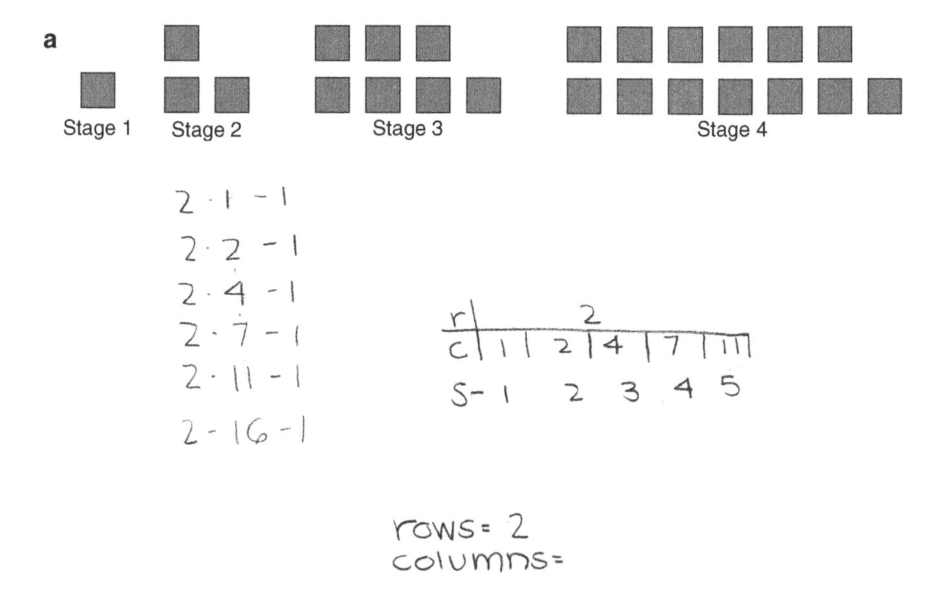

Fig. 5.19 (continued)

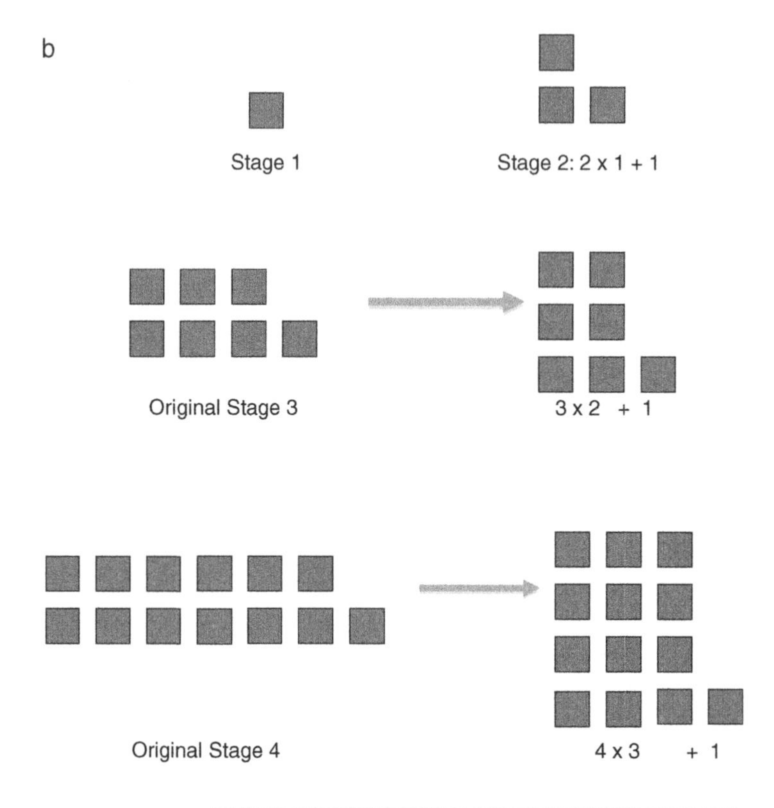

Stage Number (n)	Number of Squares (s)
1	1
2	2·1+1
3	3·2+1
4	4·3+1
5	5·4+1
6	6·5+1
n	$n(n-1)+1$

Fig. 5.19 **a** Tamara's extended pattern in relation to the semi-free construction task in Fig. 2.9. **b** A figural transformation of Tamara's pattern in Fig. 5.19a demonstrating the algebraic generalization $s = n(n-1) + 1$

Emma's situation relative to the task in Fig. 5.13b exemplifies situational discord (3). She first employed a numerical strategy in dealing with the pattern that to her was low in Gestalt goodness. She saw that stage 1 had four sticks and claimed that it was "four of stage 1." In stage 2, she counted the total number of sticks (12) and then saw it as "[*stage 2*] · 6." She then checked that her numerical reasoning also worked in stages 3 and 4. She noticed that since stage 3 had 24 sticks, it was equal to "[*stage 3*] 8." In stage 4, she claimed that 40 was "[*stage 4*] 10." She concluded that her direct formula for the total number of sticks was $s = n \cdot (2n + 2)$. When asked to explain her formula, she said that n referred to the stage number and that the numbers 4, 6, 8, and 10 were "even numbers and that to get to 10, I multiply it by 2 and add 2, so $2n + 2$." But when asked to explain how the formula might make sense in the given stages, she said, "I don't know, I don't see it in the picture." Emma's work on this task is an example of a knowledge-driven generalization with no justification because she failed to establish a structural unit.

There were students who also employed acquired knowledge in constructing an algebraic generalization despite their relative inability to establish a structural unit. However, their justifications were inconsistent and, thus, invalid. Examples of such student work are shown in Fig. 5.20a in relation to the growing chair pattern task. What these students found difficult about the pattern was how to attend to the relevant configurations or parts in each stage in a way that would lead to articulating a structural unit that could then be generalized across stages. Dung (Cohort 1, Grade 8) provided a justification in Fig. 5.20b that showed how he merely tried to fit his formula onto the stages in his pattern. Figure 5.20c shows equally valid generalizations of six students on the same pattern that represent different visual perspectives.

Like Dung in Fig. 5.20b, Frank (Cohort 1, Grade 8) also used knowledge-driven generalization in developing two equivalent direct formulas relative to the Fig. 3.9 task. He produced the same L-shaped pattern in Fig. 5.14a and constructed the formulas $S = 2x - 1$ and $S = 2(x - 1) + 1$ by first establishing a common difference of two unit squares ("there's two more than before") from one stage to the next.

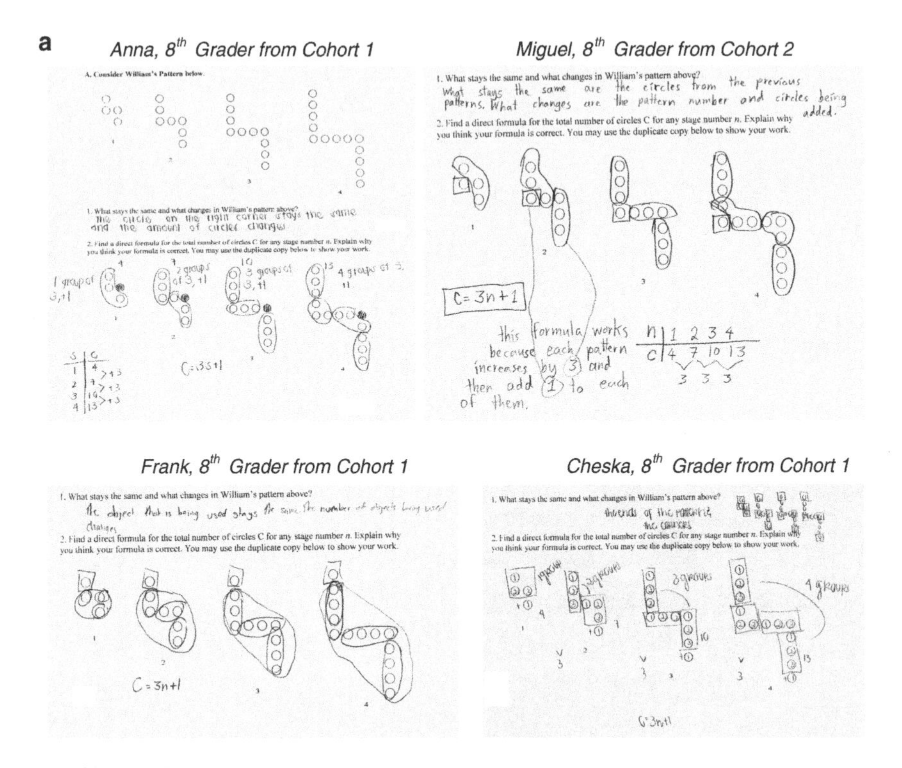

Fig. 5.20 (continued)

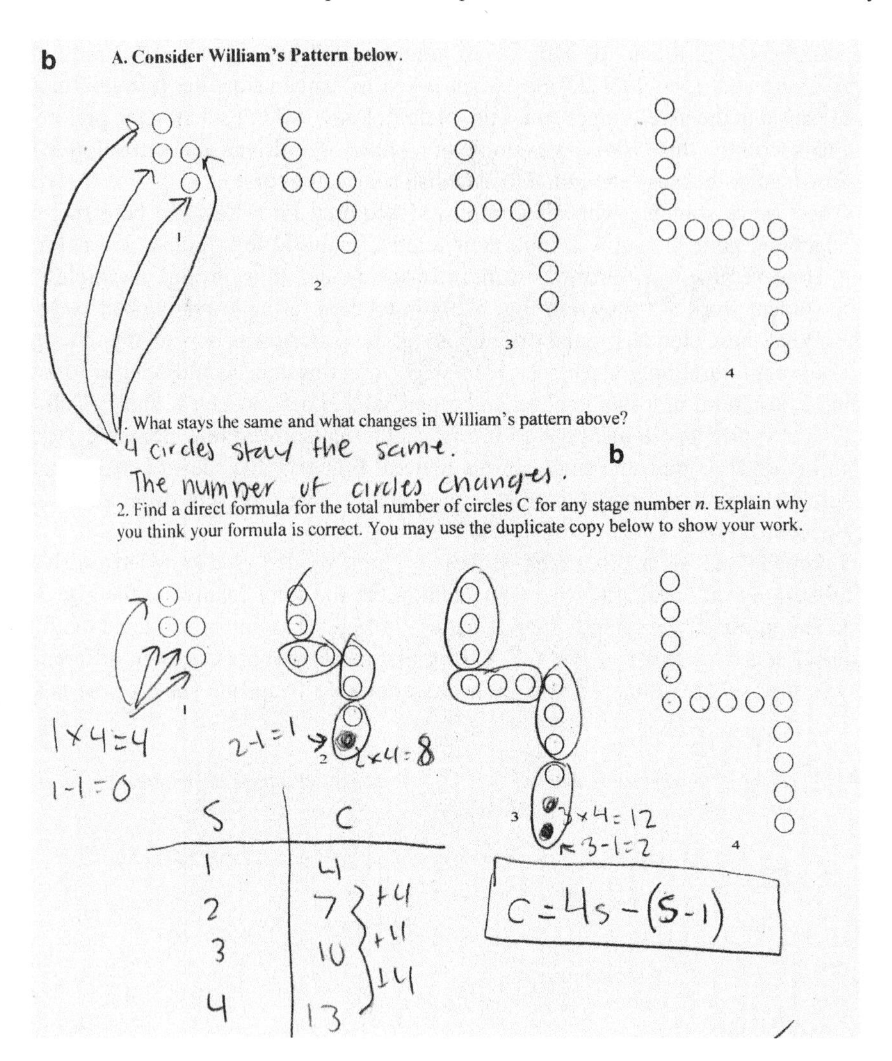

Fig. 5.20 (continued)

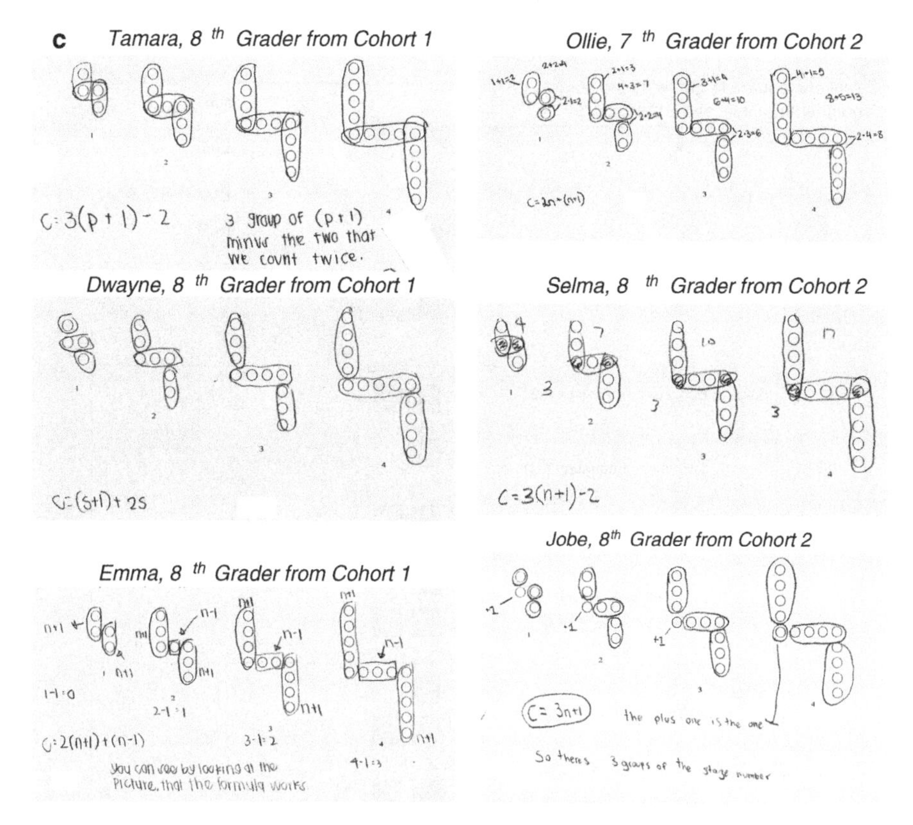

Fig. 5.20 **a** Samples of inconsistent justifications in relation to the growing chair pattern. **b** Dung's justification of his formula in relation to the growing chair pattern. **c** Samples of consistent justifications in relation to the growing chair pattern

He described his numerical strategy in dealing with the Fig. 5.14a pattern in the following manner:

> *I'd looked through the differences of each pattern, say [stages] 1 and 2 [referring to the pattern in Fig. 5.14a]. You notice that there's two more than before. So when I look at that, I wanna multiply it by the number that's different because you want to get to the next stage, right? And that's how many is needed to get to the next stage of your pattern.*

However, Frank took for granted the necessity of having a structural unit. In his justification, he tried to explain how each term in his direct formulas would assist him in constructing his stages in a step-by-step manner instead of seeing how they conveyed expressions relevant to the general structure of his pattern. Consider the following segments from his interview:

Frank's Justification of the Formula $S = 2x - 1$ (see Fig. 5.21a for a graphical illustration):

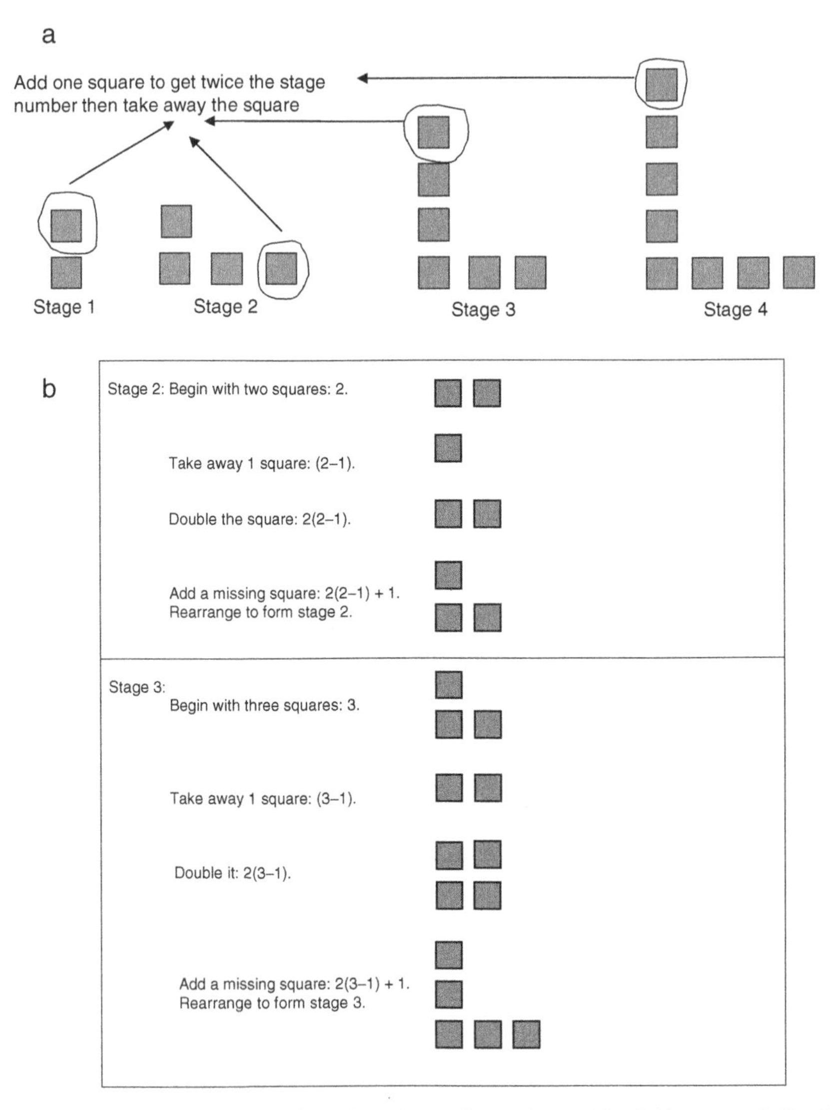

Fig. 5.21 **a** Frank's justification of his formula $s = 2x - 1$ for the Fig. 5.14a pattern. **b** Frank's justification of his formula $s = 2(x - 1) + 1$ for the Fig. 5.14a pattern

Frank: In stage 2, 3 total, so if I had another 1, that would equal 4 so subtract 1 to get to stage 3.

FDR: What about for stage 3?

Frank: You would have ahm so 2 × 3 is 6 there would be an extra so you could put wherever. [He puts the extra square at the end of the row.] But then you subtract 1 which is the extra to get to 5.

FDR: Okay, so one more.

Frank: So for stage 4, it has a total of 7. And then there's 1 more [puts the extra square at the end of the column] so you subtract that extra to get your total.

Frank's Justification of the Formula $S = 2(x - 1) + 1$ (see Fig. 5.21b for a graphical illustration):

FDR: Show me how you make sense of that formula here.
Frank: So if it's 2 minus 1, that gives you 1, multiply by 2 is 2 plus 1 gives you 3.
FDR: So where is that there [referring to the cues]?
Frank: So say, stage 2, right [starts with 2 squares]? And then 2 minus 1 is 1 [takes away 1 square] and then you multiply 1 by 2 which gives you that [puts another square] and then you add 1 there.
FDR: Hmm, so let's try that for this one [stage 3].
Frank: So it would be 3 [starts with 3 squares] for this one. And then you subtract 1 which gives you a total of two squares. 2 times 2 gives you 4 and then you're missing one so you add 1 more.

6 Structured Visual Thinking in Transformational Activity

We close this chapter by considering two issues that address functional thinking at different grade levels. In my Algebra 1 class, we deal with connections between visuoalphanumeric generalization and linear function modeling. In my Grade 2 class, we take into account the emergence of functional thinking as a result of students' experiences with figural patterns that convey structured visual representations.

In my Algebra 1 class, figural pattern generalization activity provided my students with an opportunity to engage in hypostatic abstraction and develop structure sense, both processes being relevant to issues involving generalization and symbolization that lie at the heart of algebraic reasoning (Kaput, Blanton, & Moreno, 2008, p. 20). In the Year 3 study, the students used their experiences in pattern generalization in hypostasizing about linear functions and linear function modeling.

The first activity we pursued immediately after the teaching experiment on linear pattern generalization was linear function modeling. Initially, we discussed the meaning of a function, which the students easily understood on the basis of their conceptual experiences with figural patterns. The term *function* was a hypostasized abstraction drawn from their experiences with patterns where every input corresponded to a unique output based on some explicit rule in the form of a direct formula. Also, when they were presented with the two linear modeling tasks shown in Fig. 5.22, they used their experiences with pattern generalization in dealing with the two situations. Figure 5.23 shows the written work of several students that illustrates how they successfully applied their knowledge of pattern generalization involving discrete objects to linear modeling problems whose domains were real numbers. Later, they also relied on their knowledge of increasing and decreasing

Road Test Problem

The engineers at an auto manufacturer pay students $0.38 per mile plus $250 per day to road test their new vehicles.

a. Find a function rule that relates the total daily pay P for driving x miles in one day. Explain your formula.
b. How much did the auto manufacturer pay Sally to drive 440 miles in one day? Show work.
c. John earned $545 test-driving a new car in one day. How far did he drive? Show work.

Fund Raising

The Middle Grades Math Club needs to raise money for its spring trip to Washington, D.C., and for its Valentine's Day dance. The club members decided to have a car wash to raise money for these projects. The club treasurer provided the following information:

- Charge per each car washed: The usual charge for washing one car is $4.00.
- Materials expense: The total cost of sponges, rags, soap, buckets, and other materials needed will be $117.50.

As a member of the club, your job is to produce a mathematical analysis. Do as follows.

1. Complete the table below. When done, enter the completed table in your graphing calculator and then graph.

Number of cars washed (c)	0	1	2	3	4	5	6	7
Amount of money raised (M)	−117.5	−113.5	−109.5					

2. Why is the amount of money raised with no cars washed a negative value? What happens to the club's finances?
3. Explain how the value of −$113.50 was obtained for 1 car.
4. What kind of pattern does your table above represent? How do you know for sure?
5. Find a direct formula that enables you to obtain the amount of money raised (T) for any number of cars washed (c).
6. What stays the same and what changes in the table above? How does knowing what stays the same and what changes relate to your direct formula?
7. How much money would the club raise if the members washed 12 cars? Show work below.
8. How much money would the club raise if the members washed 60 cars? Show work below.
9. How many cars must the members wash to *break even*, that is, the number of cars needed so there is no loss or no profit?
10. How many cars must the members wash to raise enough money for:
    a. the trip to Washington with an estimated cost of $8,750? Show work.
    b. the Valentine's Day dance with an estimated cost of $3,500? Show work.
11. Realistically, can the car wash raise enough money for both of these activities? Explain.

Fig. 5.22 Two linear function models

linear patterns in hypostasizing about slopes of lines, which were seen as quantitative descriptions of constant rates of change. Overall, their experiences with patterns conceptually mediated in an effective way that allowed them to easily transition and hypostasize about concepts and processes relevant to linear functions and linear function modeling.

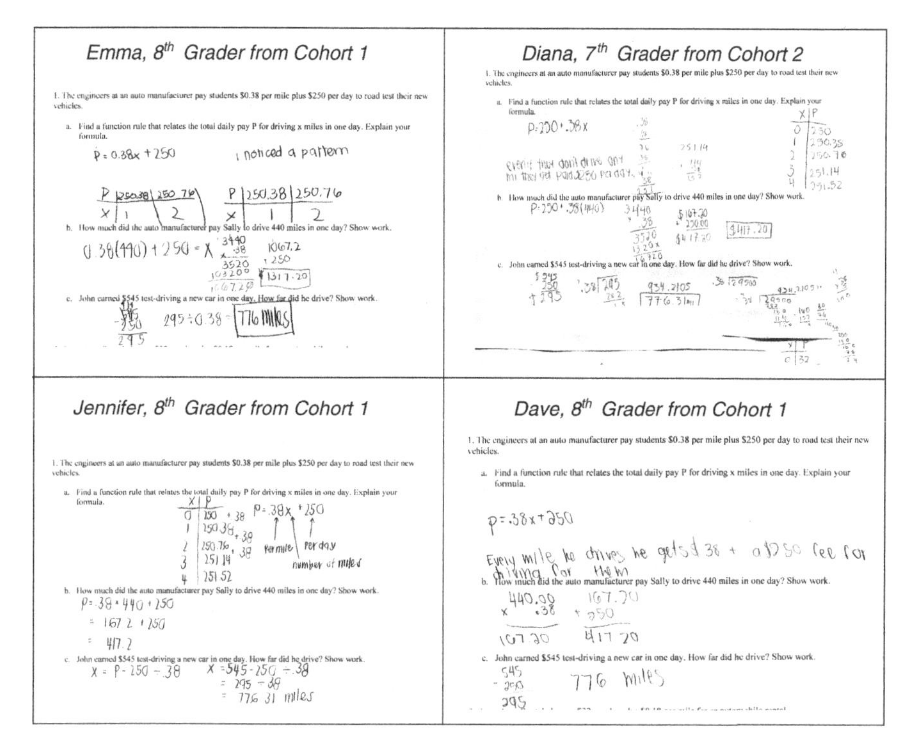

Fig. 5.23 Students' written work on the road test problem

In my Grade 2 class, when they were presented with two numerical and figural pattern tasks prior to formal instruction, none of them exhibited functional thinking relative to the numerical tasks. For example, in the case of the dogs and zebras pattern tasks shown in Fig. 5.24, they were primarily engaged in empirical counting (e.g., counting all at each stage; counting on from one stage to the next; skip counting by 2s or by 4s; combinations of count-all and count-on) with very little indication of a concern toward structural understanding. Even when they were presented with a function-based task in numerical form, shown in Fig. 5.25, none of them saw a relationship between input and output and merely inferred a relationship between two successive outputs.

However, when they were presented with the two figural tasks shown in Fig. 5.4a, b prior to formal instruction, the patterns appealed to 60% of them as having high pattern goodness with well-defined structures. In each case, they used pattern blocks to build stages 4 and 5 and drew stage 10 on the basis of how they interpreted the structure of the pattern from the known stages. Figure 5.4c, d shows samples of students' written work relative to stage 10 in each pattern task. Despite their inability to describe in words a generalization for each pattern, their visually

Dog Task

A dog has two eyes. Two dogs have four eyes. Three dogs have six eyes. Four dogs have eight eyes.

We only have four dogs on the table. So you need to imagine more dogs in your head.

A. How many eyes are there in the case of five dogs? How do you know for sure? Can you show me on paper how you were thinking about it?

B. How many eyes are there in the case of six dogs? How do you know for sure? Can you show me on paper how you were thinking about it?

C. How many eyes are there in the case of ten dogs? How do you know for sure? Can you show me on paper how you were thinking about it?

D. How many eyes are there in the case of twenty dogs? How do you know for sure? Can you show me on paper how you were thinking about it?

E. If a normal dog has two eyes, how many dogs are there if you have a total of seventeen eyes? Explain your thinking.

Zebra Task

A zebra has four legs. Two zebras have eight legs. Three zebras have twelve legs.

We only have three zebras on the table. So you need to imagine more zebras in your head.

A. How many legs do four zebras have? How do you know for sure? Can you show me on paper how you were thinking about it?

B. How many legs do five zebras have? How do you know for sure? Can you show me on paper how you were thinking about it?

C. How many legs do six zebras have? How do you know for sure? Can you show me on paper how you were thinking about it?

D. How many legs do ten zebras have? How do you know for sure? Can you show me on paper how you were thinking about it?

E. How many legs do twelve zebras have? How do you know for sure? Can you show me on paper how you were thinking about it?

F. If a normal zebra has four legs, how many zebras are there with a total of 21 legs? Explain your thinking.

Fig. 5.24 Dogs and zebras patterning tasks used with Grade 2 students in clinical interviews prior to formal instruction

driven actions, which allowed them to construct stage 10 correctly, indicate early manifestations of functional thinking. When they initially inspected stages 1–4 in each task, which the interviewer constructed for them one by one, they cognitively perceived a relationship between stage number and figural shape. Hence, figural patterns that are well defined and have a high Gestalt effect can lay the groundwork for a mathematical (symbolic) understanding of functions.

I have a magic hanky. It does something with the numbers you put under it, as follows:

A. What do you think will happen in each case below? Explain how you know.

B. Tell me what the magic hanky is doing with the numbers you put under it? How do you know?

Fig. 5.25 Numerical-based function task used with Grade 2 students in clinical interviews prior to formal instruction

7 Overview of Chapter 6

In Chapter 6, we focus on mathematical diagrams whose parts are conceptually constructed in some way. We also discuss diagrammatic reasoning, which involves distributed actions of manipulating, discerning, interpreting, and inferring relations on diagrams. While diagrams are prevalent in school geometry, which provide the conceptual tools necessary in establishing necessary and logical proofs, the examples we use in Chapter 6 are taken from school algebra and number sense. The goal, of course, is to give readers a sense of how progressive diagrammatization occurs and supports progressions in formalization, schematization, and mathematization of the corresponding alphanumeric forms. In the introduction of the chapter, we clarify the nature of diagrams in cognitive mathematical activity. In the first section, we discuss ways in which our visual system is naturally capable in seeing relationships in diagrams. The second section focuses on three fundamental issues with mathematical diagrams, which are as follows: existence, universality, and choosing between presented and generated diagrams. Having addressed the constraints in diagramming, in the third section, we discuss different diagram types and address issues surrounding progressive diagrammatization. In the last section, we explore the role of diagrammatic reasoning in mathematical reasoning.

Chapter 6
Visual Thinking and Diagrammatic Reasoning

It is difficult to decide between the two definitions of mathematics: the one by its method, that of drawing necessary conclusions; the other by its aim and subject matter, as the study of hypothetical state of things.

(Peirce, 1956, p. 1779)

Mathematics requires an inter-subjectively given object. This is supplied in modern mathematics by conceptual systems and in Greek mathematics by the diagram.

(Netz, 1998, p. 38)

Plainly the movement to accord diagrams a substantial role in mathematics is crucial to a philosophy of real mathematics.

(Sherry, 2009, p. 60)

In Chapters 4 and 5, we specifically addressed contexts in which visuoalphanumeric symbols in school algebra and number sense could be interpreted as symbolic entities that have roots in structured visual experiences. We also discussed the significance of progressive symbolization relative to intra- and inter-semiotic transitions that occur from iconic and/to indexical and to symbolic representations. In this chapter, we focus our attention on *diagrams* that are purposefully constructed to convey visual relationships and mediate in students' understanding of alphanumeric forms. Examples of such diagrams include tables in pattern generalization, graphs of functions, squares, sticks, and dots in adding and subtracting whole numbers, binary chips and number lines used in understanding integer operations, etc.

For Peirce (1976), a diagram is a relationally iconic sign whose nature is best described in terms of its relationship within a hypothetical state of things. Such a state, Campos (2007) points out, is a category of *firstness* in that it thrives in an imagined mathematical world in which pure hypotheses are as general as they are sufficiently vague and definite. In this world, a hypothetical state does not have to correspond to an actual state in reality, which is a *secondness* category. Further, everything in this world could be accessed and known primarily through reasoning and the use of logic whose propositions are either true or false. What the world then produces are mathematical truths that each "has the character of a *would-be* consequence of a hypothesis" (Campos, 2007, p. 472). One activity in such a hypothetical state involves the creative construction of diagrams, "*poietic* creations

F.D. Rivera, *Toward a Visually-Oriented School Mathematics Curriculum*, Mathematics Education Library 49, DOI 10.1007/978-94-007-0014-7_6,

© Springer Science+Business Media B.V. 2011

[that are] independent of really existing facts" (Campos, 2007, p. 473), that lead to necessary mathematical truths. Thus, Peirce's view of mathematics as "the science which draws necessary conclusions" is premised on this particular condition of the "hypothetical state of things" (cf. Dea, 2006).

Peirce describes a *diagram* in the following way:

> [It is] a *representamen* [i.e., a sign] which is predominantly an icon of relations and is aided to be so by conventions. Indices are also more or less used. It should be carried out upon a perfectly consistent system of representations, founded upon a single and easily intelligible basic idea.
>
> (Peirce, 1934, p. 341)

Mathematical activity then involves constructing, observing relations in, and experimenting with diagrams (Peirce, 1958b, p. 229). Diagrams deal with relationships among parts in objects (Stjernfelt, 2000, p. 358), "the internal structure of those objects in terms of interrelated parts, [thus,] facilitating reasoning possibilities" (Stjernfelt, 2007, p. ix). Campos (2007) captures the essential character of a Peircean diagram in the following interesting manner:

> A diagram, then, is a sign that represents in our minds the objects and relations that conform to our hypothesis. We may actualize the imagined mental diagram by way of physical models, say via graphs, drawings or equations, but the diagram is a sign that conveys a meaning to our "mind's eye" – the meaning being the form of the relations that hold according to our pure hypothesis. . . . [T]he essential act of diagramming in mathematics is not the act of drawing a geometrical figure or writing an algebraic expression; it is rather the act of imagining a representation that embodies the relations among objects that hold in our purely hypothetical world.
>
> (Campos, 2007, p. 475)

Solving an absolute value equation for the first time in my Algebra 1 class was a difficult experience for many of my students. Their primary dilemma with the algebraic form $|x| = c$ had to deal with the hypothetical condition in which the equation actually conveys seeing a number x that has a distance of c units from either the left or the right side of the origin. This complex relationship, as a matter of fact, was not articulated at all when they first learned about evaluating the absolute value of a letter as a particular object in expressions such as $|2|$ or $|-2|$. In class, they saw that such an expression meant "the distance of 2 or –2 from the origin is 2 units, so the result is always positive." However, when the letter represents a variable that could stand for any nonzero real number, that is, the general case of $|x|$, geometrically it could refer to two possible locations on a number line from the origin. It was this hypothetical possibility that my students found difficult to grasp, at least initially.

In our class, the following piecewise symbolic definition of $|x|$ made sense to them but only when they began to associate a number line diagram that showed each situation:

$$|x| = \begin{cases} x, x > 0 \\ 0, x = 0 \\ -x, x < 0 \end{cases} .$$

Figure 6.1a, for example, shows Cheska's (Cohort 1, Grade 8) number line model of the above definition. When the students later solved simple absolute value equations such as $|x| = 3$ and $|x + 2| = 5$, consistently interpreting expressions x and $x + 2$ as being 3 and 5 units away from the origin on both sides of the number line, respectively, their thinking and reasoning underwent "elaborat[ion] (while) in the process of tracing diagrams, when mind and body cooperate in the advent of the virtual" (Batt, 2007, p. 246). In other words, the number line fulfilled the "operativity of the diagram for mathematical thought" (Batt, 2007), which conveyed a type of "diagrammatic abstraction" that "preserved or highlighted" (Gooding, 2006, p. 41) the essential aspects in the definition of the absolute value.

Figure 6.1b shows the written work of Jennifer and Joko on three absolute value equations that made sense to them on the basis of how they interpreted each expression inside the absolute value as conveying two particular locations on a number line from the origin. Joko mentally perceived a number line in which he assigned two different values for the same algebraic expression inside the absolute value (i.e., 5–2x could either be 9 or –9 since | 9 | = |–9|). Jennifer's solutions transitioned from explicitly using a number line to eventually doing away with it. Nevertheless, in either visual case (mental image or evident display of a number line), their written work exemplifies how "a skeleton-like sketch of relations" (Stjernfelt, 2000, p. 358)

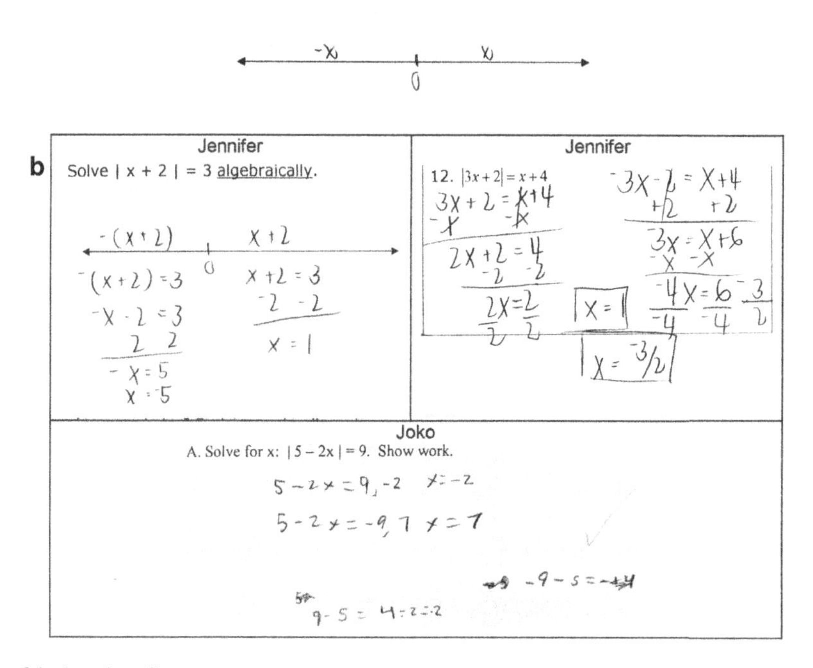

Fig. 6.1 (continued)

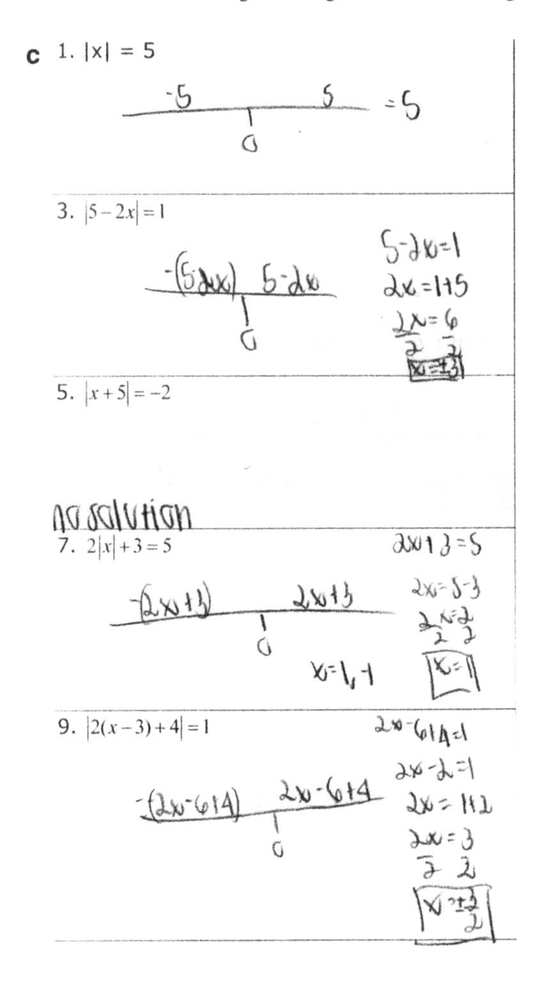

Fig. 6.1 **a** Cheska's number line model of the absolute value concept. **b** Written work of Jennifer and Joko on three absolute value equations. **c** Cheska's written work on five absolute value equations

diagram functioned in reminding them that they were considering two different kinds of situations in the case of solving absolute value equations.

Toward the end of our first classroom session on solving absolute value equations, Dave (Cohort 1, Grade 8) asked whether absolute value equations always produced two solutions. Cheska overgeneralized when she replied in the affirmative and further claimed that the two answers would always be "opposites of each other." This, in fact, was how she saw them in her diagram. Figure 6.1c shows a few of her initial solutions. So, the following day, we investigated different cases with the use of a TI 84, a graphing calculator. Graphically, the students began with two graphs that correspond to the left- and right-hand sides of a given absolute value equation (Fig. 6.2a). Then they investigated graphing situations under which an absolute value equation would yield the following: (1) no solution; (2) one solution; (3) two solutions; and (4) an infinite number of solutions (see Fig. 6.2a, b, c for examples).

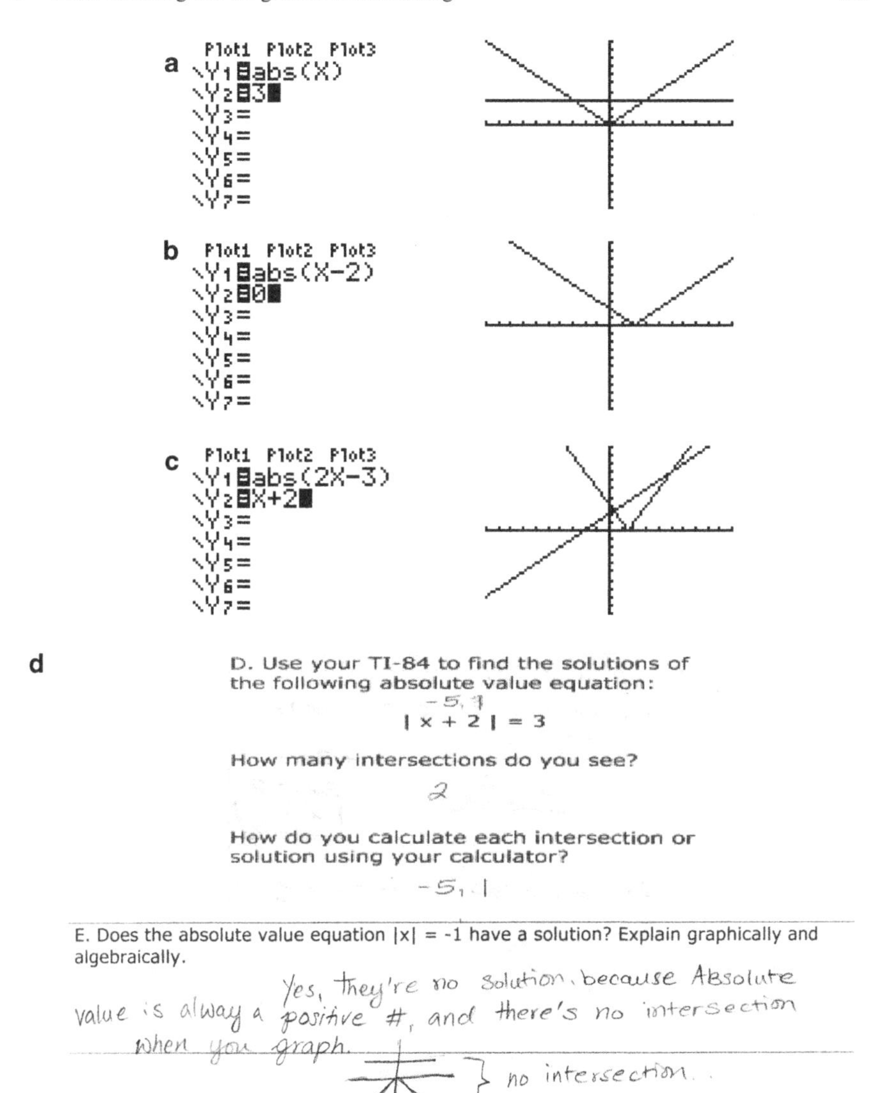

Fig. 6.2 **a** Representing an absolute value equation graphically on the TI 84. **b** Absolute value equation with one solution. **c** Absolute value equations with two solutions. **d** Ping's written work on two absolute value equations

In each case above, I purposefully asked them to present an algebraic solution with each graphical exploration so that they knew how to work on two different types of representations. For example, while the students knew that $|x| = -2$ had no solution since the two graphs

$$\begin{cases} y_1 = |x| \\ y_2 = -2 \end{cases}$$

had no point of intersection, they still solved it, at least once, in order to see why it was unnecessary to perform an algebraic procedure on a trivial case (Fig. 6.2d shows Ping's written work on a similar problem). The additional graphical experience with the calculator allowed the students to deepen their understanding of what it meant to solve an absolute value equation and how to interpret its solution/s on the basis of the relationships they visually inferred on the relevant "iconic configurations" (Batt, 2007, p. 246).

Thus, the coordinated use of diagrams in reasoning such as the number line, the alphanumeric forms that characterize an algebraic equation, and the graphs drawn from the graphing calculator provided my students with both "aesthetic and cognitive effects" (Batt, 2007, p. 246), a sense of poietic creativity in which imagining a hypothetical situation (that of the absolute value) within the ground of carefully and purposefully drawn diagrams (number line, algebraic equations, graphs) constructed the conditions for necessary mathematical reasoning. *Pace* Peirce:

> The work of the poet or novelist is not so utterly different from that of the scientific man. The artist introduces a fiction; but it is not an arbitrary one; it exhibits affinities to which the mind accords a certain approval in pronouncing them beautiful, which if it is not exactly the same as saying the synthesis is true, is something of the same general kind. The geometer draws a figure, which if not exactly a fiction, is at least a creation, and by means of observation of that diagram he is able to synthesize and show relations between elements which before seemed to have no necessary connection.
>
> (Peirce (1932, p. 383) quoted in Batt, 2007, pp. 246–247)

In this chapter, we explore the implications of diagrammatic reasoning, that is, reasoning with diagrams in formational and transformational cognitive activity. From the above situation with my Algebra 1 students, their reasoning with diagrams actually evolved over three phases. They first learned to solve absolute value equations with the aid of a number line. Then it was followed by an algebraic approach using variables. In the third phase, they engaged in a graphical analysis with the TI 84. Hence, consistent with the traveling theory that is pursued in this book, in this chapter, we demonstrate how effective diagrammatization could foster the progressive evolution of mathematical reasoning and knowledge from the initial phase of manipulating hypothetical conditions in some physical format to eventually developing hypostatic abstractions and seeing particular relations in diagrams as instantiations of more general relationships. Diagrammatization for Peirce is akin to "skeletonization" with its primary focus on the "essential relations" in a given problem situation that then enables the drawing of necessary conclusions from premises (Dea, 2006, p. 509). Bakker (2007) and Stjernfelt (2007) push this intent further in their reconceptualization of the role of diagrams in mathematics learning. For Bakker (2007), diagrammatization – which involves constructing, experimenting with, and observing/reflecting upon the results of diagrams – provides "opportunities for hypostatic abstraction [to] occur" (p. 27), a helpful step in "processes involved in learning to reason" about mathematical concepts (p. 9). In *Diagrammatology*, author Stjernfelt (2007) situates diagrams beyond their architectural significance in signage to their central epistemological role in knowledge acquisition as follows:

> The fact that diagrams displays the interrelation between the parts of the object it depicts is what facilitates its use in reasoning and thought processes, thus *taking the concept of signs far from the idea of simple coding and decoding and to the epistemological issues of the acquisition of knowledge through signs*.
>
> (Stjernfelt, 2007, p. ix; italics added for emphasis)

Four points are worth noting early at this stage. *First*, Peirce considers algebraic equations, expressions, and formulas as diagrams. Diagrams are, thus, not limited to geometric figures. However, in this section, we focus on those visually drawn diagrams that mediate in the understanding of alphanumeric forms.

Second, many of the examples that are pursued in various sections in this chapter are drawn from the algebra and number sense strands of the school mathematics curriculum, and there is a reason in doing this. While teachers take as given the visual nature of geometry, statistics, and data analysis, the objects and symbols of number sense and school algebra seem unable to escape "the cultural imprisonments of typography" (Rotman, 1995, p. 390). That is, the hegemony of alphanumeric manipulation continues to drive the content and form of much of mathematical reasoning that is characterized by a particular "phobic mode of behavior" (Rotman, 1995, p. 394) toward visual forms. This reminds me of Rotman's (1995) representational spectrum in Fig. 6.3 that shows the status of diagrams along a continuum of symbolic necessities in mathematics. While the diagram shows numeric symbols as not necessarily irreducible to writing, Rotman (1995) points out, however, the (over)valuing we tend to accord the "rigorously literal, clear, and unambiguous ideograms" over the "metaphorically unrigorous diagrams" (p. 399). Further, he notes the implications of this seeming transposition in the following manner:

> The transposition in question is evident once one puts this ranking of the literal over the metaphorical into play: as soon, that is, as one accepts the idea that diagrams, however useful and apparently essential for the actual doing of mathematics, are nonetheless merely figurative and eliminable and that mathematics, in its proper rigorous formulation, has no need of them. ... [t]his alphabetic prejudice is given a literal manifestation: linear strings of symbols in the form of normalized sequences of variables and logical connectives drawn from a short, preset list determine the resting place for mathematical language in its purest, most rigorous grounded form.
>
> (Rotman, 1995, p. 399)

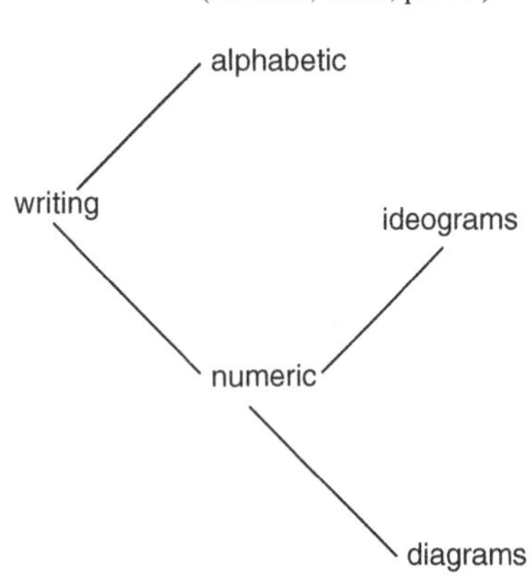

Fig. 6.3 Rotman's (1995) representational spectrum (p. 399)

Third, Sherry (2009) has surfaced the metonymic issue of insufficiency of diagrams (because they are drawn as empirical objects) and diagrammatic reasoning in dealing with both the abstract nature and abstract properties of mathematical objects. The matter, of course, is confounded by the fact that there are constraints in our perceptual ability to gain access to the relevant abstraction on the basis of ambiguities and inexactness of the diagrams we use to reason with. Sherry (2009) offers a provocative (Wittgensteinian) response that we pursue in some detail that hopefully allows us to fully understand the nature and context of diagrammatic reasoning (perhaps beyond Peirce) and further enrich our understanding of diagrams in mathematical activity that oftentimes deals with the hypothetical state of things. Sherry's (2009) provocative perspective that we view diagrammatic reasoning as an instance of applied mathematical reasoning resonates well with progressive mathematization and formalization that we have discussed in earlier chapters in this book.

Fourth, recent research studies empirically demonstrate a relationship between type of visuospatial representations employed and success in mathematical problem solving. We briefly discuss a few of these studies below.

Hegarty and Kozhevnikov (1999) distinguished between pictorial imagery ("constructing vivid and detailed visual images") and schematic imagery ("representing the spatial relationships between objects and imagining spatial transformations") with the latter type falling under the Peircean notion of a diagram. Based on their work with 33 sixth-grade male students, they found that diagrammatic and pictorial representations are positively and negatively, respectively, correlated with mathematical problem solving. Diagrams include relationships that are relevant to solving a problem unlike pictures that contain details irrelevant to a problem solution.

Booth and Thomas (2000) provided anecdotal evidence that shows older students (ages 11–15 years) with low visuospatial skills preferring pictures over diagrams in solving arithmetical problems perhaps because of a psychological need to "bring" the pictures "nearer to their perception of reality" (p. 185) that unfortunately distracted them from completely and correctly solving the problems.

Van Garderen's (2007) multiple probe design work with three eighth-grade students (one female, two males) with learning disabilities (LD) provided evidence of improved problem-solving performance on tasks that involve one- and two-step computational word problems due to a successful shift in thinking from pictures to diagrams. Van Garderen's (2006, 2003) empirical work in the case of a larger sample of students grouped by problem-solving ability (66 participants consisting of students with LD, average-achieving, and gifted) produced findings that were consistent with those found by Hegarty and Kozhevnikov (1999) above, including the following important facts: (1) the gifted used diagrams (conveyed by drawing, gesturing, or talking aloud) more significantly than did the other two groups; (2) the LD used pictorial representations more frequently than did the gifted; and (3) diagram- and picture-use, respectively, positively and negatively correlated with spatial visualization ability, which is correlated with mathematics achievement and is important in learning geometry and in solving complex word problems.

Zahner and Corter (2010) statistically analyzed the "spontaneous (unprompted) use of external inscriptions" (p. 178) of 34 adult students (mean age of 28 years) who solved six different problems based on different probability topics in the context of a clinical interview. The inscriptions the novices produced were classified as follows: schematic diagrams, which depict relationships; pictorial or iconic figures, which illustrate physical appearances of the objects involved in a problem; and other "forms of spatial organization and tabulation of problem information" (p. 180). Zahner and Corter concluded that there is an association between (1) the use of inscriptions and specific probability topics; (2) the choice of inscription and rate of success in solving a probability problem; and (3) particular types of inscriptions and stages in probability problem solving. To illustrate (1), the novices used an outcome listing method in dealing with a combinatorial problem or a Venn diagram in solving a compound event problem. In the case of (2), those who chose to, say, use a schematic diagram or a tree in making sense of a problem involving the fundamental principle or combinations and conditional probability, respectively, obtained significantly higher rates of solution success than the overall mean performance rate for the problem.

Recently, I read a blog entry of a high school mathematics teacher who shared with interested readers how he taught his students to factor simple quadratic trinomials of the form $x^2 + bx + c$ using a popular method shown in Fig. 6.4. On close inspection, nothing in the numerically driven diagram reflects the iconic configurations that might be conceptually relevant in factoring such trinomials unlike the tic-tac-toe diagrams shown in Fig. 2.4a–c. That is, his students used the diagram in Fig. 6.4 basically as a tool for generating the factors and nothing more. When I emailed the teacher and commented that the method the students acquired did not appear to at least resemble the mathematical process involved in factoring quadratic trinomials such as the ones shown in Fig. 2.4a–c, his response below reflected the use of a diagram that has been drawn from alphanumeric activity.

> My students don't have experience in algeblocks (or any manipulatives) from algebra 1, so I hadn't based the method off of them. But it is probably worth doing. I'll design something for my classes to try next week to see if it helps their understanding.
>
> (Teacher X, 2009, email communication)

The point is, in Rotman's (1995) words, "the distorting and reductive effects of the subordination of graphics to phonetics [alphanumeric symbols in the case of algebra] and [how they] have made it their business to move beyond this dogma" (p. 390). In this chapter, we explore the implications of how thinking diagrams in algebra and number sense would allow students to visually experience the "intricate interplay of imagining and symbolizing [that is] familiar to mathematicians within their practice" (Rotman, 1995, p. 392). This way, abstraction and generalization in algebra and school mathematics, more generally, are seen as having "material, empirical, embodied, [and] sensory dimension[s]" (Rotman, 1995, p. 391).

This chapter has four sections. Section 1 deals with four psychophysical and cognitive perspectives on diagrams. Section 2 addresses three fundamental dilemmas with diagrams, namely: problem of existence; concern for universality; and choosing between presented and self-constructed diagrams. In Section 3, we discuss different types of diagrams and implications of progressive diagrammatization.

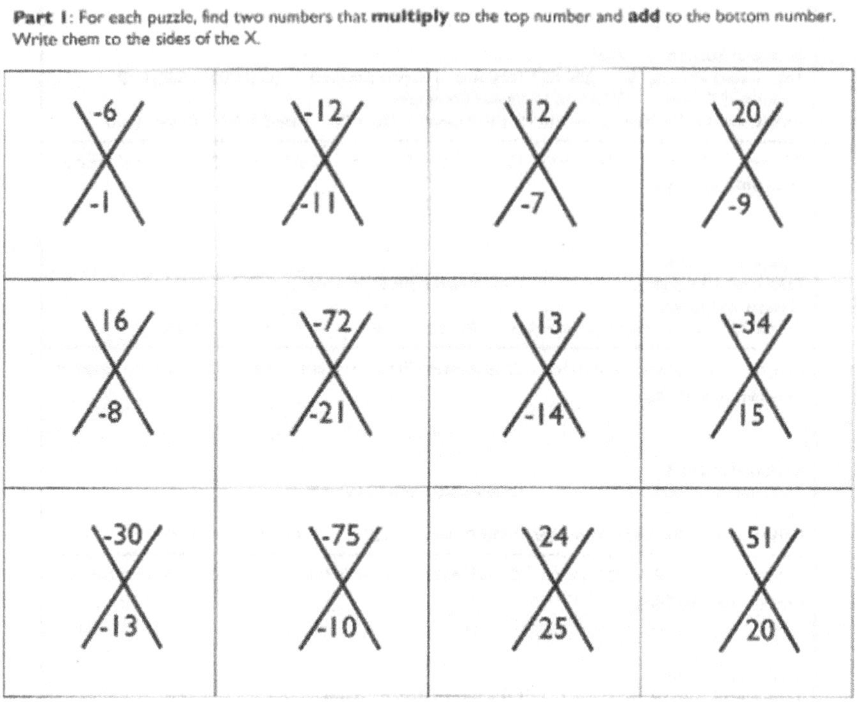

Part I: For each puzzle, find two numbers that **multiply** to the top number and **add** to the bottom number. Write them to the sides of the X.

-6 / -1	-12 / -11	12 / -7	20 / -9
16 / -8	-72 / -21	13 / -14	-34 / 15
-30 / -13	-75 / -10	24 / 25	51 / 20

Fig. 6.4 Sample of factoring quadratic trinomials of the type $x^2 + bx + c$ by the x method (Teacher X, 2009)

Section 4 explores in some detail Sherry's (2009) point about diagrammatic reasoning as an instance of applied mathematical reasoning.

1 Psychophysical and Cognitive Perspectives on Diagrams

Pylyshyn (2006) notes at least four interesting points about the role of the human visual system in diagrammatic activity. *First*, our vision has natural mechanisms that enable us to detect relationships and construct abstractions in diagrams. Adults, in fact, tend to visually attend to structures unlike infants who initially focus on objects (cf. Leslie, Xu, Tremoulet, & Scholl, 1998).

Two examples from my classes illustrate the above point. In my Algebra 1 class, the graphing calculator activity in Fig. 4.20 allowed the students to primarily use their vision in establishing a relationship between the y-intercept B in $y = Ax + B$ and the straight line by visually attending to lines that either went up, down, or crossed the origin depending on whether $B > 0$, $B < 0$, or $B = 0$, respectively. In fact, the visually drawn inference became the basis in which they formed hypostatic abstractions relevant to types of slopes and intercepts of graphs. Further, what they saw consistently across several examples allowed them to state their generalizations about the role of the y-intercept C in the case of $y = x^2 + C$ and much later in the

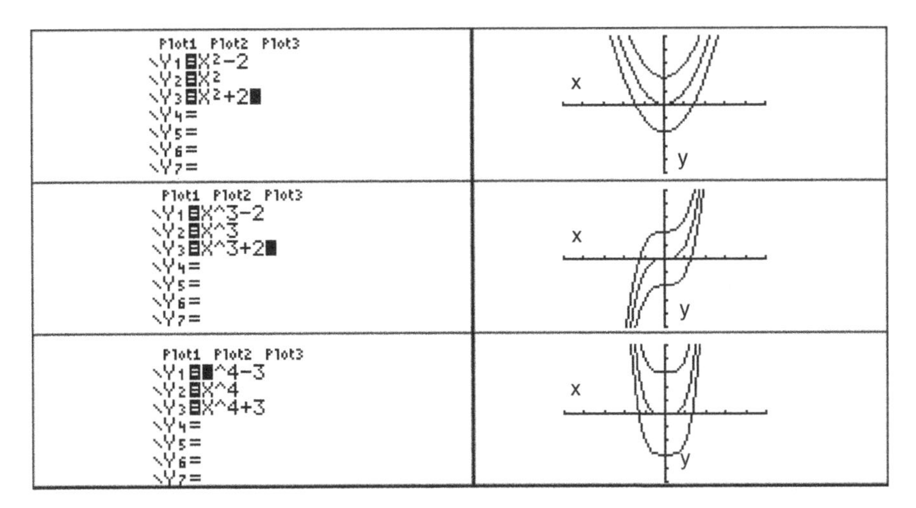

Fig. 6.5 Power functions on the TI 84+ using transformation graphing application

general case of power functions of the form $y = x^n + C$, where n is a whole number (Fig. 6.5). In my Grade 2 class, 60% of the students naturally exhibited structural awareness relative to the two tasks shown in Fig. 5.4a, b prior to formal instruction on pattern generalization. Figure 5.4c, d provides samples of their work on two far generalization tasks.

Second, Pylyshyn (2006) points out how steps in a diagram actually function in facilitating some kind of a concrete justification in visual form. For example, the diagram in Fig. 2.1 visually assisted my Algebra 1 students in establishing the steps needed to factor a quadratic binomial involving the difference of two squares. Also, the visual steps in Figs. 3.14 and Fig. 3.15 provided them with three models of a visual explanation of the Pythagorean theorem. In my Grade 2 class, the students used their diagrammatic experiences (e.g., Figs. 1.1 and 3.1) in understanding the numerical algorithm for adding and subtracting whole numbers with regrouping.

Third, even when our vision is predisposed to noticing relationships in a diagram, it also depends on what we visually attend to that influences what and how we see. In other words, when we glance at a diagram, some property, attribute, or relationship may take more effort and time to acquire than others that would then influence the steps we construct in relation to the diagram. For example, Emma had a difficult time abducing a plausible relationship in the case of the figural stages in Fig. 5.13b unlike Diana, who immediately saw each stage as the union of a growing perimeter of sticks ($4x$) and two symmetrical sets of (x–1) rows of x sticks in the interior of each square. Dung saw each stage in Fig. 5.13b in a much more complicated manner. Figure 6.6 is a visual illustration of the steps he employed in his visual thinking. He initially saw n disjoint rows and counted the total number of sticks per row. In counting the number of sticks per row, he once again saw n disjoint squares for a total of $4n$ and then subtracted the overlapping sticks, n–1. He then saw that

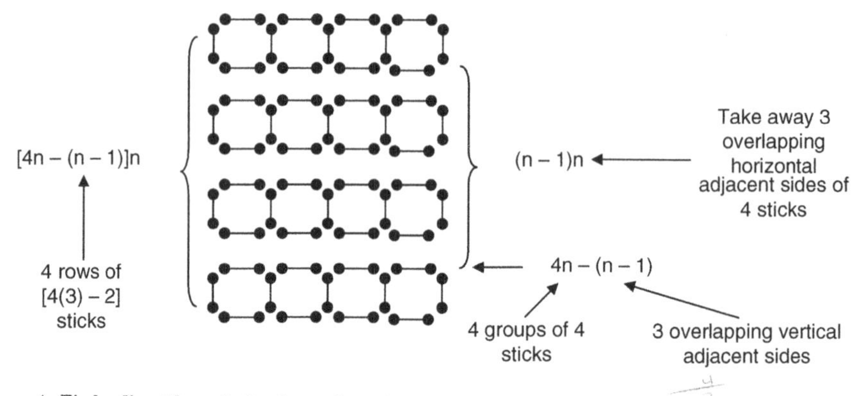

Fig. 6.6 Dung's construction and justification of his formula for the pattern in Fig. 4.13b

there were $(n-1)n$ overlapping horizontal rows (sides) in the interior of the larger square.

In all reported accounts relevant to pattern generalization activity, in particular, what is not oftentimes pointed out when students' thinking transition from particularizing to generalizing is the qualitatively drawn observation that certain visual mechanisms are operating behind the scenes that somehow assist them in constructing an algebraic generalization. For example, Dung's visual mechanism in Fig. 6.6 had him parsing each complex stage into subconfigurations of separate rows of squares and separate smaller squares per row. The diagrammatic reasoning he employed biased the manner in which he interpreted each known stage in the pattern, which he then projected onto the unknown stages. The relative absence or ineffective use of the visual mechanism of parsing and biasing[1] could help explain why Emma was unable to justify the numerically driven formula she produced for the Fig. 5.13b task. When she found it difficult to parse and bias the pattern stages in a way that surfaced a structural unit, she resorted to using a numerical strategy that

[1] The process of figural parsing and biasing shares many of the characteristics that Gal and Linchevski (2010) identified in relation to the operative apprehension of geometric figures (Duval, 1998), which involves engaging in an appropriate and purposeful "dimensional deconstruction of figures" that leads to "infer[ring relevant] mathematical properties in axiomatic geometry" (Gal & Linchevski, 2010, p. 180).

bypassed such mechanisms altogether. The same explanation could be claimed in those cases when students resorted to superficially (versus structurally) iconic generalizations. It was evident that they either did not use diagrams effectively (implicitly or explicitly) or simply avoided them.

Fourth, the use of the number line diagram in Figs. 6.1a, b, including the diagram in Fig. 6.7 that Jackie (Cohort 2, Grade 7) used when she solved word problems,

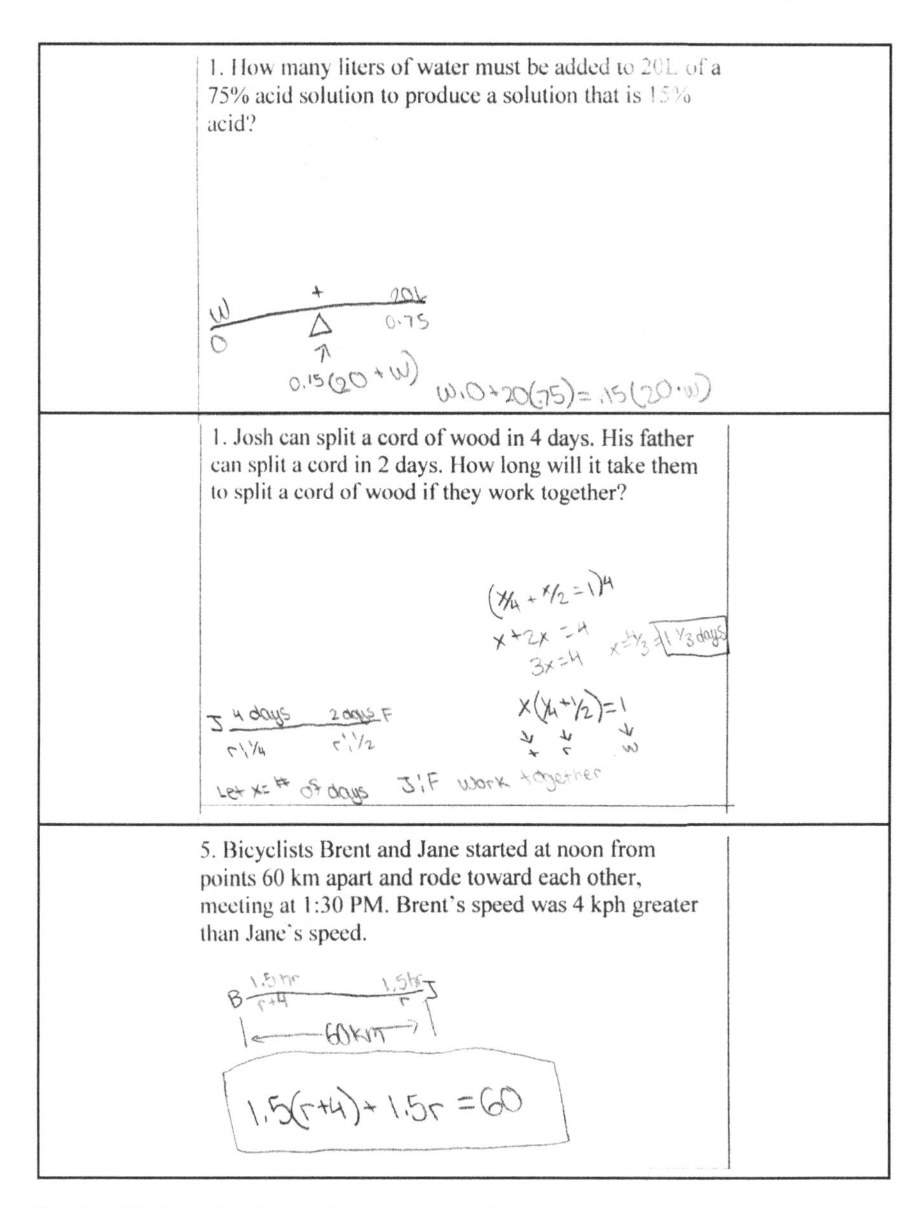

Fig. 6.7 Ollie's word problem-solving strategy by diagramming

conveys the view that a diagram "not only lessen(s) the burden on memory, but also encourag(es) certain assumptions and ways of thinking about the problem(s)" (Pylyshyn, 2006, p. 455). The diagrams shown in Figs. 6.1a, b and 6.7, in fact, biased the manner in which the students dealt with such mathematical objects in a way that could not be inferred on the objects themselves. They also assisted the students when they had to hypostasize and generalize the processes involved and deal with a variety of similar problem situations.

"Diagrams and other abstract models," Novick (2006) points out, and rightly so, "have a long history of guiding human activities, from the intellectual to the practical to the leisure" (p. 1826). Table 6.1 lists examples of diagrams that have been developed over many years by individuals in everyday and mathematical settings. Across the different forms, diagrams address basic issues such as *computational offloading* and *re-representation* (Ainsworth & Loizou, 2003, p. 670; Zahner & Corter, 2010, p. 196). Gemiliano (second grader) in Fig. 3.1 consistently and very quickly used sticks and dots in helping him deal with all subtraction problems with regrouping. The actions he performed on the diagrams he constructed effectively enabled him to make the necessary changes and interpretations in the digits involved in the symbolic aspect (i.e., the subtraction algorithm). The hypostasized term "regrouping" for Gemiliano and his classmates, in fact, cued them to use their stored visual image in guiding their symbolic actions on the relevant digits in any subtraction problem.

In problem-solving situations, when texts are available for comparison, diagrams seem to provide more efficient and enhanced analyses that could also significantly reduce cognitive effort. For example, Fig. 6.8b shows a combined visual–numerical inductive-structuring table diagram that could help students develop an algebraic generalization for the classic *frog problem* shown in Fig. 6.8a. Lewis (2009) notes that when the same problem was presented to a sixth-grade class, the students used a similar diagrammatic approach and noticed that the solution in each case resembled

Table 6.1 Examples of diagrams (Novick, 2006, pp. 1826–1827)

Geometric diagrams
Route maps
Multiplication tables
Abstract charts resembling networks
Abstract charts for illustrating patterns and navigation schemes
Origami diagrams for paper folding
Networks
Abstract matrices and networks used in game boards
Graphic-language expressions on blackboards
Scientific drawings
Tables
Corporate logos
Iconic diagrams such as photographs and line drawings of objects
Schematic diagrams or conceptual graphs displaying abstract concepts
Circuit diagrams, hierarchical trees, and Venn diagrams as schematic diagrams
Charts and graphs such as line graphs, bar graphs, and pie charts
Cartesian graphs

a palindrome. Further, the organized presentation of arrows and numerical values in Fig. 6.8b enabled them to infer with relative ease that the total number of jumps and slides in each case involves a perfect square and an even number, respectively. Thus, diagrams used in this manner allow learners to conveniently search for information that otherwise would be difficult to accomplish with a solely textual approach (cf. Larkin & Simon, 1987).

Diagrams can also "trigger the retrieval processes" relevant to "extracting knowledge from memory" (Zhang, 1997, pp. 180–181). For example, my Algebra 1 students sustained their ability to factor a difference of squares $a^2 - b^2$ throughout the school year each time they were reminded of the generic diagram shown in Fig. 2.1. My Grade 2 class successfully added and subtracted whole numbers having two and three digits with and without regrouping throughout the school year because they could easily recall the visual processes they developed with squares, sticks, and dots (Figs. 1.1, 2.13, 2.20, and 3.11a). One powerful diagram that my Algebra 1 students always referred to whenever they had to solve quadratic equations of the type $ax^2 + bx + c = 0$ is the visual model shown in Fig. 3.12b, which reminded them of the alphanumeric steps needed to solve such equations. In this particular situation, it was interesting how the diagram had more conceptual impact than the quadratic formula that helped them solve such equations. Figure 6.9 is a popular 10×10 unit square grid approach that teachers have found effective in assisting their students to understand the addition of two decimal fractions. When Suh, Johnson, Jamieson, and Mills (2008) implemented the grid approach to decimals in a fifth-grade class, the students (1) understood decimals as "an extension of the base-ten number system and not a different system"; (2) established "generalizations about the relationship

a

A male-and-female couple sits on two chairs separated by a chair in the middle. They can switch places in three moves, as shown below.

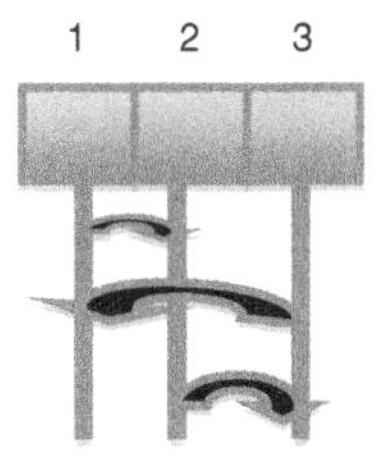

On a row of seven chairs, three couples sit so that the three men M sit on the three chairs on one end beginning from the left and the three women F sit on the other chairs starting from the right. The middle chair is left vacant. What is the minimum number of moves needed to interchange the men and the women under the following conditions:

1. The men can only move from left to right and the women only from right to left.
2. Each of them can *either slide* onto an adjacent chair *or jump* over one other person to the vacant chair immediately after him or her.

Fig. 6.8 (continued)

b

Case	Visual Solution	Number of Jumps	Number of Slides	Number of Moves
1 pair	S-J-S (see figure below)	1	2	1 + 2 = 3
2 pairs	S-J-S-J-J-S-J-S (see figure below)	4	4	4 + 4 = 8
3 pairs	S-J-S-J-J-S-J-J-J-S-J-J-S-J-S (see figure below)	9	6	9 + 6 = 15
4 pairs	S-J-S-J-J-S-J-J-J-S-J-J-J-J-S-J-J-J-S-J-J-S-J-S	16	8	16+ 8 = 24

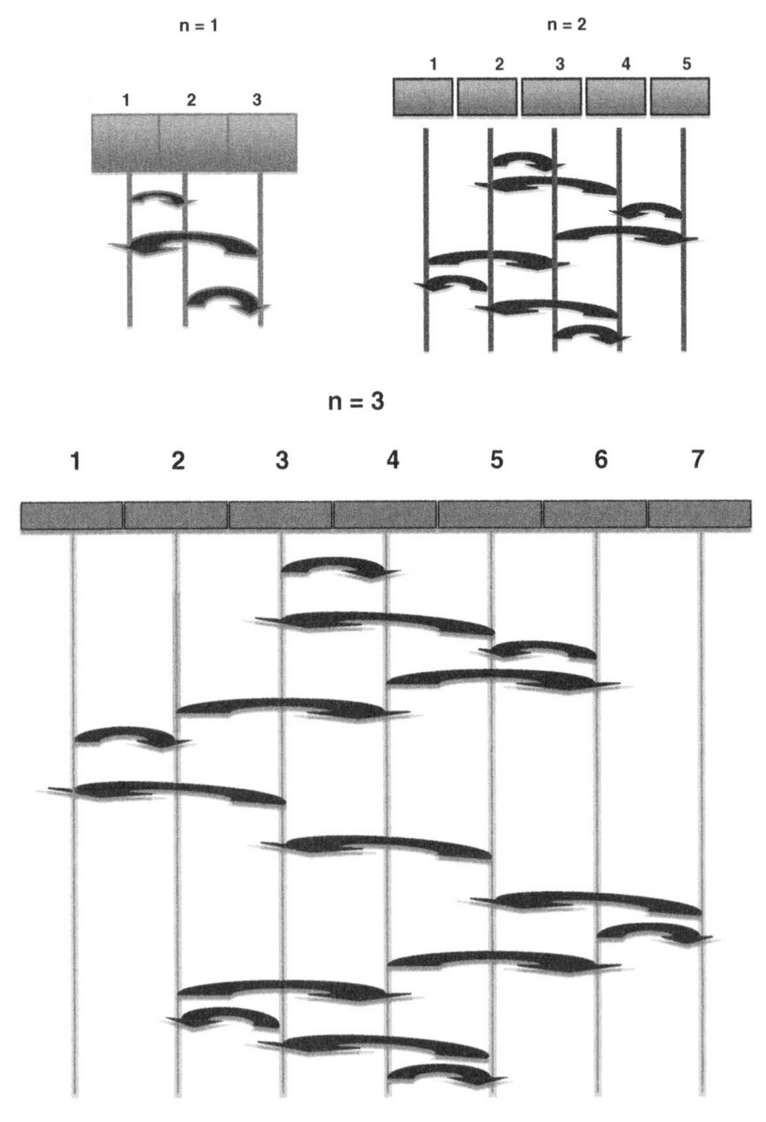

Fig. 6.8 **a** The frog problem (Lewis, 2009, p. 420). **b** A combined visual-and-table solution to the Fig. 6.8a task (Humphries, 2010; Lewis, 2009, p. 422)

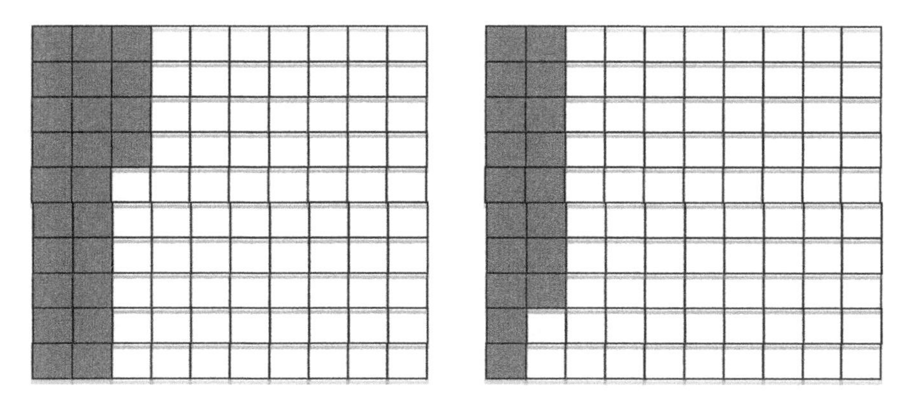

Fig. 6.9 Fifth-grade student's work (Suh et al., 2008, p. 48)

of numbers and their place value by stating that the value of the digits was determined by the place"; and (3) "helped [them] calculate the decimal sums, state the decimals in words, and relate them to fractions" (p. 49).

2 Three Basic Issues with Diagrams

In this section, we deal with three fundamental dilemmas with diagram use, meaning, and construction in mathematics. The first two issues are concerned with matters involving existence and universality. The third issue deals with effectiveness of presented and self-constructed diagrams.

Issues with existence deal with how we know a priori that our target object, the subject of diagramming, actually exists. For Kulpa (2009), the easy answer is that if it could be drawn, then it exists. Also, if it cannot be drawn, then it does not exist. Apparently, in the case of, say, logic tables, Venn diagrams, and graphs used in statistics and discrete mathematics, the problem of existence is a nonissue. In cases involving other objects that have a continuous domain, however, something more needs to be assumed. For example, a diagram of a segment that contains all the real numbers in [0, 1] would be difficult to distinguish from a segment consisting of all the rational numbers in the same interval. In relatively similar situations, it makes sense to perceive a constructed diagram only as a physical representation of the ideal diagram that exists in the mind. Because our minds are usually able to distinguish between the two diagrams, we tend to actually reason with the ideal diagram itself.

Kulpa (2009) also notes that with existence comes the issue of precision. Imprecision errors occur when there is no harmony between the ideal and the physical diagram, which is, in fact, a consequence of the "physical limitations of the diagrammatic medium" (p. 80). For example, in my Algebra 1 class, we confronted this matter when we had to model irrational numbers in which case the physical metric devices we used could only support rational approximations that made the task more difficult and less convincing for beginning students. One interesting visual activity we did in class is shown in Fig. 6.10. In the activity, the students had to determine why it seems to be the case that rearranging the first diagram to form the second diagram actually produced two different areas. The activity reinforced the importance of the constant nature of slope for any pair of points on a straight line, which should explain the imprecise manner in which the second diagram in Fig. 6.10 has been drawn.

Issues with existence are not limited to constructing iconic diagrams that preserve properties of their target objects. In many situations, students also construct diagrams that retain some of their iconic properties but are used primarily to facilitate concept or process attainment and development in either formational or transformational context. Such diagrams exist that fulfill, in Dörfler's (2001a) words, those "cases where in principle the mathematical notion is not amenable to diagrammatic methods and one has to stick to a kind of conceptual reasoning based on linguistically or metaphorically prescribed properties" (p. 1; see also Dörfler, 2001b). With such diagrams, the relevant issue is not one on precision but in seeing to it that the diagram does not mislead students to produce unnecessary misconceptions.

For example, when my Algebra 1 students learned to simplify rational algebraic expressions later in the school year, I used their experiences with fraction strips (Fig. 3.4) in helping them understand the significance of what I refer to as a "unit diagram" in simplifying variable expressions. When they manipulated the strips, they developed the view that fractions could be constructed by taking multiples of

Cut up an 8 by 8 square and rearrange the pieces to make a 5 by 13 rectangle as shown. What is the area of each figure?

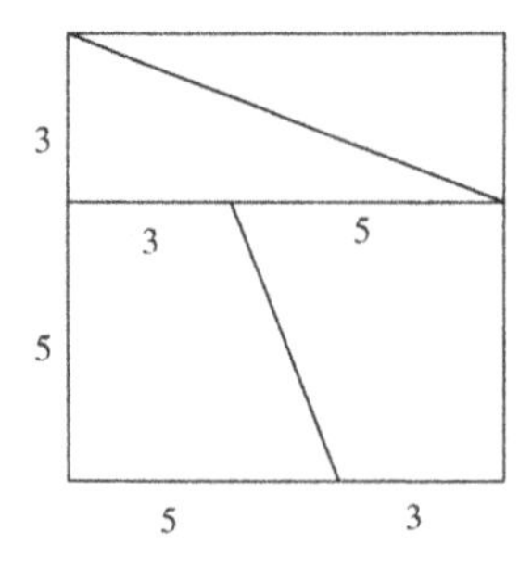

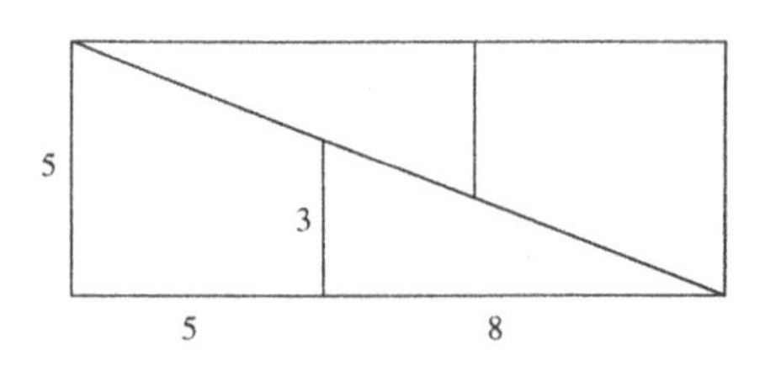

Fig. 6.10 Area problem (Teaching Committee of the Mathematical Association, 2001, p. 7)

some unit fraction of the form $(1/n)$, where n is a counting number. So (1/4) is a half copy of (1/2), (3/5) means three copies of (1/5), 2(1/3) has seven copies of (1/3), etc.

When the students added and subtracted fractions and rational expressions that had the same denominator such as the ones shown in Fig. 6.11a, I had them initially stating an appropriate *common unit*. Doing this emphasized the central role of *unitizing* in performing the two operations. When we repeated the same process several more times, the students began to draw an empty box consistent with their experiences with fraction strips. Then they filled in the box with the correct common unit. Figure 6.11b illustrates how they used the unit diagramming in dealing with fractions that had different denominators.

Figure 6.12 illustrates how they used the unit diagram approach in making sense of multiplying fractions and rational expressions. The diagram, in fact, played a significant role that allowed them to see why the rule for multiplication worked that way and not some other way.

Finally, we made sense of division not by working through examples but by thinking about it in the general context of rule construction. As shown in Fig. 6.13, we first established the appropriate unit diagram, transformed each fraction appropriately, and then focused on the numerators leading to the construction of a rule for division.

Thus, in the case of learning to simplify fractions and rational algebraic expressions, the students manipulated a unit diagram that assisted them in making sense of the four operations in a more coherent manner. Two related points are worth noting. *First*, the students' initial experience with unit diagramming took place when they learned to add and subtract polynomial expressions with algeblocks. For example, Fig. 6.14 shows an algeblocks activity that involves adding polynomials. In performing the operation, they initially determined the terms that were similar before adding the relevant coefficients. Their understanding of similar terms relied on the general concept of a unit. *Second*, consistent with their experiences with polynomials and rational expressions, the students also used unit diagramming in making sense of adding (and also subtracting) radical expressions (Fig. 2.9b).

Issues involving universality deal with the general and representative nature of constructed diagrams. One concern is universality drawn from specializing – that is, when we use a diagram to explain a proposition, how do we justify the generalizability of the diagram "to a wide (usually infinite) class of configurations" (Kulpa, 2009, p. 81)? Our first response below is taken from Kulpa (2009), who suggests employing a reasoning strategy he refers to as domain splitting. A second response is given in Section 4 in relation to Sherry's notion of diagrammatic reasoning as an instance of applied mathematical reasoning.

Kulpa (2009) suggests placing constraints on the relevant domain, splitting the domain over two or more subsets, or restricting the property that we want to prove. For example, in school geometry, when we investigate the claim that "the altitudes of all triangles intersect at a single point inside a triangle," we may require constraining the set of triangles to, say, acute triangles so that the following statement is true: Altitudes of all acute triangles intersect at a single point inside a triangle.

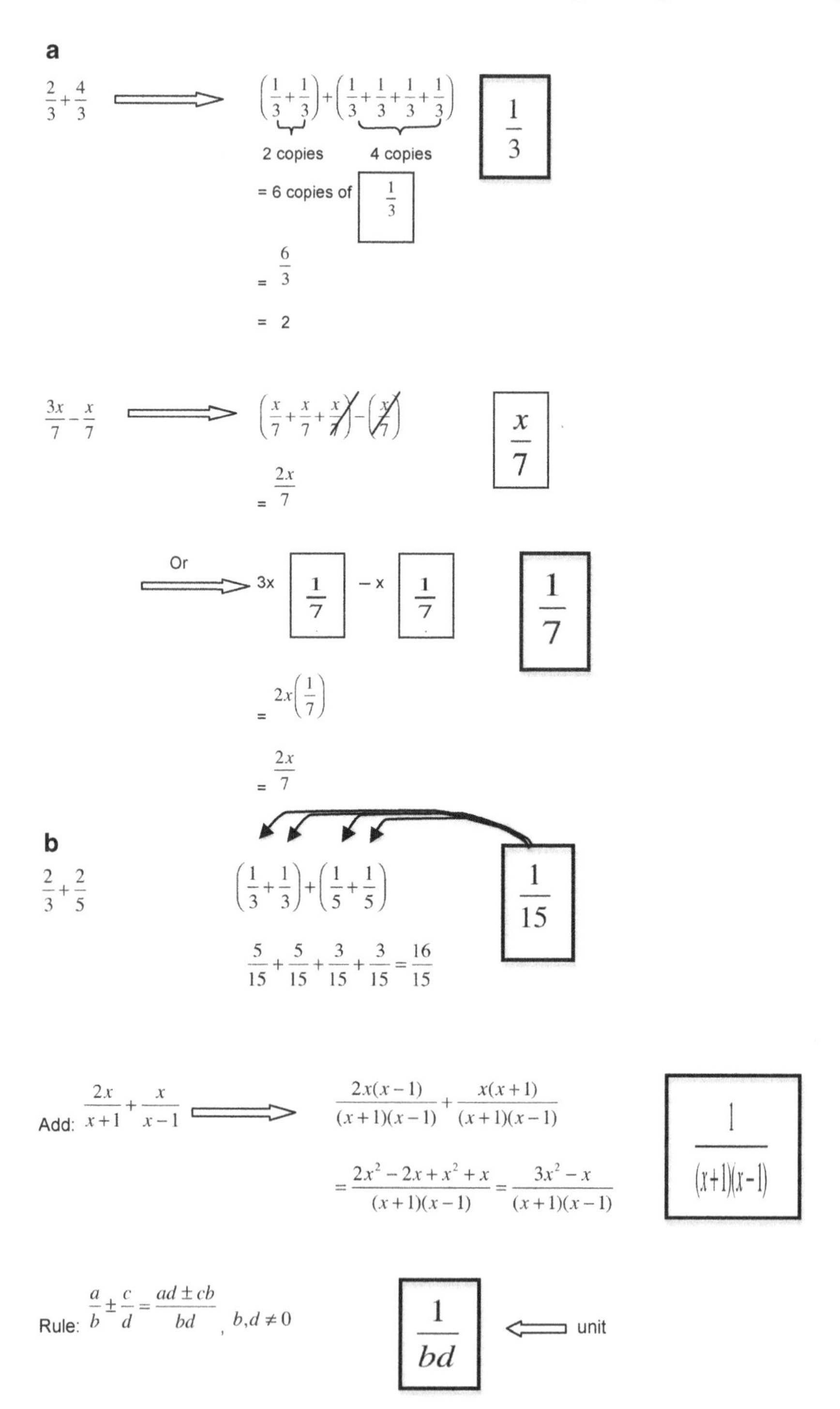

Fig. 6.11 **a** Adding and subtracting fractions with the same denominator by unitizing. **b** Adding and subtracting fractions with different denominators by unitizing

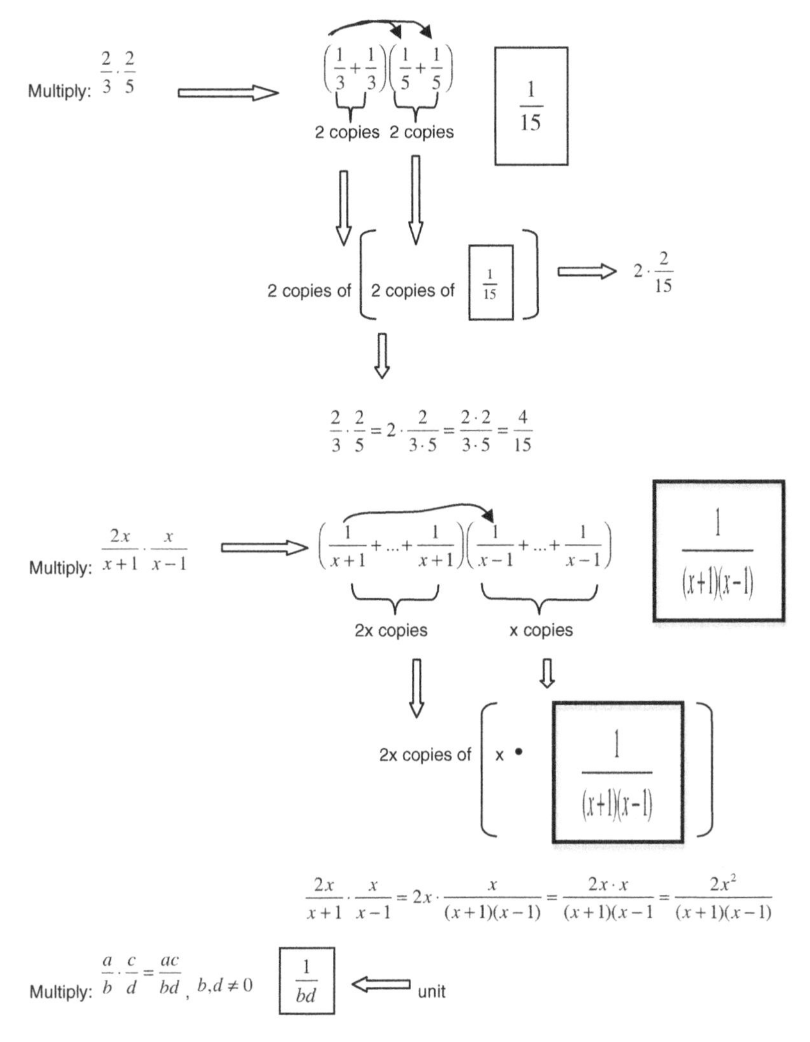

Fig. 6.12 Multiplying fractions by unitizing

Or we may choose to further restrict its property as follows: Altitudes of all triangles intersect at a single point. A second illustration is taken from school algebra and number sense. Diagram construction is difficult to accomplish when the nature and the context of variables transition from the particular to the general case. For example, the multiplicative expression $a \cdot b$, which conveys a (equal) copies of b or a (equal) groups of b, is limited to counting numbers relative to the variable a. However, there is no diagrammatic interpretation for $a \cdot b$ when both variables are, say, irrational numbers. Hence, constraints have to be imposed on the domains in most cases that involve variables.

$$\frac{a}{b} \div \frac{c}{d} \Longrightarrow \frac{a}{b} \div \frac{c}{d} = \frac{ad}{bd} \div \frac{bc}{bd} = ad \div bc = \frac{ad}{bc} = \frac{a}{b} \cdot \frac{d}{c}$$

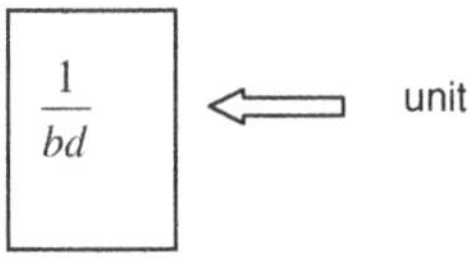

$$\boxed{\frac{1}{bd}} \Longleftarrow \text{unit}$$

Fig. 6.13 Explaining division of fractions by unitizing

A third issue with diagrams and diagrammatic activity in school mathematics deals with the situation that Pantziara, Gagatsis, and Elia (2009) refer to as *presented diagrams* versus *self-constructed diagrams*. The authors note that learners in problem-solving situations perceive tasks with and without diagrams to be different despite having the same basic structural characteristics. Cox and Brna (1995) made a similar claim in the more general case of external representations in problem-solving contexts. A recent study by Barmby, Harries, Higgins, and Suggate (2009) conducted with two cohorts of elementary school students of upper ability in mathematics [20 students in Year 4 (ages 8–9) and 14 students in Year 6 (ages 10–11)] shows that their students successfully generated a variety of counting strategies that enabled them to determine the total number of circles in a presented array representation involving the multiplication of two whole numbers of up to two digits long. However, they also saw that the array format "encourage(d) the over-use of inefficient counting strategies, in particular for the younger pupils, although this was less prevalent for the older pupils" (p. 235).

The results of Pantziara et al. (2009), Cox and Brna (1995), and Barmby et al. (2009), in fact, empirically illustrate a phenomenon that Zhang and Norman (1994) refer to as *representational effect* in which "different isomorphic representations of a common formal structure can cause dramatically different cognitive behaviors" (Zhang & Norman, 1994, p. 88). Drawing on their study with 194 Grade 6 participants on six problem-solving tasks, Pantziara et al. (2009) saw that "(i)n the problem solving condition with diagrams, students needed to interpret and interact with the presented diagrams, whereas in the condition without diagrams students had to search for a solution strategy, such as constructing their own diagram" (p. 16).

In my own studies, however, it seems that a more fundamental issue with diagrams is not about assessing which representational format actually produces the most significant learning or the most meaningful cognitive behavior but how students use them, presented and constructed, in a lifeworld-dependent context with a distributed nature. The 20 students that participated in a study by Ainsworth and

Lab 4-2: Combining Like Terms on the Basic Mat

Remember that terms with exactly the same variables are called ***like terms****. These terms can be combined by adding the coefficients of the variables.*

Example: Simplify: $x^2 - 4x + 2x^2 + 2x$

Step 1.	Step 2.	Step 3.	Step 4.
Model the entire expression on the mat.	Combine like terms. Remove zero pairs.	Read the mat. 3 of x^2 and $^-2$ of x	Record. $3x^2 - 2x$

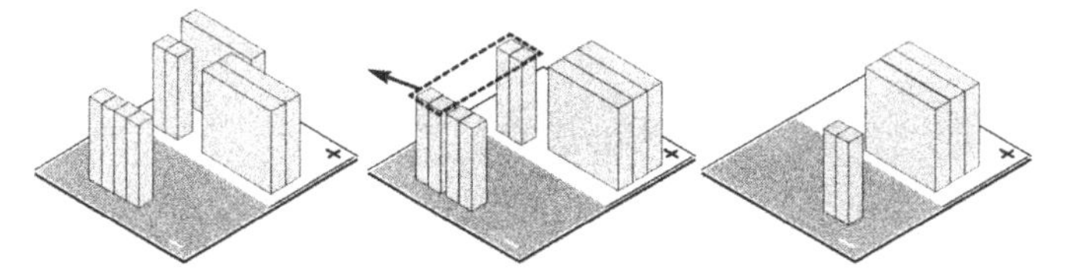

Try It

1. In the example, 2 positive x blocks and 2 negative x blocks were removed from the mat. What do these blocks represent? ____________________
2. In the example, what terms are combined? ____________________

Practice

Use Algeblocks and the Basic Mat to combine like terms and simplify the expression. Sketch each step of your work, and write the final expression.

3. $3x + 4 - x - 1$ **4.** $2x^2 + 4xy - 4x^2 - xy$ **5.** $y^2 - 2x - 3y^2 - x + 3$

6. $5 - 2x + 3 - 3x$ **7.** $y + 3xy - 4y - xy$ **8.** $3x^2 - xy + 4 - 5xy$

Simplify each expression. Sketch your final answer and write the expression.

9. $xy - 2y - 2x + 4xy$ **10.** $2y^2 - 3x - 4y^2 + 7$ **11.** $^-4 - 3x^2 + 2 - 2x^2$

12. $5x - 3 + 2x - 3xy$ **13.** $y + 3x - 4x - y$ **14.** $3y^2 - 2y - y^2 + 4y$

15. $3y^2 - x^2 + 2xy - 4 + 2y$ **16.** $4x^2 + 2x - x - 1 - 3x^2$ **17.** $4xy - 3 + 2xy - 5 - 2y$

Fig. 6.14 Adding polynomials with the algeblocks (© ETA/Cuisenaire®, 2003, p. 53)

Loizou (2003), for example, produced more self-explanations in presented pictorial diagrams than in presented text-alone diagrams, which significantly improved their learning and, thus, aided learning more effectively. With my participants, many individual- or group-generated diagrams became presented diagrams for others and many presented diagrams became the basis in developing self-generated ones.

As a further illustration of function over choice, Ponce (2008) has suggested using the concept maps shown in Figs. 6.15 and 6.16a in helping middle school students to recall and visually understand the algebraic process relevant to solving absolute value equations and inequalities. The presented map in Fig. 6.15 is similar to the one I used in my Algebra 1 class. In Ponce's (2008) case, as well as mine, the presented diagrams provide a powerful visual context that students can actually use

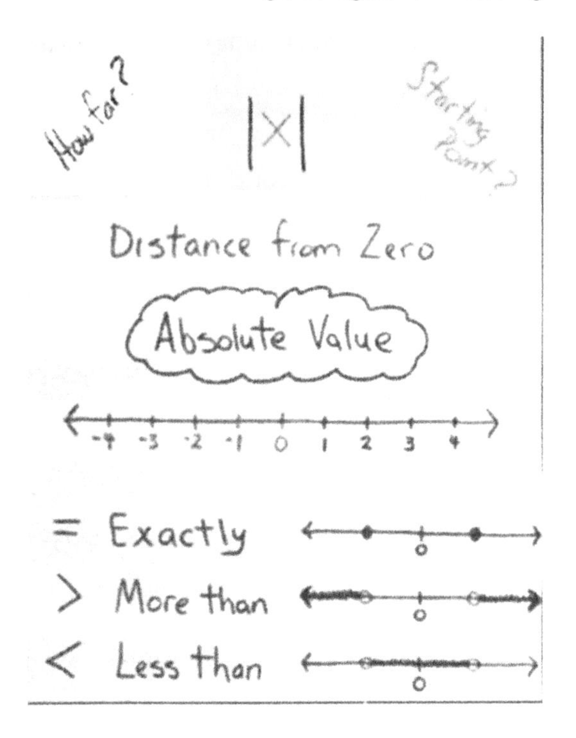

Fig. 6.15 Concept map for solving absolute value equations and inequalities (Ponce, 2008, p. 237)

whenever they find it necessary to construct their own diagrams either physically or mentally in order to solve a particular absolute value equation. Figure 6.16b shows the written work of Jing (Cohort 2, Grade 7) on a problem involving an absolute value inequality. Jing's work reflect an appropriation of a number line diagram in making sense of the inequality that allowed her to successfully obtain the correct solution set.

It, thus, goes without saying that the usefulness, meaningfulness, and representational power of either presented or self-constructed diagrams actually depend on how students acquire them in classroom activity. In the school geometry curriculum, students are always given presented diagrams as visual scaffolds that help them produce a conjecture whose necessity they will eventually prove or disprove. In the statistics and data analysis curriculum, students frequently generate diagrams such as scatter plots, histograms, and bar graphs in order to have a visual sense of, say, the behavior of a data set leading to statistical claims, generalizations, and meaningful inferences. The point is that in either presented or generated context, what matters significantly more than the diagrams themselves has to do with what students do with them. Zahner and Corter (2010) expressly articulate the need for diagrams to have a "facilitative effect" on individual learners especially in situations when they are confronted with complex or nonroutine problems. Certainly, the facilitative effect could be taught. Pantziara et al. (2009), for example, emphasize the crucial role of effective teaching strategies in helping students effectively use diagrams. Novick (2006) argues along similar lines, that is, students' ability to

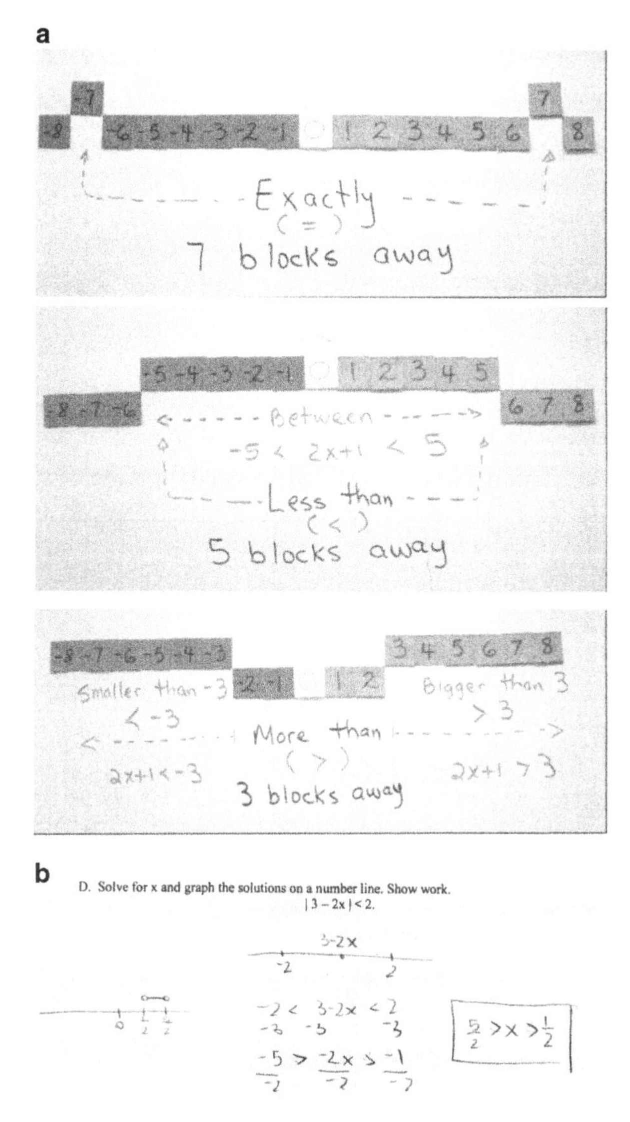

Fig. 6.16 **a** Examples of absolute value equations and inequalities using the concept map in Fig. 6.15 (Ponce, 2008, pp. 238–239). **b** Written work of seventh grader Jing on an absolute value inequality

"be effective diagrammatic reasoners" critically depends on instruction that helps them see "applicability conditions that are most important for selecting an appropriate diagrammatic representation" (p. 1850). For example, Fig. 6.17b, c shows the generated diagrams of two pairs of middle school students in Tarlow's (2008) study in relation to the pizza problem shown in Fig. 6.17a. The work of each pair of students models diagrammatic transitions or progressive "instrumental reifications"

a

A Pizza Problem

A local pizza shop has asked us to help design a form to keep track of certain pizza choices. The shop sells a plain pizza, topped with only cheese and tomato sauce. A customer can then select from the following toppings: peppers, sausage, mushrooms, and pepperoni.

1. How many different choices for pizza does a customer have?

2. List all the possible choices.

3. Find a way to convince one another that you have accounted for all possible choices.

4. Suppose a fifth topping, anchovies, was available. How many different choices for pizza does a customer now have? Why?

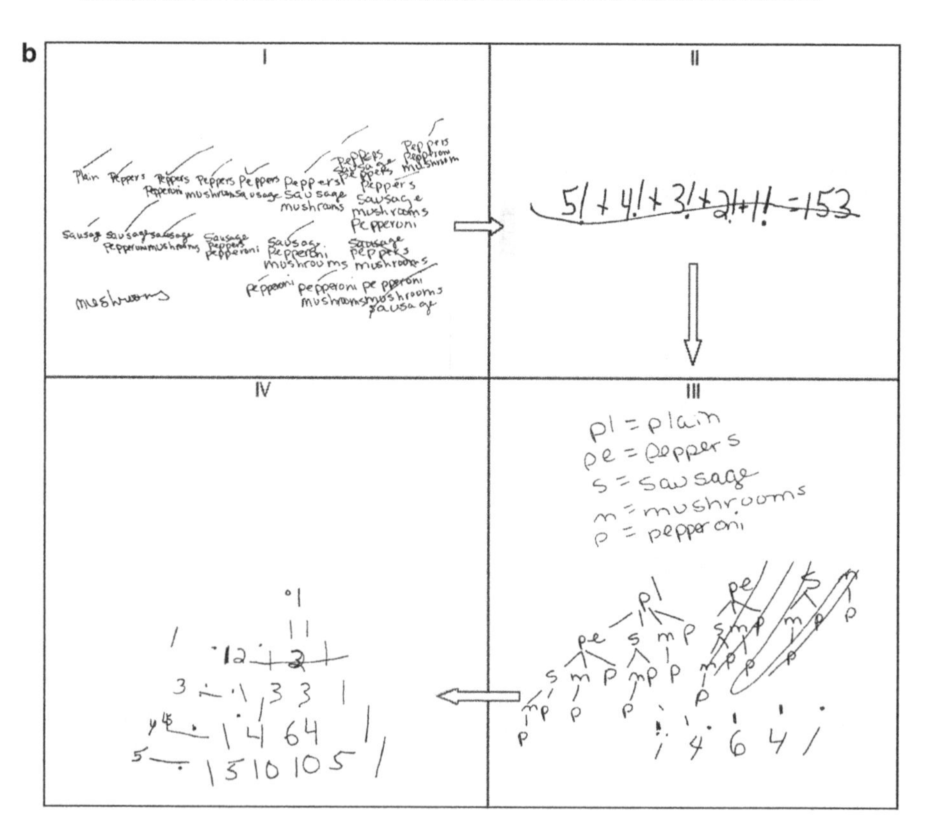

Fig. 6.17 (continued)

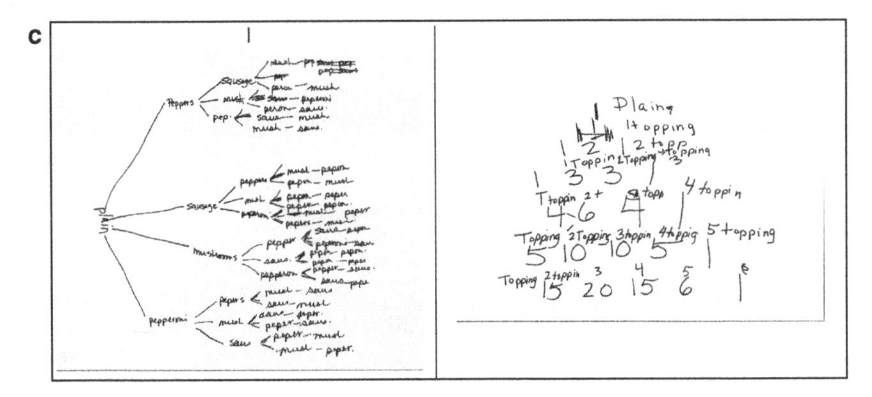

Fig. 6.17 **a** Pizza problem (Tarlow, 2008, p. 485). **b** Progression of diagrams in the written work of Shelly and Stephanie in relation to the Fig. 6.17a task (Tarlow, 2008, pp. 486–487). **c** Progression of diagram in the written work of Angela and Magda in relation to the Fig. 6.17a task (Tarlow, 2008, p. 488)

(Dea, 2006, p. 509) from the informal (i.e., personal) to the formal (i.e., increasingly symbolic and abstract) that Tarlow made possible when she provided her students with an opportunity to "develop their own meaningful representations so as to have the foundation necessary to make sense of mathematical abstractions" (p. 488). In Tarlow's teaching account, she effectively used her role in supporting her students' "progression of [diagrammatic] representations," from "find(ing) a solution to the Pizza problem, to provid(ing) a justification for their solution, and to develop(ing) combinatorial reasoning" (p. 488).

I also underscore the significance of effective curriculum sequencing and instructional orchestration of diagrams (Ng & Lee, 2009). For example, the recommended diagrammatic sequence in Fig. 6.18a to teaching multiplication of whole numbers at the elementary level if implemented on a large scale would predictably foster inefficient counting strategies since the presented diagrams mainly encourage more exploratory and less structured visual thinking. Further, the suggested teaching sequence would actually take more time than, say, the diagrammatic sequence offered in the California-adopted Singapore elementary mathematics textbooks that consistently employ place-value diagramming. In Fig. 6.18b, the place-value diagram already conveys the use of the grid method, which is then linked to the standard numerical algorithm for multiplying two whole numbers of up to three digits.

Referring to the study of Booth and Thomas (2000) concerning the issue of pictorial versus diagrammatic representations, their recommendation of *cognitive integration* should help overcome the cognitive phenomenon of differing representational effects of presented and self-generated diagrams and, more generally, any two opposing forms of representations. Cognitive integration involves having "two different modes of thinking" or "two qualitatively different hemispheric processors" to interact that results in "an integrated whole with two connected, distinct, simultaneous modes of operation" (p. 172). For example, in my 3-year study on pattern generalization at the middle school level, I discussed in several places in this book differences between alphanumeric versus visuoalphanumeric generalizations in the case of linear patterns in terms of numerical versus visual modes of representations,[2] respectively. The numerical approach involves constructing a *differencing*

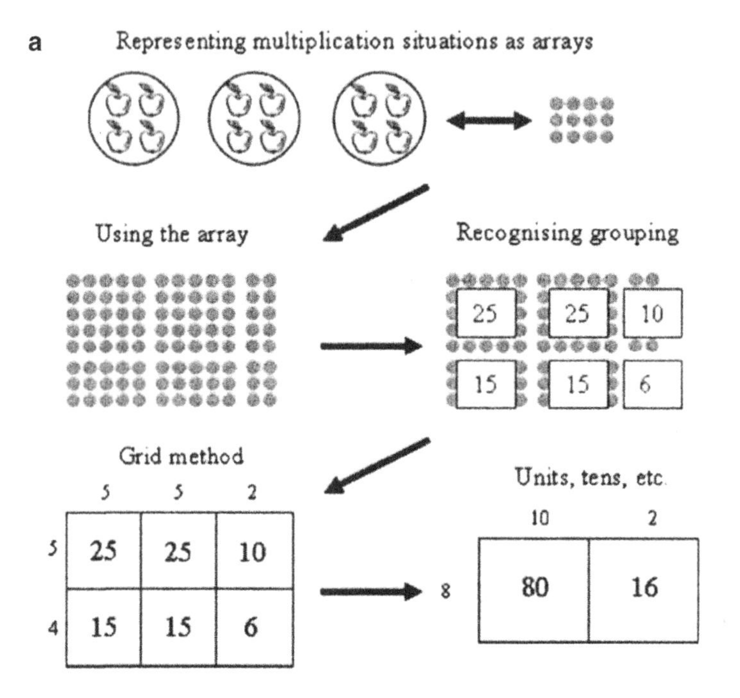

Fig. 6.18 (continued)

[2]Swafford and Langrall (2000) found differing representational effects between presented and self-generated tables of values in patterning activity among 10 sixth-grade students prior to formal instruction. They note:

> Tables seemed to be more useful when students constructed them to make sense of the problem. However, when the interviewer provided a table for the student to complete or to examine, the table seemed to be more of a distraction than an aid, diverting students' focus from the context of the problem to a string of numbers" (pp. 106–107).

In my 3-year study with my middle school students, however, those student-constructed tables that merely show common difference (e.g., Fig. 5.3b) were problematic because they funneled my students to a particular, narrow form of numerical strategy that encouraged mostly constructive standard generalizations with no room for more creative and other complex forms of generalization (e.g., constructive nonstandard; deconstructive). The proposed inductive-structuring tables (e.g., Fig. 5.3a) as an alternative diagrammatic representation help overcome issues that Swafford and Langrall (2000) identified as unproductive and ineffective actions relevant to table use ["hinder(ing) their abilities to recognize and describe the relationship between dependent and independent variables implicit in the situation;" "cloud(ing) rather than clarify(ing) the students'

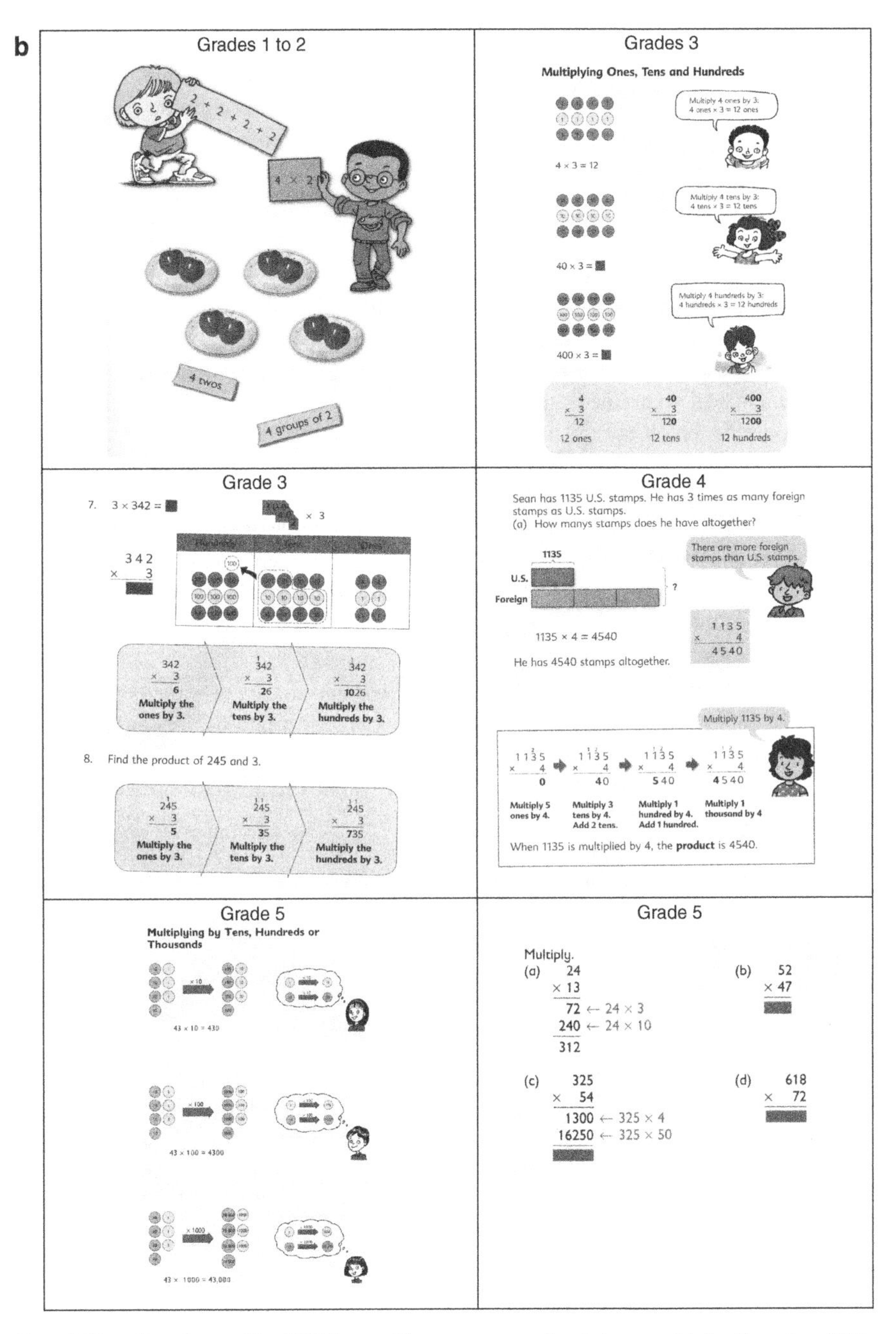

Fig. 6.18 **a** Barmby et al.' s (2009) teaching sequence involving multiplication (p. 237). **b** Singapore approach to multiplication of whole numbers in the California-adopted version (Curriculum Planning and Development Division, 2008, pp. 51 & 99; Copyright belongs to the government of the Republic of Singapore, c/o Ministry of Education, Singapore, and has been reproduced with their permission)

table (e.g., Fig. 5.3b), while the visual approach involves implementing the 3D template model shown in Fig. 5.15. A cognitive integration approach involves using an *inductive-structuring table* (e.g., Fig. 5.3a), which involves a simultaneous process of visually establishing and justifying a structural unit and setting up a structuring numerical table of values that takes into account all the information acquired from the structural unit.

3 Diagram Types and Progressive Diagrammatization

Following Peirce, diagrams for Dörfler (2001a) are inscriptions that have the quality of permanence in some medium (paper, virtual, etc.). They are either geometric or "less geometric [such as] arithmetic or algebraic terms, function terms, fractions, decimal fractions, algebraic formulas, polynomials, matrices, systems of linear equations, continued fractions and many more" (p. 2). Good diagrams, according to Cheng (2002), exemplify "systems with limited abstraction, being neither too abstract nor too specific" (p. 686).

Dörfler (2001a) provides additional characteristics of diagrams along several dimensions as follows:

- *They are structural and relational.* Objects in diagrams are arranged and parts or components are related in some way. They express relationships and, thus, are relational than figurative.
- *They possess internal meaning.* There are rules on how to operate on and transform them.
- *They possess external or referential meaning.* There are rules that allow them to be used and interpreted within and outside mathematics.
- *They are generic.* They convey a kind of visual generality that encompasses individual instances of the same type.
- *They are perceptual and material.* The relevant operations are performed in a perceptual and material context.

Dörfler (2001a) also points out that considering the many different ways in which we use diagrams, it is *how they are used* to solve a problem that captures the meaning of *diagrammatic reasoning* (DR). Examples of DR in this book are as follows: the presented diagram in Fig. 2.1 that helps algebra students understand why $a^2 - b^2 = (a - b)(a + b)$; the presented diagrams in Fig. 2.11a–c that illustrate the use of a bar representation in solving arithmetical problems; the student-constructed

recognition of a relationship between the independent and dependent variables;" "draw(ing) students' attention to the recursive relation between consecutive values of the dependent variable instead of the relation between the independent and dependent variables"; "forc(ing) an artificial relation between the numbers in the table with no regard for the context of the situation" (pp. 105–106)].

graphs in Fig. 1.5c that help precalculus students obtain the correct intervals corresponding to the solution of the stated polynomial inequalities; the presented pebble arithmetic diagram in Fig. 3.16 that shows why the sum of even numbers should be even; the student-generated diagrams in Fig. 4.14 that visually explain integer sums and differences; the unit diagrams from Fig. 6.11a through 6.13 that help students understand the rules involving rational algebraic expressions; and the presented diagram in Fig. 6.14 that explains how algeblocks could be used to perform addition of relatively simple polynomials. Bakker (2007) notes the following three important steps in employing DR: (1) construct a diagram that represents the necessary and significant relationships; (2) experiment with the diagram based on actions that are permitted by the representational system that enabled its construction in the first place; and (3) observe and reflect upon the results in order to articulate important relationships (pp. 17–18).

Using DR to solve a mathematical problem involves getting rid of nonessential information and transforming the problem in some abstract analogue, that is, skeletonizing it so that it surfaces only the essential elements needed in the (necessary) solution. With the aid of DR, the focus switches to detecting, constructing, and establishing regularities and invariant relationships that eventually take the shape of concepts and theorems that are themselves diagrams in some other format. There is, thus, a sense of *progressive diagrammatization* that is involved and necessary in using diagrams for DR. For example, Tarlow's (2008) experiences with her students in relation to the Fig. 6.17a combinatorial task exemplify a DR situation in which her students' diagrams (Fig. 6.17b, c) underwent instrumental reification and evolution from situated to mathematical.

One use of diagrams that for Dörfler (2001a) does not convey DR is when they are used primarily to represent abstract objects and models and their structures and nothing else; "they are then more kind of a methodological scaffold possibly unavoidable but to be dismissed when successful" (p. 2). In light of this distinction, Dörfler (2001a) articulates what he considers to be the essence of DR as follows:

> This [i.e, the representational view of diagrams] is diametrically opposed to DR where the focus is on the diagrams themselves as the objects of study and of operations and not on their doubtful mediation with virtual objects. In this representational view mathematics is a predominantly mental activity supported by diagrams whereas mathematics as DR essentially is a material and perceptual one. And this does not reduce mathematics to meaningless symbol manipulations since the diagrams have meaning through their structure, their operations and transformations and of course via their applications. This holds for all diagrams as considered here in a way completely analogous to how geometric figures can have meaning.
> (Dörfler, 2001a, p. 2)

Thus, Dörfler's (2001a, 2001b) account of DR focuses on the integral and central role of diagrams play in assisting students to perceptually and materially establish abstract relationships primarily by manipulating what they are able to visualize or see in empirical form. Those diagrams that basically do nothing else but represent do not convey DR. For Dörfler (2007), and Peirce (1976) before him, diagrams are

> not just a means for investigating an object different from them. The diagrams turn into the very objects of investigation and this latter one is carried out by manipulating and (inventing) diagrams and detecting recurring patterns within these actions with diagrams. Those patterns can then be formulated as theorems or by formulas which again are diagrams.
>
> (Dörfler, 2007, p. 853)

For example, inductive-structuring tables (Fig. 5.3a) demonstrate how tables in pattern generalization activity could be effectively utilized as a tool for DR (Rivera, 2009). Dörfler's (2007) interpretation of DR above should help us understand the following description of DR offered by Peirce himself:

> The first things I found out were that all mathematical reasoning is diagrammatic and that all necessary reasoning is mathematical reasoning, no matter how simple it may be.
> By diagrammatic reasoning, I mean reasoning which constructs a diagram according to a precept expressed in general terms, performs experiments upon this diagram, notes their results, assures itself that similar experiments performed upon any diagram constructed according to the same precept would have the same results, and expresses this in general terms. This was a discovery of no little importance, showing, as it does, that all knowledge without exception comes from observation.
>
> (Peirce, 1976, pp. 47–48)

In the school mathematics curriculum, students deal with either pure or hybrid diagrams. *Pure diagrams* utilize only one representational medium. For example, in school geometry, paper-folding activity involving diagrams alone can be used to explore construction of geometric figures or to establish area formulas for circles and certain quadrilaterals. In a second course in algebra, graphs of conic sections can be introduced using a focus–directrix graphing diagram (Fig. 6.19) that consists of parallel lines that are spaced one unit apart and concentric circles spaced in a similar manner. A line takes the role of a directrix, while the common center of the circles is the focus. A focus–directrix diagram allows students to investigate and construct conic forms that are generated when they manipulate a particular eccentricity (value), a concept that is central to the definition of a conic section.

Hybrid diagrams, presented and self-generated, can be either static or dynamic depending on the context in which they arise. Algebraic expressions are combinations of alphabet (or variable) and ideogram that together exemplify alphanumeric representations. Other hybrid diagrams combine some other types of diagrams, say, pictorial and textual labels, with the texts used to demonstrate further the reasoning that is being targeted (Kulpa, 2009, p. 91). The goal of DR in a hybrid diagram is to establish a correspondence between the two diagrams and the relevant proposition that is being modeled. Figure 6.20 is the Nicomachus version of the left diagram shown in Fig. 4.17. In such diagrammatic situations, Kulpa (2009) notes "the variable is actually the whole diagram, parametrized by its size ... varying over the set of naturals" (p. 91).

Networks, matrices, and hierarchies exemplify what are oftentimes labeled as spatial or schematic diagrams. They also have a hybrid character (Hegarty, Carpenter, & Just, 1991; Novick, 2006). Novick (2006) points out how "each diagram is optimized to serve a different representational function" (p. 1851). At the very least, matrices store relational information between item pairs in two sets,

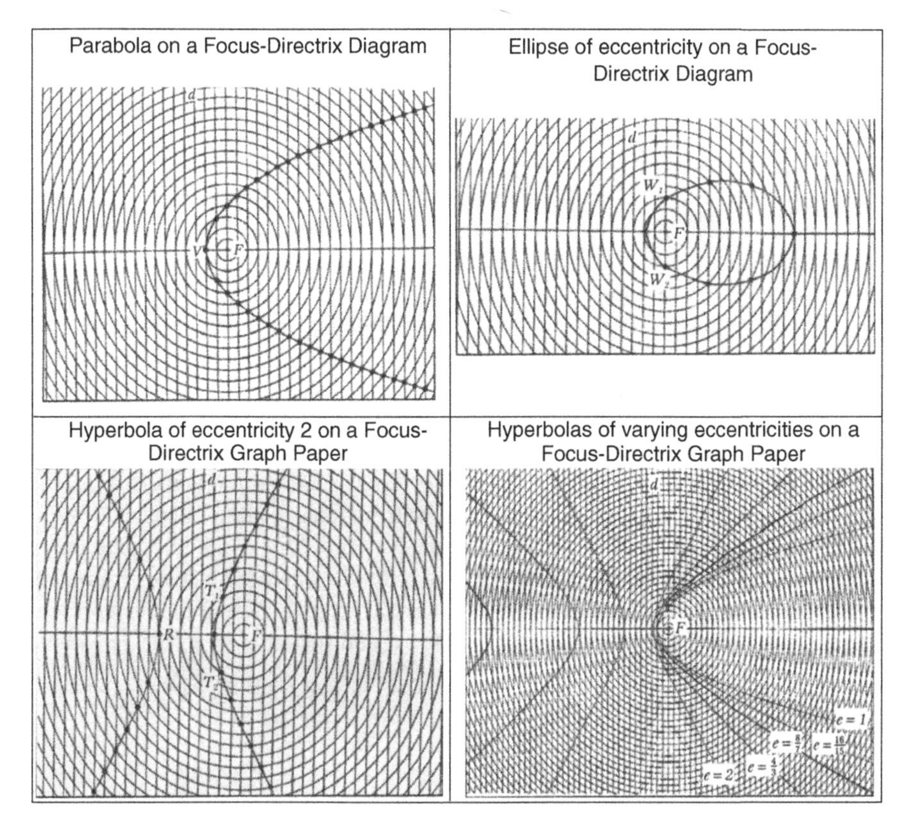

Fig. 6.19 Exploring conic sections on a focus–directrix graphing paper (Brown & Jones, 2006, pp. 322–327)

networks show local and global connections and routes between nodes, and hierarchies model power structures and precedence relations among items in a set. Figure 6.21a–c shows examples of a network, a matrix, and a hierarchy, respectively, drawn from particular areas in school mathematics.

Recent advances in graphing technologies enable students to construct hybrid dynamic diagrams. Figure 6.22 shows dynamic transformations of

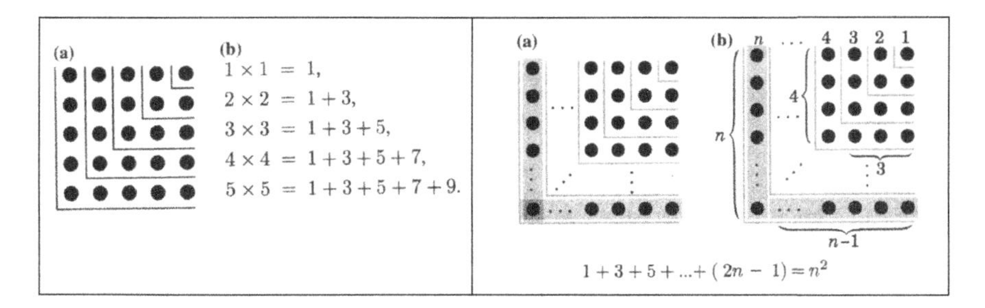

Fig. 6.20 Nicomachus version of the Fig. 4.17 diagram on the left (Kulpa, 2009, pp. 90–91)

$f_2(x) = af_1(b(x-c))+d$, where $f_1(x) = x^2$, with the aid of a graphing calculator that has a built-in computer algebra system with a slider (Zbiek & Heid, 2009, p. 543). The same task can, in fact, be accomplished on a regular graphing calculator using a transformation application.

In either pure or hybrid context, the dynamic component of the diagrams enables students to experience and establish the abstract relationships involved by providing them access to the changes that are taking place in the graphical representations on the same screen of their calculators. Also, while the foregoing discussions echo the central role of diagrams in mathematical reasoning, Lomas (2002) points out how the perceptions we draw from "the shape properties of concrete diagrams" act like "a *surrogate* for conscious awareness of shape properties of abstract geometric objects depicted in the diagrams" (p. 210).

Dörfler's point above about the mere representational versus diagrammatic quality of diagrams (following Peirce) underscores the need for progressive diagrammatization. The distinction suggested by Dörfler about diagrams is the same issue we tackled in Chapter 4 about transitions from iconic and/to indexical to symbolic representations. I have pointed out in the earlier sections in this chapter and in preceding chapters as well instances in my own classes where students appropriated, constructed, and internalized visual representations progressively from the representational to the diagrammatic, from the metrical and quantitative to the schematic

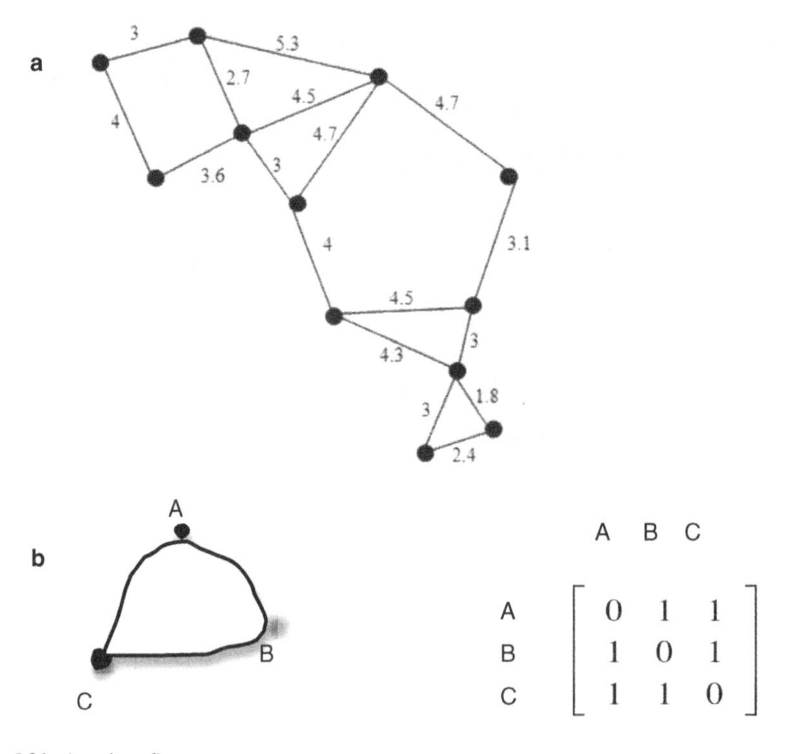

Fig. 6.21 (continued)

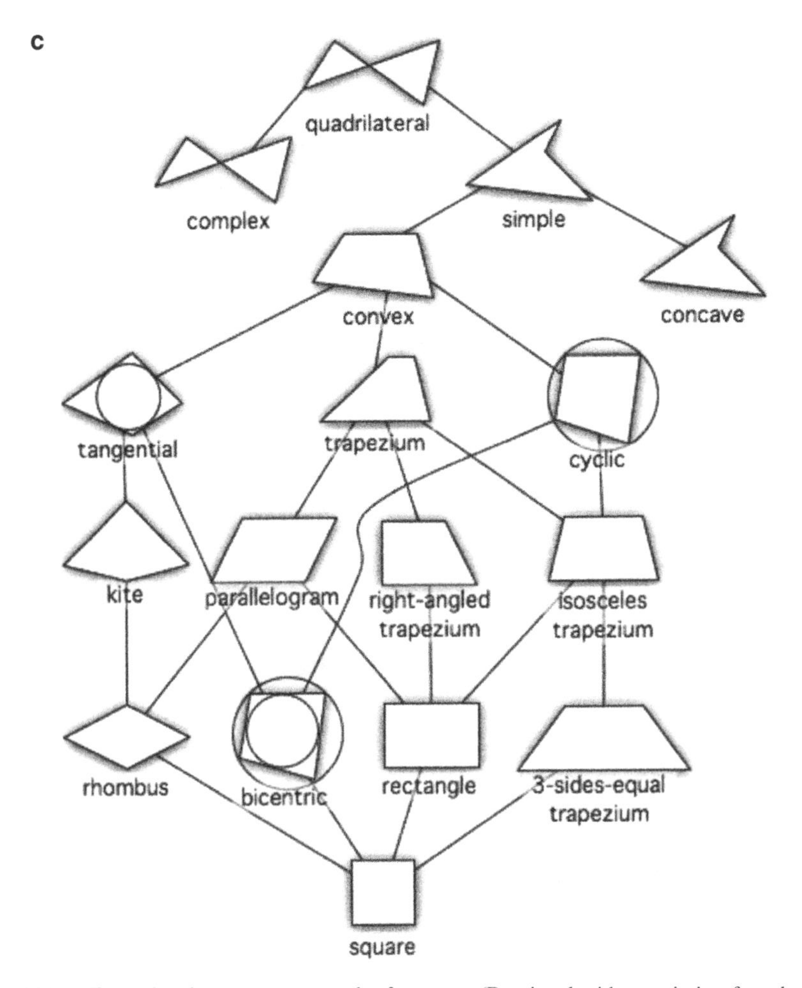

Fig. 6.21 **a** Example of an energy network of cameras (Reprinted with permission from http://faculty.kfupm.edu.sa/SE/salamah/operations_research_I/Khobar_intersections%20network.jpeg by Muhamaad Al- Salamah). **b** Example of an adjacency matrix of a graph. **c** Hierarchy of quadrilaterals (http://commons.wikimedia.org/wiki/file:Quadrilateral_hierarchy.png)

and qualitative, and from the material to the imagined symbolic or abstract. In other words, the diagrams in this book, especially those that are used in teachers' classes and in school mathematics textbooks, encompass all types that actually fall along a continuum of progressive diagrammatization. Here I am reminded of Netz's (1999) interesting cognitive historical analysis of the nature and context of deduction in Greek mathematics, which had him considering diagrams (and language) as central to the emergence, development, and success of the deductive method in ancient Greece. For example, in shaping the condition of generality in both form and practices, diagrams were constructed initially to develop a mathematical argument and convey specific relationships in some material form as lettered

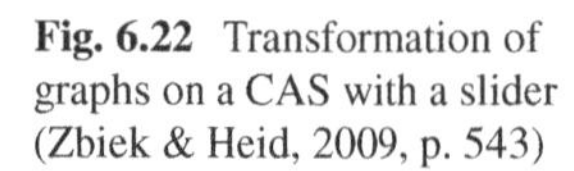

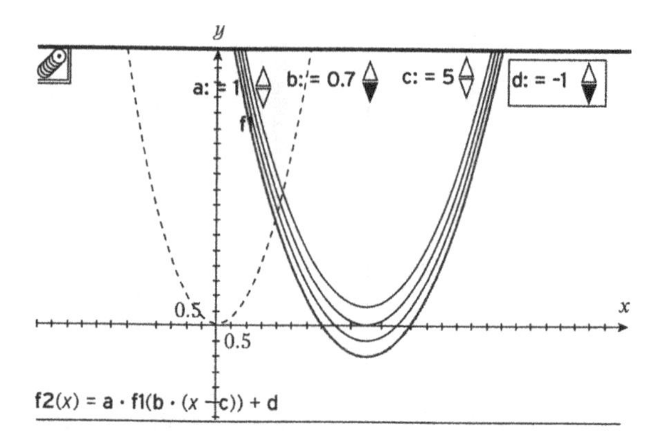

Fig. 6.22 Transformation of graphs on a CAS with a slider (Zbiek & Heid, 2009, p. 543)

or labeled diagrams, which were then repeated, permuted, and later assessed for structure and invariance. Among the Greek mathematicians, in fact, the use of a diagram came with it a "whole set of procedures for argumentation based on the diagram" (p. 57).

4 Extending DR Beyond Establishing Regularities and Constructing Mathematical Knowledge: Supporting Growth in Mathematical Reasoning Through DR

Since the hypothetical state of things in the mathematical world consists of relations and structures with invariant ones considered to be the most valued, constructing diagrams is one powerful way of reifying them. Diagrams metonymically capture their essence in some visible form that could then be manipulated in order to yield, in some cases, hypostatic abstractions and, in other cases, more powerful (abstract) diagrams in the form of, say, theorems and meaningful propositions. They "suggest definitions and proof strategies [and] function as 'frameworks' in parts of proofs" (Carter, 2010, p. 8). In many cases, they do not even have to be drawn precisely. *Contra* Kulpa's (2009) concern for precision in diagrams, Sherry (2009) thinks that poorly drawn or inexact diagrams do not "generally affect proofs" despite the possibility that they "may affect our success at guessing theorems" (p. 63). A haphazard construction of partitioned squares in the case of Fig. 3.5a, which visually demonstrates the series $\sum_{n=1}^{\infty}(1/4)^n$, is likely to prevent students from developing the conjecture that its sum is (1/3). In school geometry that assumes Euclidean principles, a poorly constructed diagram of a convex polygon with n sides might deter students from seeing and inferring that the sum of its n external angles equals 360°. However, their symbolic proofs are not at all influenced by the perceptions drawn from their respective diagrams.

For Sherry, what matters significantly more than constructing a precise diagram is the target mathematical knowledge that is, in fact, "nowhere to be seen" (Sherry, 2009, p. 64); it does not reside in the diagrams themselves. This means to say that when students are unable to establish the knowledge, it is not so much because of a deficiency in the diagrams but in their "inability to grasp" the relevant "conceptual relations" (Sherry, 2009, p. 68).

Sherry (2009) offers a description of DR that further enriches our understanding of the role of diagrams in mathematical reasoning beyond, but still compatible with, prevailing views of diagrams in the Peircean context. From an instrumental point of view (in the sense of Verillon and Rabardel), when we let students "treat a diagram as an instance of" a mathematical concept and encourage them to "deduce various conclusions about the elements of the diagram" (p. 65), they acquire inferential habits that enable them to gain access to many of the characteristics associated with mathematical reasoning. Here, *inference* is distinguished from *implication*. While an implication refers to a logical relationship between propositions, an inference is "a type of mental act, whose outcome is a possible change in belief . . . [that is,] a transition between two (personal-level) mental states" (Norman, 2006, p. 18). With changes in our beliefs, there are bound to be changes in our conceptions. Thus, greater competence in manipulating diagrams and their parts means developing better mathematical knowledge as a result of the inferences that are constructed. Sherry (2009) writes:

> We learn more about a diagram just as we learn more about physical object, viz., by applying mathematical concepts and drawing inferences in accordance with mathematical rules. Recognizing that a diagram is just one among other physical objects is the crucial step in understanding the role of diagrams in mathematical argument.
>
> (Sherry, 2009, pp. 65–66)

To illustrate, in my Algebra 1 class, we used the transformation activity shown in Fig. 6.23a with a TI 84, a graphing calculator. The activity took place near the end of the school year when the students have already acquired some knowledge of the basic graphs (i.e., linear, quadratic, absolute value, and power functions). Initially, they used their calculator and graphed the function ($y_1 = x^2$). Then they graphed Maria's function, $y_2 = x^2 + 1$, on the same viewscreen in which case they knew that all the points on the original graph would vertically shift one unit up along the dimension of the y-axis (Fig. 6.23b). In obtaining the reflection graph of y_2 about the x-axis, I anticipated that they would enter the function $y_3 = -x^2 + 1$ instead of $y_3 = -(x^2 + 1)$. When most of them did, I used the overhead viewscreen, graphed the two functions (Fig. 6.23c), and asked them to compare the two graphs. When they saw that the two graphs were different, they decided to repeat the task all over again in which case they concluded that Maria's graph was represented by the function $y_3 = -(x^2 + 1)$ or $y_3 = -x^2-1$, while Lina's graph took the form $y_4 = -x^2+1$. They then used their experience to explore situations involving different y-intercepts and different graphs that enabled them to eventually infer the generalization that the commutative property does not hold in the case of a vertical translation and reflection about the x-axis. That is, in visuoalphanumeric form, the action sequence

$$y = (f(x) \rightarrow y = -f(x) \rightarrow y = -f(x) \pm c.$$

a Use your TI-84 to graph $y = x^2$.

Maria initially performed a vertical shift of 1 unit followed by a reflection about the *x* axis. Lina initially performed a reflection about the *x* axis followed by a vertical shift of 1 unit.

1. Would they yield the same graph? Why or why not?
2. Does commutativity hold in the general case of a vertical shift of *c* units and a reflection about the *x* axis?
3. Explore other graphs you know and investigate commutativity involving vertical shift and reflection about the *x* axis.

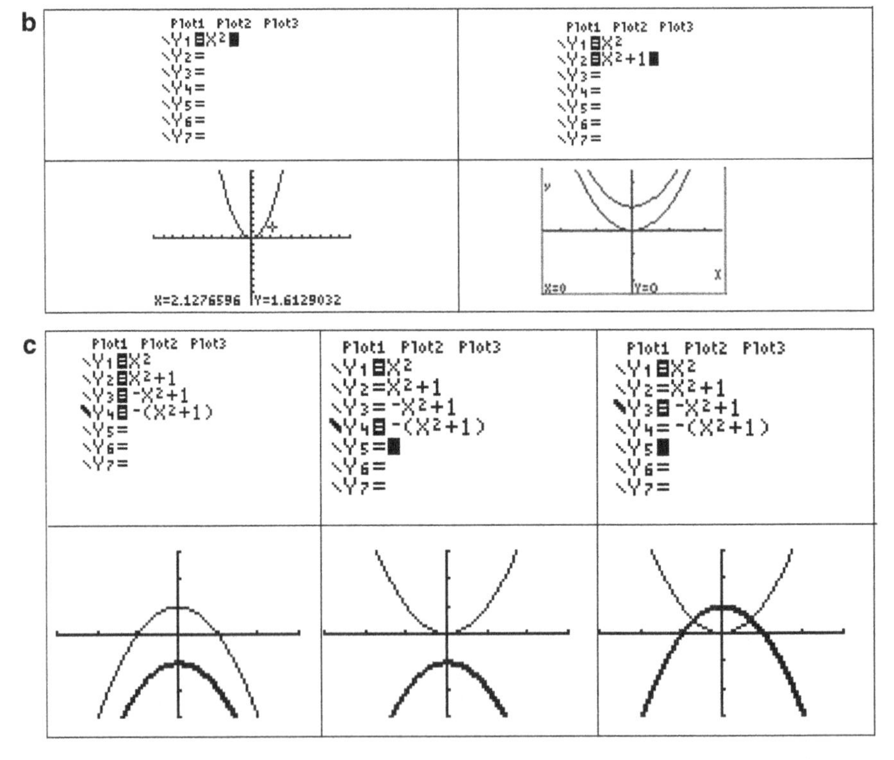

Fig. 6.23 **a** A transformation task involving simple graphs. **b** Graphs of $y = x^2$ and $y = x^2 + 1$ on a TI 84. **c** Graphs of $y = x^2$, $y = x^2 + 1$, $y = -x^2 + 1$, and $y = -x^2 - 1$ on a TI 84

was, in fact, different from the action sequence

$$y = f(x) \rightarrow y = f(x) \pm c \rightarrow y = -(f(x) \pm c) = -f(x) \mp c$$

In the above diagrammatic activity involving transformations of graphs, the students established generalizations and acquired knowledge relevant to inference making (e.g., repeated abduction and induction of an empirical regularity) on the basis of manipulating graphs that they treated like everyday objects. Their beliefs and conceptions about the two graphing actions transitioned from one mental state (i.e., not paying attention to graphing action) to another (i.e., being aware of noncommutative graphing action). Later, with more examples, their inferences transitioned from being reasonable (explaining with a few specific examples) to necessary (explaining at the visuoalphanumeric symbolic level).

Thus, the act of constructing a diagram, presented and self-generated, is not so much about establishing an exact figure (in a metric sense) as it is more about developing inferences from the "topological properties like incidence relations" (Sherry, 2009, p. 65) rather than the physical or material nature of such diagrams. They provide a basis for constructing explanations on the route to the operative and abstract and, when applicable, deductive knowledge. For example, when my Algebra 1 students used a number line to demonstrate their understanding of the absolute value of a real number x, it mattered less to them whether they were a tad sloppy in locating x and $-x$ relative to the origin as they were more focused on visually demonstrating a conceptual fact relevant to the absolute value symbol.

Sherry's version of DR is as follows:

> [D]iagrammatic reasoning recapitulates habits of applied mathematical reasoning. On this view, diagrams are not representations of abstract objects, but simply physical objects, which are sometimes used to represent other physical objects. . . .. For obtaining mathematical knowledge . . . consists not in gaining information about abstract objects, but in constructing inference rules. Mathematical proof is the construction of such rules, and oftentimes the construction involves a diagram.
>
> (Sherry, 2009, p. 67)

DR as conceptualized above takes us to the basic activity of mathematical reasoning, which involves learning inference rules that enable us to construct, establish, and justify abstract propositions that are mediated by, and extracted from, diagrams. In a Peircean context, when students effectively manipulate a diagram, their mathematical knowledge is (hopefully) an account of progress from the concrete to the abstract. In a Sherrian context, however, their experience with the diagram should provide them with an opportunity to see the mutually determining relationship between inferential rule construction and the development of mathematical knowledge, especially the role of mathematical explanation in establishing and justifying propositional claims.

Consistent with his view that DR involves habits of applied mathematical reasoning, Sherry situates the origin of inference rule formation in pragmatism, that is, in a lifeworld-dependent social context that allows the phenomenon of "hardening into a rule" to occur:

> [Wittgenstein] describes mathematical propositions as empirical propositions that have hardened into rules . . .. An empirical proposition has hardened into a rule once it has come to be used by a community for inferring one empirical proposition from another. . . . In order

> for an empirical proposition to harden into a rule, there must be overwhelming agreement among people, not only in their observations, but also in their reactions to them.
>
> (Sherry, 2009, p. 66)

For example, in my Algebra 1 class, the students' diagrammatic experiences relative to the Fig. 6.23a activity demonstrate how the inferential technique of establishing an empirical regularity and generalization were both initially drawn from their visual experiences with the graphs on their calculator. Then repeated similar and noncontradictory experiences across different values of the y-intercept and different types of graphs allowed them to infer something about the noncommutative nature of the two transformation actions they were investigating. Following Wittgenstein, Sherry (2006) notes that such empirical propositions harden into rules "when one no longer has a clear concept of what experience would correspond to the opposite" (p. 496), which then leads to a necessary inference.

One last point about DR deals with its productive role in what Cellucci (2008) refers to as *explanatory reasoning by the analytic method*. The analytic method, Cellucci points out,

> is the method by which, to solve a mathematical problem, we formulate a hypothesis that is a sufficient condition for its solution. The hypothesis is obtained from the problem, and possibly other data, by some non-deductive inference: induction, analogical, and so on. ... The hypothesis must not only be a sufficient condition for the solution of the problem but must also be plausible, that is, compatible with the existing data, in the sense that, comparing the reasons for and the reasons against the hypothesis on the basis of the existing data, the reasons for prevail over those against. Plausibility is distinct from probability, as it appears from the fact that induction from a single instance often leads to hypotheses that are plausible but whose probability is zero when the number of possible instances in infinite.
>
> (Cellucci, 2008, p. 205)

When students construct diagrams and develop DR relevant to a task, they initially produce plausible hypotheses (individual or shared) that are capable of undergoing modification when more information and data are known. The hypotheses have explanatory power when they assume an essential role in solving the task (Cellucci, 2008, p. 206). For example, the number line diagram contains useful hypotheses about the absolute value notations that my Algebra 1 students used to solve equations and inequalities successfully. Thus, mathematical explanations behind diagram construction and interpretation involve establishing hypotheses that have "heuristic value" and are effective "as a means of discovery" (Cellucci, 2008, p. 207), which then support mathematical proofs that necessitate the use of a combination of analytic and synthetic (axiomatic) methods. Figure 6.24 provides another interesting and rigorous DR-based mathematical explanation of the Pythagorean theorem that is based on Cellucci's explanatory analytic framework.

Problem. Explain the Pythagorean Theorem.

Given a right triangle, construct the square on its shorter leg. Make a copy of the triangle and construct the square on its longer leg. Put the two resulting figures together as shown below.

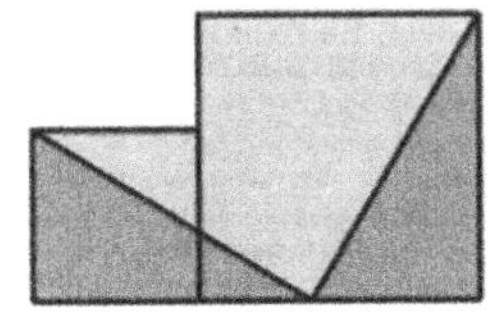

Then slide up both the triangle on the left and the triangle on the right, forming a quadrilateral as shown below. Such quadrilateral equals the sum of the squares on the two legs of the triangle.

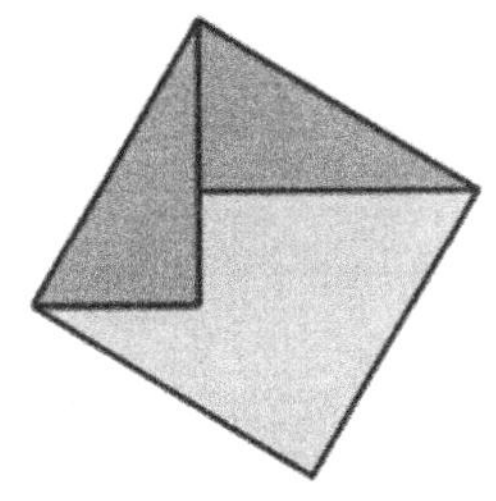

Therefore, to show that the square on the hypotenuse equals the sum of the squares on the two legs, we need only show that the quadrilateral is the square on the hypotenuse. To show that we state the following hypothesis:

(A) The three interior angles of a triangle are equal to two right angles.

Hypothesis (A) is capable of solving the problem since it implies that the two non-right angles of the right triangles add up to a right angle. Thus, each angle of the quadrilateral is a right angle, so the quadrilateral is a square – the square on the hypotenuse of the right triangle. Since hypothesis (A) plays an essential role in solving the problem, it explains why the square on the hypotenuse equals the sum of the squares on the two legs.

Fig. 6.24 A DR-based mathematical explanation of the Pythagorean theorem (Cellucci, 2008, pp. 207–208)

5 Overview of Chapter 7

In Chapter 7, we discuss cultural influences and blind-specific issues relevant to visual attention, thinking, and performance in mathematics. Under cultural influences, we talk about differences in how mathematical objects and their images, including mathematical proofs, are perceived, captured, and interpreted as effects of sociocultural orientation and intentionality that yield a sense of shared experiences with them in activity. A materialist account of mathematical objects is also pursued that links visual forms of mathematical knowledge to socially shared feelings and practices. Under blind-specific matters, we discuss the nature of image construction and processing among blind learners in order to point out the need to broaden our prevailing understanding of the sources and nature of visual representations to include all aspects of our sensory modalities. We close the chapter with the notion of multimodality learning in mathematics.

Chapter 7
Cultural and Blind-Specific Issues and Implications to Visual Thinking in Mathematics

Every culture, with its privileges or taboos of seeing, shapes a certain way of thinking, as it is in turn shaped by norms or vetoes of looking.

(Beltin, 2008, p. 189)

It is, of course, the norm rather than the exception that more than one sense modality contributes to the perception of an object. We hear, see, smell, and feel the same object. That is not because all senses provide the same information, but because the contributions from different modalities converge and overlap sufficiently to be felt as "same" rather than as "different."

(Millar, 1994, p. 47)

Overall, neuropsychological reports, neuroimaging and behavioral findings support the view that visuo-spatial processes and numerical representation are intimately related. This relationship may constitute an early and fundamental link which, although partially shaped by our cultural constraints, remains an essential component of our cognitive architecture.

(de Hevia, Vallar, & Girelli, 2008, p. 1370).

In this chapter, we explore cultural and blind-specific issues and their implications to visual thinking in mathematics. Cultural issues refer to those patterns of knowledge and skills, tool use, thinking, acting, and interacting that are favored by, and specific to, groups that support them. It is interesting to consider the possibility that students' developing explicit knowledge in a lifeworld-dependent context interacts with macroconditions such as taken-as-shared cultural practices, which are likely to influence the way they perform visualization without them being aware of them. In a constrained Baudrillardian perspective, the social sign-mediated context that Radford (2002) has usefully inserted in semiotic activity could also be interpreted as conveying particular forms of sociocultural conditions and practices that effectively "exclude and annihilate all symbolic ambivalence[1]" through their structures that "designate, abstract, and rationalize" (Baudrillard, 1981/1972, p. 149). That is,

[1] For Baudrillard (1981/1972), the rationality of signs is seen in its ability to distinguish between signifier and signified and

F.D. Rivera, *Toward a Visually-Oriented School Mathematics Curriculum*, Mathematics Education Library 49, DOI 10.1007/978-94-007-0014-7_7,

© Springer Science+Business Media B.V. 2011

such conditions and practices assist individuals by cueing them – their interpretant (in a Peircean sense) and modes of signification (in a Wittgensteinian sense) – to look at situations in a particular way.[2] Ferreira (1997) expressly articulates the (subtle) influence of the cultural standpoint in the following manner:

> Different worldviews – the socially constituted world and its cosmological foundations – and the everyday experience of active individuals account for the diversity of strategies of mathematical reasoning. In other words, different cultures, and individuals within any given culture, proceed differently in their logical schemes in the way they manage quantities and, consequently, numbers, geometrical shapes and relations, measurements, classifications, and so forth.
>
> (Ferreira, 1997, p. 135)

Blind-specific issues are narrowly confined to neurophysiological and neuropsychological factors that influence the manner in which totally or partially blind learners construct and process images, which may yield useful implications in our developing understanding of the role of visual thinking in mathematics. For example, based on an extensive review of literature, Cattaneo et al. (2008) found that blind and visually impaired individuals tend to use different cognitive mechanisms to compensate for their loss of sight (in varying degrees). While many of the documented mechanisms appear to be visual-like or similar to those used by nonblind subjects, the authors underscore the necessity of multisensory capacity and performance among blind subjects especially when they perform various cognitive tasks that test their competence in processing visual and spatial information.

> functions as the agent of abstraction and universal reduction of all potentialities and qualities of meanings that do not depend on or derive from the respective framing, equivalence, and specular relation of a signifier and signified. This is the directive and reductive rationalization transacted by the sign – not in relation to an exterior, immanent "concrete reality" that signs would supposedly recapture abstractly in order to express, but in relation to all that which overflows the schema of equivalence and signification; and which the sign reduces, represses, and annihilates in the very operation that constitutes it. . . . The rationality of the sign is rooted in its exclusion and annihilation of all symbolic ambivalence on behalf of a fixed and equational structure.
>
> (Baudrillard, 1981/1972, p. 149).

I see social and cultural practices as providing conditions that engender the possibility of reduction, repression, and annihilation of sign ambivalence. The well-defined character of signs used in formal and institutional mathematical knowledge indicates this necessary elimination of ambivalence.

[2]The term *culture* has many meanings and is certainly more complicated than the one that I describe in this chapter [cf. De Abreu, 2000; Lerman, 2001; Nunes 1992; Rivera & Becker, 2007b; special issue of the *Journal of Intercultural Studies* on culture in mathematics education edited by FitzSimons (2002)]. Byrne et al. (2004), for example, identify six dimensions of culture that merit individual analysis – that is, culture as pattern (social transmission processes); a sign of mind (solidified cognitive habits drawn from social transmission); a bonus (knowledge and practices resulting from social transmission); inefficiency (negative consequent actions resulting from social transmission); physical products (cultural actions and tools); and meaning (rituals). My characterization of culture is pragmatic; it has useful elements that allow me to explain the ideas I pursue later in relation to cultural forms of seeing in mathematics.

The examples we discuss in some detail in this chapter in relation to the issue of culture and visualizing have been drawn from both past and recent studies that basically foreground the idea that visualizing mathematically is an *interpretive reproduction*[3] of particular cultural practices. In the case of totally and partially blind students, having a sense of how they acquire knowledge allows us to further enrich our understanding of what it means to visualize mathematical objects, concepts, and processes other than what has already been discussed in the preceding chapters with an eye on understanding the epistemic significance of a physiologically driven sensuous dimension in mathematical learning. Millar's (1994) thoughts in the opening epigraph provide an attractive source of insight and characterization in this emerging idea of the sensuous mode in mathematical knowing. Drawing on her longitudinal work with preschool and adolescent blind subjects, Millar shares the view regarding the necessity of supplementation from a variety of modality sources in the extreme case of loss of one modality (say, sight), however, with the additional condition that convergence toward the same information should take place since "redundancy facilitates processing" (p. 47).

Eight sections comprise this chapter with the first five sections addressing the cultural issues. In Section 1, we begin with a very interesting interpretive historical account that could explain subtle differences in visual attention and the construction of concepts among students in diverse classrooms. We also discuss the theory-laden and socialized nature of scientific knowledge construction. Both macro-contexts provide sufficient reflection of the powerful influence of sociocultural practices in lifeworld-dependent conditions that shape how students visualize objects, concepts, and processes in school mathematics. In Section 2, we discuss interpretive historical accounts of differences in the way objects and their images were perceived, captured, and interpreted as a consequence of sociocultural modes of seeing that influenced the nature and type of mathematical content that was valued and produced. Section 3 discusses the implications of a materialist view of objects in mathematics, which sees visual forms of mathematical knowledge as evocations of shared social and cultural feelings and practices. Section 4 pursues a cultural analysis of the nature of mathematical proofs in terms of particular ontological perceptions of objects, which then influence the manner in which truths and the empirical world are constructed. In Section 5, we deal with two interesting empirical studies that illustrate the effectiveness of culture-specific visual practices in early childhood mathematics education, which could assist us in developing a model for visual instruction in mathematics at the upper grade level.

The remaining three sections highlight findings from research conducted with totally and partially blind subjects. In Section 6, we discuss perspectives drawn from a few studies that describe the nature of image construction and processing among them. The main point that is addressed deals with the need to broaden our prevailing understanding of the sources and nature of visual representations to include all

---

[3]See footnote 4 in Chapter 2 (p. 55).

aspects of our sensory modalities. Section 7 explores three implications for effective visual representation on the basis of studies that compared blind subjects with their nonblind counterparts. Section 8 closes the chapter with a discussion on the significant role of multimodality learning in mathematics with a particular focus on learning tools that target several different modalities in simultaneity.

1 Cultural Preferences Relative to Visual Attention and the Construction of Concepts

The work of Nisbett, Peng, Choi, and Norenzayan (2001) provides a useful starting point in talking about cultural differences in perceptual processes. Based on a synthesis of research, including their own, Nisbett et al. claim that the cultural differences between ancient China and ancient Greece in relation to their views and practices in science, mathematics, and philosophy could be traced to differences in how they lived their social lives and employed cognitive procedures in acquiring some knowledge of the world around them.

The ancient Greeks valued personal agency and freedom. Their sense of curiosity led them to discover rules and establish causal models, which Nisbett et al. characterize as early manifestations of analytic thought that

> involv(es) detachment of the object from its context, a tendency to focus on attributes of the object to assign it to categories, and a preference for using rules about the categories to explain and predict the object's behavior. Inferences rest in part on the practice of decontextualizing structure from content, the use of formal logic, and avoidance of contradiction.
>
> (Nisbett et al., 2001, p. 293)

The ancient Chinese valued collective agency and community-driven practices. Their advances in technology and science, far more superior than those of the ancient Greeks, were actually drawn from a combination of practicality, intuition, and empiricism, which Nisbett et al. characterize as early manifestations of holistic thought that

> involv(es) an orientation to context or field as a whole, including attention to relationships between a focal object and the field, and a preference for explaining and predicting events on the basis of such relationships. Holistic approaches rely on experience-based knowledge rather than on abstract logic and are dialectical, meaning that there is an emphasis on change a recognition of contradiction and of the need for multiple perspectives, and a search for the "Middle Way" between opposing propositions.
>
> (Nisbett et al., 2001, p. 293)

Current Western and Eastern findings seem to suggest the continued prevalence of holistic and analytic thought with the former manifested in contexts such as associative thinking and computing on the basis of similarity and contiguity and the latter in terms of a predisposition towards constructing symbolic representational systems and structured or rule-based computations (Nisbett et al., 2001, p. 293). Table 7.1, which summarizes cultural differences between the ancient Chinese and

Table 7.1 Cultural differences between the ancient Chinese and ancient Greeks in relation to various aspects of science, mathematics, and philosophy (Nisbett et al., 2001)

Notions	Ancient Chinese/Eastern standpoint	Ancient Greeks/Western standpoint
Orientation of the body	The self is oriented toward the social environment (collectivist), that is, attends to the field or the situation relative to the object	Capacity of the self to know an object (individualist), that is, attends to the object based on his or her goal
Object categorization	Predisposition toward relational knowledge and family resemblances: preference for understanding part–whole relationships in order to know something about an object	Rule-based: predisposition toward isolating an object in order to derive its attributes
Cognitive processing	Preference for an explanation of an event relative to an object's field; preference for a dialectical approach that involves "accepting, transcending, and reconciling apparent contradictions"	Preference for decontextualization and separation of an object from its field; use of formal systems of logic
Continuity versus discreteness	Valuing of the interconnectedness of object and matter (content)	Predisposition toward the singularity of objects that have universal qualities which make categorization possible
Sensitivity to the field	Field dependent: wholes over parts; one comes to know an object through a "field of forces"	Field independent: parts over wholes; one comes to know an object through its parts
Relationships and similarities versus categories and rules	An object is known in the context of some event; the whole exists with parts defined in a relational context	An object is known independent of context
Epistemological processes	Experience-based knowledge resulting from direct perception	Logical and abstract knowledge with a focus on reason over senses

the ancient Greeks in most aspects of thinking, provides some context that might explain differences in visual performance and attention in mathematics in diverse classrooms. Certainly, the intent is not to put individual learners in narrow boxes but to explain possible sources of differences in visual attention and visual processing of mathematical objects, concepts, or processes.

In the context of professional and scientific disciplines, Kuhn (1970) also contributed significantly to what is now accepted as the theory-laden and socialized nature of scientific observation and practices. Scientists oftentimes rely on the taken-as-shared practices of their communities as an analytic way of coping with complexity brought about by disordered data. Thus, observations and facts are always interpreted in the context of the techniques, methods, and tools that are shared in communities (cf. Löwy, 2008). Necessarily, individual visual experiences are mediated and framed within one's cultural frame of seeing and sensing that then influences his or her perspective of what counts as objective, impartial, and abstract/de-referential knowledge. This perspective is frequently more asserted in various historical accounts of discovery and invention in the sciences than in mathematics.

In school mathematics, students who use manipulatives in a lifeworld-dependent situation are provided with an opportunity to become involved in particular acts of selective attention and visualization. In fact, effective use of mathematical objects necessitates the acquisition of appropriate conceptual relations and actions on them. For example, recent graphing technologies relevant to graphic and virtual objects provide powerful dynamic visual contexts in which both visualizing and thinking are organized from the point of view of the mathematics community that values particular ways of thinking about relationships, say transformations of graphs.

In an important sense, the subtle sociocultural influences, espoused in the works of Nisbett et al. (2001) and Kuhn (1970) above, foreground the Wittgensteinian perspective about one possible source of mathematical practices – that is, in Watson's words:

> Mathematics, says Wittgenstein, derives not from the material world, but from *our particular ways of systematizing* as we bring the physical world into the social world. Wittgenstein gives us powerful tools for lifting the mantle of myth and mystery with which mathematics cloaks itself. He urges us to look for a non-discursive rationality, in introducing the notions of "language-game" and "forms of life." These two concepts are tied up with what it is to grasp a rule without discursively interpreting it – that sometimes momentary, but unshakeable conviction that "doing it like *this* is the right way to do it."
>
> (Watson, 1990, p. 285; italics added for emphasis)

Thus, a valued aspect of visualizing a mathematical object, concept, or process involves making sense of the forms seen and/or produced and (re)organizing them either by negotiating or conforming with institutional ways of seeing and interpreting (i.e., that which comprises the language game). Certainly, this does not mean stifling creative visualizing since visualizing is still an interpretive phenomenon. For example, visual tools such as manipulatives by themselves are rendered useless unless an interpretive structure is constructed, shared, and negotiated. But an interpretive action on the tools is to some extent influenced by institutional practices that provide intentional use and ascertain the validity of what is visually sensed. Referring to the work of the Polish mathematician Leon Chwistek, who developed the theory of multiple realities, Löwy (2008) writes:

> Each theoretical/scientific truth, [Chwistek] added, is relative, that is, dependent on a given system. We regard some truths – those that repeat themselves in all the known systems – as absolute, but this is merely a result of the generalization of conventions. The truth of the lay

> experience is to some extent absolute because we do not doubt it, but it is in fact grounded in widely shared beliefs, many of which will not withstand a truly critical examination.
>
> (Löwy, 2008, p. 379)

2 Sociocultural Influences in Seeing and Imaging from a Mathematical Perspective

In this section, we discuss sociocultural influences in the development of mathematical ways of perceiving objects and their images in early history. We begin with Belting (2008), whose work zeroes in on differences in the manner early Arabs and Westerners thought about images. While the Arabs in early history were primarily interested in visual theory and the science of optics, which explains their interest in geometry, the Westerners were more into pictorial theory in relation to their interest in art. Consequently, differences emerged in the manner each group perceived the role of mathematics, including "modes of seeing and their aesthetics" (Belting, 2008, p. 184). Ibn An-Haitham (Alhazen), an Arab theorist in the eleventh century who wrote the *Book of Optics* that was translated in the Latin under *Perspectiva* or *De Aspectibus*, lived in an Islamic society whose ethos favored an "an-iconic culture" that enabled him "to dismantle the ancient authorities and to break with their dependence on bodies and idols in visual perception" (Belting, 2008, p. 184).

Belting's (2008) analysis of the translation of Alhazen's book in English shows a radical departure in terminology from the ancient Greek version. At the very least, Alhazen was more interested in geometry and light, which to Western thought was considered as abstractions, than "with anything mimetic and pictorial" (Belting, 2008, p. 185). For example, Alhazen concerned himself with matters involving reflection and refraction properties of light and did not pay attention to the relevant iconic images that were also being formed. Belting (2008) sums up the early Arab view of the intricate relationship between image construction and geometry as follows:

> Geometry in Arab culture serves as the equivalent of what pictures are in Western culture. At the same time, however, it functions quite differently from pictures, if we consider that one of its main goals was to protect the eye from all sensuous distraction. . . . Geometry served as a medium for purifying the world of the senses while at the same time representing the supreme reign of light in the world. The light is not just there to illuminate bodies, as it does in Western art, but has a superior existence. For Alhazen, mathematics in turn had a superior beauty, as it is based on calculation, a beauty not just seen but "read" in a cognitive act of perception. Alhazen often stated that he wanted to bridge physics and mathematics, the one being material and the other abstract, and we may observe the same tendency in Islamic art where geometry transforms the physical reality of its objects and buildings into the mathematical beauty of tile patterns that, like a skin, obscure or eclipse the corporeality of vessels or buildings underneath.
>
> (Belting, 2008, p. 187)

Bier's (2006) research on Islamic and Western art complements Belting's (2008) views. While Western art was preoccupied with figural and pictorial images (e.g., human portraits, images in coins), Islamic art tended to be more planar (two-dimensional) and pattern-driven (hence "ornamental"). The Islamic approach to art and architecture, in particular from the eighth to the eleventh centuries, could be seen as "visual analogues to Islamic ideas or ways of thinking that were central to the philosophical debates of the time" (p. 261). Among artists during this period, the prevailing interest was on the mathematics of space and form in two and three dimensions, "about surfaces and the plane, about units and repeats, about properties of the circle and the nature of space" (p. 269) that were all new to them. From the point of view of religion and metaphysics, the notion of multiplicity was seen as important and derived from the unit viewed as "indivisible" with number, more generally, seen in the context of learning something about the soul and God.

How medieval Islamic artists came to construct and establish patterns from various ornamental designs could be explained in terms of their close relationship with "mathematicians who taught [them] practical geometry" and "played a decisive role in the creation [and design] of those patterns" (Özdural, 2000, p. 171). Based on Özdural's (2000) textual analysis of available historical documents concerning the relationship between mathematics and the arts in medieval Islam, records indicate that regular meetings between geometers and artists took place quite frequently that enabled the former to teach the latter mathematical matters, resulting in the creation of more consistent and intricate pattern configurations. Artists relied on mathematicians who provided them with practical solutions such as the use of the method of dividing and assembling (cut-and-paste methods) in, say, constructing a larger square from two, three, five, and six smaller squares, which they used in designing wooden door and tile patterns that appear in mosques and other architectural media.

Panofsky (1991) has also argued about cultural differences in thought between the Renaissance and antiquity periods on matters involving the concept of perspectives. That the differences in the pictorial representation of three-dimensional space could be attributed to the manner in which each period perceived the world around them. In particular, the structures that shaped different works of art in each period were reflections of particular cosmologies, perceptual practices, and visual experiences. For example, in antiquity, perspective length size (i.e., the retinal image) was seen as curved depending on its proportional relationship to an angle of view (i.e., conformal view); in the Renaissance, such a line was perceived to be straight (i.e., collinear view). The following thought sums up Panofsky's (1991) view of differences in spatial representations on the basis of differing cultural practices: "[I]t is essential to ask of artistic periods and regions not only whether they have perspective, but also which perspective they have" (p. 41).

3 Materialist Versus Formalist Views on Mathematical Objects, Concepts, and Processes

Bier (2000) offers the following provocative explanation regarding differences in the notion of symmetry between artists and mathematicians:

> Although mathematicians treat symmetry as an ideal, in nature all symmetry is approximate. The study of patterns in Oriental carpets may lead one to suppose that in art, as in nature, it is in the approximation of symmetry, rather than in its precision, that beauty is to be found. . . . The study of patterns and pattern formation in Oriental carpets provides insights into the nature of beauty, which relies upon the beauty of nature in the realm of human choice.
>
> (Bier, 2000, p. 132)

Her thoughts about patterns being approximate and rooted in human choice exemplify a materialist, grounded perspective that we explore in some detail in this section.

Gerdes's (2004) interpretation of the mathematical structures underlying Tonga baskets illustrates a basic materialist point concerning the significance of "starting-points that will allow learning to develop more easily" (p. 119) among students. For Gerdes, artifacts and tools used in communities provide one useful source that should motivate learners to make sense of mathematical objects, concepts, and processes that are frequently claimed and supported on universal grounds. For example, Fig. 7.1 shows a basket made by the Tonga weavers in the southeastern part of Mozambique and its corresponding mathematical structure. In his analysis of several weaving patterns, Gerdes (2004) traces their structural differences on social, cultural, demographic, and periodic changes that took place in the evolving lived experiences of the Tonga weavers. That cultural variability enabled weavers to produce more complex design patterns and develop shared practices of weaving. Form, regularity, order, and symmetry, Gerdes (2004) notes, are not a priori but are "formed in productive activity" (p. 114), a reflection of "the societal experience of production" (p. 115). Further, he writes:

> The capacity to recognize order and regular spatial forms in nature has been developed through active labor. Regularity is the result of human creative labor and not its presupposition. It is the real, practical advantages of the invented regular form that lead to the growing awareness of order and regularity.
>
> (Gerdes, 2004, p. 114)

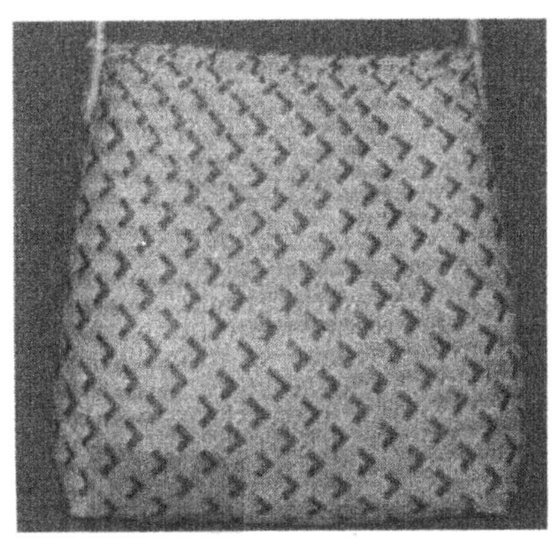

Bag with a [2/3, 5] pattern

Plaiting texture of a [2/3, 5] pattern

Fig. 7.1 An example of a Tonga handbag (note: [2/3,5] refers to the class of plane patterns that uses an over-two-under-three twill and has period 5)

Thus, the initial ability to perceive geometric forms and patterns could be interpreted as having been derived from the context of practical and useful activity that produces visual forms of knowledge. When such an activity transitions to reflections about the corresponding art and shape, mathematical thinking emerges in ways that allow the possibility of creatively imagining forms and patterns that may or may not have material existence. "Early mathematical thinking," Gerdes (2004) points out, "started to liberate itself from material necessity: form becomes emancipated from matter, and thus emerges the concept and understanding of form. The way is [then] made free for an intra-mathematical development" (p. 115).

While it is an accepted fact that concrete objects for counting in elementary mathematics classrooms provide sufficient context that assist students to understand the value of knowing mathematics, a materialist approach would have them see visual processes relevant to, say, recognizing coin and paper money as being related to institutional conceptions and social practices that people have about them. For example, Zebian (2008) investigated the role of artifact use on the everyday numeracy (i.e., orally and paper-based) practices of Lebanese micro-business owners from the southern and Bekka regions using a combination of cognitive ethnography and an experimental reaction–time study. He found that among the adult group, artifact use

> influences numerical recognition and conceptualization in a way that suggests tight linkages between the visuo-spatial processes involved in recognizing numerals embedded in cultural artifacts and the semantically based processes involved in the conception of these numerals. (Zebian, 2008, p. 359)

In his study, Zebian (2008) employed an ecological[4] approach to explore a possible relationship between currency use and concept of number. He initially utilized a social constructivist framework to assess the extent to which "socially situated artifact use and higher order cognition" among adult Lebanese sellers and nonsellers influenced their cognitive representation and processing of numbers (p. 361). He then used cognitive scientific methods to determine whether their understanding of Arabic numerals was either notation independent or notation dependent. In particular, he focused on "whether the visuo-spatial properties of currency based numerals, such as patterns and color of the bill" actually had an impact on how the Lebanese owners internally represented currency-based magnitudes (p. 361).

The micro-business numeracy practices in Lebanon fall into two categories, namely: oral- and paper-based. Oral-based practitioners (OBPs), on the one hand, pride themselves in their memory skills despite having access to paper. All purchases are stated orally. Transactions are oftentimes lengthy and involve repeatedly enumerating items that are bought. OBP numbers are seen in context, that is, in

[4]Zebian's (2008) "ecologically-informed approach to numeric cognition" involves coordinating social and cognitive approaches. The cognitive approach "offers a way of thinking about how surface notation is recognizable and decoded and how these processes are related to semantically-based processes" (p. 362). The sociocultural approach "is acutely sensitive to the socially situated cognitive demands of numeric practices" (p. 362).

terms of the currency bills used with little to no attention paid to the corresponding numerals. Paper-based practitioners (PPB), on the other hand, use cash registers, books, logs, ledgers, invoices, and many others to record all of their transactions. Further, numbers are seen in the context of numerals without referring to particular currency bills.

Zebian's findings indicate that while adults continue to rely on artifacts in thinking, it seems that a more pertinent concern is not the assessment of whether dependence on artifacts would diminish over time but to understand "what dimensions of socially situated artifact use will impact which kinds of cognitive processes" (p. 380). For Zebian,

> artifacts do not merely amplify thoughts but rather call upon cognitive processes which enable the coordination of different representations in a way that achieves or approximates socially situated task goals.
>
> (Zebian, 2008, p. 380)

Zebian's (2008) materialist perspective would have us view differences in numeracy practices in culture-specific situations as fundamentally determining the manner in which artifacts are used to influence thinking. Here we are reminded of a recent study done by Wood, Williams, and McNeal (2006) about primary children's mathematical thinking in different classroom cultures. While artifact use and visual strategies were not addressed in their work, it is reasonable to assume that differences exist in the role and use of artifacts and visual strategies in each of the following classroom cultures that seem to favor particular mathematical practices relevant to arithmetical learning: (1) conventional textbook-driven classes; (2) conventional problem-solving classes; (3) reform strategy-reporting classes; and (4) reform inquiry/argument classes.

Were's (2003) materialist perspective involving object construction in particular cultures shares many of the same ideas espoused by Gerdes (2004), Ferreira (1997), Zebian (2008), and other scholars who work within a situated cognitive view of mathematical practices. However, for Were (2003), there is also a need to explain the "role of objects as agents that activate mathematical thought" as a way of further enriching our understanding of the "object world as a tool in the learning of mathematical concepts" (p. 25). By object world, Were refers to the surrounding contexts that help explain how individuals construct their mathematical knowledge.

Three studies are worth noting briefly in order to highlight Were's (2003) materialist perspective. Toren (1990) made an observation that individuals in Fiji acquire their notion of hierarchy in the context of a *kava* drinking ceremony, a social practice in which hierarchy is associated with spatial reference points. Further, they acquaint themselves spatially through reference points, say, within buildings. Saxe (1991) found that the numerical processes of the street children in Brazil are tied to the tools they use and their everyday experiences with problems as confectionary sellers. Consequently, successful performance on school mathematics tasks with this group of children seems to correlate with situations they encounter in their everyday work as sellers. Jürg Wassmann (1994) observed that while Western conceptions associated with space tend to be egocentric, the Yupno of Papua New Guinea uses three different ways of spatial perception (i.e., body-centered or relative reference system; field names or local landmark system; absolute reference system).

In all three studies above, Were (2003) interprets the constructed mathematical knowledge as a consequence of the relevant social and cognitive processes without the authors fully taking into account the "material quality of objects" (p. 27). By materiality of an object, it means the associated "synthesis of thought process" (p. 28) that is evoked once individuals in particular cultures use the object in some way. For example, patterns and knots as objects convey complexes of ideas, feelings, relations, and social and cultural practices that are recalled and felt when used in an activity. Hence, instead of dwelling on contexts (the object world) to explain mathematical thinking, Were (2003) focuses on relationships between the materiality of an object and mathematical thinking.

For example, we frequently perceive patterns used in the mathematics classrooms as reflective of artifacts drawn from particular cultures. However, in Papua New Guinea, for instance, recent curriculum changes in the elementary mathematics curriculum reflect a shift in thinking from a language-based to an object-centered learning in which patterns play a central mediating role in the emergence and development of mathematical knowledge. At the very least, geometry and numeracy are taught in the vernacular that also use pattern artifacts drawn from local craft and art. Were (2003) writes:

> [C]ertain properties of objects can harness mathematical thoughts, and it is these thoughts that can be used fruitfully as a vehicle for learning. So for example, concepts of numeracy and spatial reasoning are tapped directly from the mathematics embedded in various phenomena with shape, symmetry, time, pattern, color, set theory, number, angle length, and capacity illustrating some of the specific mathematical concepts the curriculum program harnesses. The program also utilizes mathematics through the performative aspects of culture: traditional dance (drumbeats and movement); traditional music (rhythm and timing); and traditional arts and crafts in the teaching of pattern, shapes, and designs. The focus on visual and material culture in the curriculum emphasizes the concreteness of the learning experience utilizing students' familiarity with objects and materials, concepts, and cultural practices.
>
> (Were, 2003, p. 30)

Figure 7.2 is an example of a patterning activity for elementary students, while Fig. 7.3 is an example of a curriculum guide on patterns for teachers. Patterns as material objects are used to learn relevant mathematical concepts and processes. Children use them to develop spatial reasoning; create local arts and crafts; learn about number sequences and color patterns, seasons, and time; learn about their central role in musical performances and dance routines; explore their environments, name shapes and associate them with the local language; help them construct rules; and so on.[5]

---

[5]Were (2003) provides a thorough and engaging discussion of the object of *kapkap*, a patterned-shell ornament in the Western Island Melanesia, which he uses to demonstrate the rigor of mathematical thinking that comes with understanding its social and cultural significance. While the object conveys to these people certain "everyday and ritual performances," it is the "translation from mental to material form that mobilizes mathematical thinking and spatial reasoning" (p. 42).

1. Look at the walls of the houses in your community. What patterns do you see? Draw them.
2. Look at the material of the clothes you are wearing, what patterns can you see? Draw them if you can.
3. Look for patterns in the things made in your community. Make a list of the things and draw the patterns. If you can you should collect items with different patterns on them.
4. Were all pigs, which are now adult, baby pigs once?
5. Were all butterflies baby butterflies once?
6. How many different life cycle patterns can you think of?
7. How many different patterns can you see in the things around you?
8. Numbers can also be arranged in patterns. The numbers in counting order form a pattern made by adding one to the last number. Write down the pattern made by adding two to the last number.
9. Does it matter which number you start with?

Fig. 7.2 Example of a patterning activity at the elementary level in Papua New Guinea (Were, 2003, p. 32)

Who in the community has knowledge of the traditional patterns and designs?
What are the traditional patterns and designs?
Why do we use traditional patterns and designs?
When do we use traditional patterns and designs?
Where do we use traditional patterns and designs?
How do we draw traditional patterns and designs?
Here are two examples from Ovatoi village in the Trobriand Islands, Milne Bay Province.

Example 1
Who? Sobububwaluwa.
What? Traditional carving designs.
Why? Sobububwaluwa uses these designs in his carvings because they are attractive and catch the people's attention.
When? These patterns are used when new canoes are cut and on new carvings to be sold to tourists.
Where? They are traditionally used on the prow boards and on the rear of the village canoes. These patterns are also used on the yam houses, and the chief's dwellings.
How? These are carved with special carving knives and chisels. Another special tool is called the ligogu in the Kiriwina language.

Fig. 7.3 Example of a teacher's guide involving patterns at the elementary level in Papua New Guinea (Were, 2003, p. 31)

4 Cultural Views on What Constitutes a Mathematical Proof

In 1979, Bishop talked about his research on the "visual and spatial aspects of mathematics" (p 136) using data drawn from 12 male first-year undergraduate students in Papua New Guinea. Acknowledging cognitive differences between his participants and their counterparts in the UK, he provided evidence of cultural practices in language, drawing, and visualizing in Papua New Guinea that were not aligned with Western practices, which certainly had implications in how and what mathematics was learned in those communities. For example, processes relevant to generalizing had to take into account the linguistic structure in Papua New Guinea, which then appeared to be drawn toward the use of specific rather than general terms.

Also, comparing quantities was difficult or almost impossible to accomplish in most cases since there appeared to be more adjectival and subjective than invariant and objective units that were used in everyday activity. Bishop's thoughts below about the nature of conventions capture the sense in which we talk about cultural views pertaining to the nature of a mathematical proof.

> Conventions are of course learnt, as are the reasons for needing them, *and* the relationship between the pictures and the reality that are conventionalizing. The hypothesis is therefore provoked: perhaps much of the found difficulty with spatial tasks lies in understanding their conventions, and that if these *are* known by those people, from both non-Western and Western cultures, who are supposedly weak spatially, then perhaps they would not appear to be quite so incapable.
>
> (Bishop, 1979, p. 138)

Raju's (2001) interesting work attempts to explain differences between Eastern and Western practices in proof in mathematics in terms of the value accorded to the empirical content of proof. The Western view favors a (neo)Platonist perspective that sees mathematics to be consisting of knowledge that is located "between the gross empirical world and the higher Platonic world of ideals" (p. 329). Platonists believe that we can acquire knowledge of ideal objects only through the "shadows" and the reflections they produce, say, in some medium. This explains why the early Greeks considered mathematical activity (in particular, geometry) as a form of "spiritual exercise ... which turns one's attention inward, away from sense perceptions and empirical concerns, and 'moves our souls toward Nous' (the source of the light that illuminates the objects)" (p. 326). Further, contemporary Western thought is skeptical of the empirical world because the truths that are produced are at best contingent and not necessary.

Raju (2001) notes that while the West has accepted the nonabsolute status of mathematical truths, however, it does not mean jettisoning "necessary" truths but "shift(ing) the locus of this 'necessary truth[s]' from theorems and axioms to proof" (p. 326). That is, in mathematical proofs, one establishes "tautological relation(s) between hypotheses and conclusions," which is not the same in physics that focuses on "the empirical validity of the hypotheses/conclusions" (p. 327).

Certainly it is also reasonable to assume that some cultures do not subscribe to the above Western views concerning the contingency of the empirical world and the nonempirical status of necessary truth. For example, all the major schools of thought in India between the seventh and early sixteenth century such as the Lokayata believe in the opposite view – that is, they see the empirical as the main source of validation and inference as fallible.[6] An example of an empirical proof of the Pythagorean theorem in shown in Fig. 7.4. The content of the proof relies heavily on an empirical, visual demonstration that shares the same intent as the ones shown in Figs. 3.14 and 3.15 (in the context of a visual discovery explanation approach).

[6]Raju (2001) also notes that Buddhists value inference as a valid means of validation but reject authoritative testimony, while Naiyayikas value all three (empirical, inference, authoritative testimony) means, including analogy (p. 328).

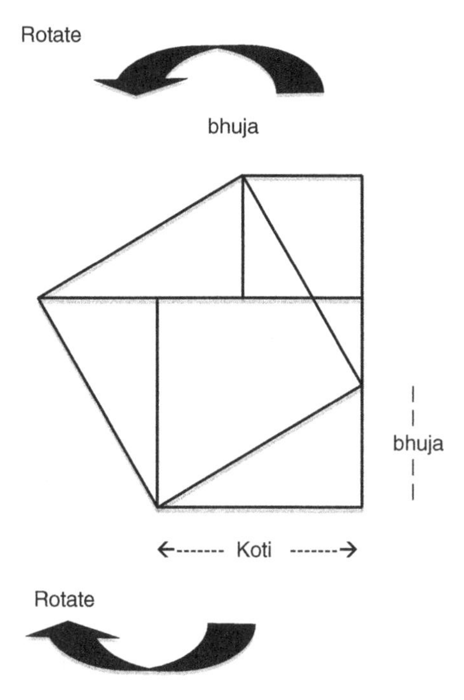

Rationale for the sine rule in the Yukubhasa. The square corresponding to the smaller side (bhuja) is drawn on a palm leaf and placed on the square corresponding to the bigger side (koti), as shown. The bhuja is measured off from the SE corner of the larger square, and joined to the SW corner of the larger square and the NW corner of the smaller square. Cutting along the joining lines and rotating gives the square on the hypotenuse. This simple proof of the "Pythagorean" "Theorem" involves (a) measurement and (b) movement of the figure in space.

Fig. 7.4 Ancient Indian proof of the Pythagorean theorem (Raju, 2001, p. 328)

For Raju, the Western disavowal of empirical content in favor of a nonvisual synthetic and abstract approach, especially during the time of Hilbert, influential German mathematician in the nineteenth and early twentieth centuries, was a consequence of an emerging industrial civilization that was at the time exploring the full power and capabilities of machines. "So it is no surprise," Raju writes,

> that Hilbert's view of mathematics was entirely mechanical ... a proof had to be so rigidly rule-bound that it could be mechanically checked – an acceptable proof had to be acceptable in *all* cases.
>
> (Raju, 2001, p. 330)

Those mathematical propositions that involved measurement and were established empirically did not reach the status of theorems in Hilbert's terms. For example, Hilbert thought that the famous side-angle-side (SAS) theorem in Euclid's *Elements* would have to be regarded simply as a postulate since the proof offered by the Greeks was visual and empirical (using transformations). "Hilbert," Raju points out:

> reflected the then-prevalent Western view that doubted the role of measurement and the empirical in mathematics. Picking and carrying involves movement in space, and it was thought that movement in space may deform the object .... The avoidance of picking and carrying in the proofs of the subsequent theorems in the *Elements* was interpreted, by the twentieth century, as an implicit expression of doubt about the very possibility of measurement.
>
> (Raju, 2001, pp. 330–331)

Also, the notion of *coincidence* was seen as empirical that explains why the SAS proposition would have the character of a postulate and not a theorem (which should possess necessary truth). German philosopher Schopenhauer writes:

> *[C]oincidence* is either mere tautology, or something entirely empirical, which belongs not to pure intuition, but to external sensuous experience. It presupposes in fact the mobility of figures; but that which is movable in space is matter and nothing else. Thus, this appeal to coincidence means leaving pure space, the sole element of geometry, in order to pass over to the material and empirical.
>
> (Schopenhauer quoted in Raju, 2001, p. 331)

Among early Indian mathematicians, however, there was never a dichotomy between (1) the sensible and the intelligible, (2) pure and applied, (3) mathematical and physical, (4) spiritual and practical needs, (5) the abstract and empirical, and (6) proof and calculation. They viewed mathematics as a means to some practical end. Mathematical calculations were valued for their practical value in commercial transactions and their usefulness in matters involving, say, astronomy. Finally, since mathematical proofs were seen in the context of providing "rationale for methods of calculation" (Raju, 2001, p. 333), they were simply taught and not recorded.

5 The Role of Artifacts and Visual Language in Early Mathematical Development

Findings obtained from recent cross-cultural studies foreground the influential role of artifacts and visual language in early mathematical development. Ginsburg, Lin, Ness, and Seo (2003) investigated the everyday mathematical behavior of 4- and 5-year-old American and Chinese children during free play. By everyday mathematics, Ginsburg et al. refer to

> whatever mathematics children acquire in their ordinary physical and social environments. It may be formal and intuitive; it may be based in social and cultural experience; and, to a small extent, it may even be formalized.
>
> (Ginsburg et al., 2003, p. 236)

Ginsburg et al. observed that the children in both cultures who participated in their study were actually engaged in similar types of everyday mathematics during free play. They did not exhibit any preference for particular objects and worked through tasks such as patterns and shape, magnitude, and enumeration quite comfortably with tools such as Legos and blocks.

Ginsburg et al. (2003) developed some interesting hypotheses in relation to particular aspects of "cultural differences in the frequency and complexity of everyday mathematics" (p. 255) between the two groups. Overall, while they found that the two groups did not differ in complexity of mathematical thinking, the Chinese group engaged more frequently than the American group in activities that involved patterns and shapes. One hypothesis that could explain why the Chinese group appeared to be doing more everyday mathematics than did the American group could be biological, that is, the former has the biological disposition toward mathematical activity compared with the latter. While the remaining hypotheses that the authors raised were environmental,[7] two are worth noting in light of their relevance to the ideas that are pursued in this book.

The first hypothesis concerns exposure to types of objects used by the children during free play. While the participating Chinese preschools offered very few playing objects due to limited physical space, which included Legos and blocks that support activities involving patterns and shape, the participating American preschools had a larger number of playing objects that consequently distracted the children from fully engaging in mathematical activity.

The second hypothesis deals with what Ginsburg et al. refer to as "useful intuitive foundations of mathematics" that foreground the significance of a visual orientation in mathematical thinking as follows:

> Whatever its origins, the greater frequency of mathematical activity among Chinese children particularly if it persists over time, can provide a useful intuitive foundation in mathematics. Extensive exploration of pattern and shape may result in visual intuitions that can provide a sound basis for many mathematical ideas. Thus, a picture of a staircase may more vividly capture the idea of a linearly increasing number series than may a numerical equation. To the extent that intuitions of mathematics can usefully be comprised of visual images, experience with pattern and shape may enhance mathematical thinking. Further investigations of the role of visual imagery in Chinese children's later mathematical superiority may be informative. Further, does the need to learn the intricacies of characters make Chinese children more visually oriented (and hence more mathematically oriented) than Americans?
>
> (Ginsburg et al., 2003, p. 256)

Hence, a strategic use of visual objects could support growth in visual intuitions and visual knowledge, which could enhance mathematical thinking in the long term.

It is interesting to peg the first environmental hypothesis above on a study on manipulatives by Uttal, Scudder, and DeLoache who also note that the use of

> many different kinds of bright, beautiful manipulatives may push children's attention toward the objects themselves and away from where it needs to be – on the relation of the symbol to what the children are supposed to learn.
>
> (Uttal et al., 1997, p. 50)

---

[7] Ginsburg et al. (2003) also suggested the hypothesis that the teachers involved in the study influenced what the children were doing during free play. But this perspective was discounted due to the fact that the teachers observed were not seen interacting with the children during free play.

Uttal et al. cite a study by Stevenson and Stigler (1992) who saw that

> Japanese teachers ... use the items in the math set repeated throughout the elementary school years ... [while] American teachers seek variety. They may use Popsicle sticks in one lesson, and marbles, Cheerios, M&M's, checkers, poker chips, or plastic animals in another. The American view is that objects should be varied in order to maintain children's interest. The Asian view is that using a variety of representational materials may confuse children, and thereby make it more difficult for them to use the objects for the representation and solution of mathematical problems.
>
> (Stevenson and Stigler quoted in Uttal et al., 1997, p. 50)

With respect to the second hypothesis above, a study by Miura, Okamoto, Vlahovic-Stetic, Kim, and Han (1999) provides empirical evidence of the influence of language characteristics in mathematical thinking and performance. Miura et al. assessed first- and second-grade Croatian, Korean, and US students in relation to their fraction understanding prior to classroom instruction. They saw that culture-specific visual language practices actually assisted their participants in understanding mathematical concepts such as place value, counting, and fractions.

For example, in Korea, the term *fraction* is linguistically interpreted in the context of part–whole relation. The fraction 1/3, spoken as *sam bun ui il*, literally means "of three parts, one," which is not the same in the USA in which case the fraction is called a "third." Miura et al.'s study shows that while the performance of the Korean, US, and Croatian children that was tested did not differ in relation to the task that asked them to associate numerical fractions with their pictorial representations, the Korean students performed significantly better than the two other groups by the end of first grade. Also, at the start of second grade, 75% of the Korean students were more capable than the other groups in correctly identifying fractions that would not be typically used at the grade level. Miura et al. then points out that

> (a)ssuming that exposure to fractions and fraction terms for children in Grade 1 is similar across countries, Korean children may be better able to make sense of the relation between fraction terms and their visual representations because the Korean language supports the connection between the two. The part–whole relation is an integral part of the linguistic term. Once the unit fraction (e.g., ¼ or "of four parts, one") is understood, that knowledge can be extended to more complex fractions.
>
> (Miura et al., 1999, p. 363)

6 Multisensory Convergence on Imagery Tasks in Totally and Partially Blind Subjects

In this section, we discuss the nature of image construction and processing in totally and partially blind subjects. One basic issue confronting research in this field is the extent to which the images they form are visual (pictorial) or nonvisual (e.g., spatial). Further, in those cases when a sensory substitution device is used to help blind subjects acquire information via the auditory or the tactile route, more research data are needed to assess whether the information is processed the same as the original input or is transformed into visual data. "There is convincing evidence" however,

Ward and Meijer (2010) note, "from both tactile and auditory sensory substitution devices that users of these devices recruit regions of the brain that are normally specialized for vision" (p. 492).

Many congenitally blind subjects have been observed to have a tendency to process and code imagery tasks more spatially than visually. For example, in a reported study by Kerr, Foulkes, and Schmidt (1982), the congenitally blind subjects described their dream settings in terms of spatial features that share many basic features common to visual images such as "the size and shape of a room and the location and orientation of objects and people in the environment" (Kerr, 1983, p. 266). Landau, Gleitman, and Spelke (1981) describe the case of a congenitally blind 2½-year-old child who successfully developed and employed innate metric properties of Euclidean space in accomplishing a locomotor task that tested knowledge of space and spatial relationships among objects. In a tactual mental rotation task, Marmor and Zabeck (1976) observe that the imagery strategies of their blind subjects did not use visual representations [see, also, Carpenter and Eisenberg (1978) who inferred the same conclusion but in the context of a letter orientation task]. They claim, however, that having a visual experience "might enhance the speed and accuracy of [their] performance" (Kerr, 1983, p. 266). A similar claim has been offered on matters involving perspective taking and perspective structure (Heller & Kennedy, 1990; Rieser, Guth, & Hill, 1986).

We point out two reasons why this section and the two that follow are included in this book. *First*, Kerr (1983) has echoed the "need for a broader definition of imagery" as a result of her work with blind subjects. Indeed, it is an important matter to understand the characteristics of effective and meaningful image construction, internalization, and processing in blind subjects and use such knowledge to develop better visual instruction for sighted learners. Kerr (1983) writes:

> In an effort to avoid visual metaphor in describing the imagery of blind subjects, it is tempting to substitute a metaphor based on another, non-visual, sensory modality. ... [B]lind subjects are not limited to modality-specific image processing but are capable of a broader range of imagery experience, including imagery that is spatial in nature but without specific ties to one sensory modality. ... It has not been easy for sighted researchers and theorists to dissociate themselves from visual metaphor and analogy in thinking about images and imagery. The visual component of the imagery experience is surely its most salient aspect for most sighted persons. Yet, ... [we] underscore *the need for a broader definition of imagery than one that is specifically tied to the visual processing system*. Kolers and Smythe (1979) offered as one definition of an image, "a mental event that seems to preserve the spatial, figural, chromatic, textural, and like properties of objects imaged" (p. 159). If one omits the word "chromatic," the definition characterizes precisely this kind of imagery apparently experienced by the blind subjects who participated [in our study].
>
> (Kerr, 1983, p. 275; italics added)

Tinti and Galanti, for example, observe that blind subjects effectively employ auditory imagery in helping them enhance their memory and learning. The authors also recommend the need to encourage them "to use a visual imagery strategy in learning tasks" but with an extra caution that "not too many items be included in a single image" (Tinti & Galanti, 1999, p. 583). Cattaneo et al. (2008) also note the same observation with blind subjects based on a number of studies that show their predisposition to employ "different cognitive strategies" when confronted with imagery tasks. That is, they tend to "rely more on verbal/semantic, haptic, or purely spatial (i.e. without a visual content) representations" (p. 1347). Cattaneo et al. note:

> In fact, it is likely that blind individuals compensate for the lack of vision both at a perceptual level, by enhancing their auditory capacities . . ., and at a higher cognitive level, by developing conceptual networks with more acoustic and tactile nodes . . ., thus contradicting the view that semantic networks are less elaborate in congenitally blind individuals.
>
> (Cattaneo et al., 2008, p. 2347)

Second, findings from studies with partially and totally blind individuals will further enrich our understanding of the nature and context of visualization in school mathematical learning, including the many purposes we assign to representational tools that we use to teach mathematical objects, concepts, and processes. Kerr (1983), in particular, notes that blind people are capable of producing visual images. What this means is perhaps best captured in the following sentences below by De Beni and Cornoldi (1988), who questioned the commonly accepted view that "visual imagery is based on visual experiences" and that those who have "not previously had visual experiences should not have visual images" (p. 650).

> In our opinion, the weak point of this argument is the presupposition that visual images are necessarily the products of visual experiences; rather, they may be representations that are based on information collected through different sensory modalities and therefore may maintain some of the properties of visual objects. We assume that people go beyond the limitations imposed by the properties of single sensory channels and construct representations that integrate information coming from different sensory modalities.
>
> (De Beni & Cornoldi, 1988, pp. 650–651)

Ward and Meijer (2010, for example, document the visual-like experiences (e.g., "seeing sounds," "perceiving edges, depth, movement, and color") of their two participants who became blind at ages 21 and 33 years. At the time of the study, the blind subjects were already considered expert "users of a sensory substitution system that converts visual images into auditory signals" (p. 498). The sensory substitution tool allowed them to encode visual information, which could be seen as a consequence of "experience-dependent learning of sensorimotor contingencies (in sensory substitution) augmented by non-learned multi-sensory associations" (p. 499). Farah argues along similar lines as De Beni and Cornoldi as follows:

> (I)magery [among blind subjects] is not visual in the sense of necessarily representing information acquired through visual sensory channels. Rather, it is visual in the sense of using some of the same neural representational machinery as vision. That representational machinery places certain constraints on what can be represented in images and on the relative ease of accessing different kinds of information in images.
>
> (Farah, 1988, p. 315)

The above remarks provide a complementary neurophysiological dimension to the psychologically derived distributed context of visualization that we have discussed in various sections in this book. In the psychologically derived view, visualization is tied to an assemblage of factors that shape the process of visualization (external environment, individual learner, task, etc.). That visualizing could be seen in terms of using and integrating various multiple sensory modalities, which then enable the reification of an image and give the appearance and feel of being visual. Here we are reminded of the blind French topologist Morin,[8] whose famous work involves the homotopy of a sphere – the eversion of a sphere (how to turn it inside out) – that he initially constructed out of clay models by "touch alone ... to communicate to the sighted what [he] sees so clearly in his mind's eye" (Johnson, 2002, p. 1246). Morin's spatial ability and imagination provide an example of a blind subject who does not need a visual experience prior to developing spatial understanding. His spatial imagination, in fact, enables him to visualize objects

> from outside to inside, or from one "room" to another ... [which] seems to be less dependent on visual experience than on tactile ones. "Our spatial imagination is framed by manipulating objects," Morin said. "You act on objects with your hands, not with your eyes. So being outside or inside is something that is really connected with your actions on objects." Because he is so accustomed to tactile information, Morin can, after manipulating a hand-held model for a couple of hours, retain the memory of its shape for years afterward.
> (Johnson, 2002, p. 1248)

Johnson shares another story of Morin, who was asked how he was able to compute a sign in relation to a thesis that actually required a colleague to employ extended calculations involving determinants. His following response echoes the foregoing points raised by De Beni and Cornoldi: "I don't know – by feeling the weight of thing, by pondering it" (quoted in Johnson, 2002, p. 1248). Further, Johnson points out,

> Morin believes there are two kinds of mathematical imagination. One kind, which he calls "time-like," deals with information by proceeding through a series of steps. This is the kind of imagination that allows one to carry out long computations. "I was never good at computing," Morin remarked, and his blindness deepened this deficit. What he excels at is the other kind of imagination, which he calls "space-like" and which allows one to comprehend information all at once.
> (Johnson, 2002, p. 1248)

7 Benefits of Visual Representations: Three Lessons from Blind Studies and Implications for Sighted Learners

Both blind and nonblind learners experience similar difficulties when they deal with, say, geometry tasks that involve spatial representations and perspective taking. Several researchers who have worked with blind subjects point out that the crucial issue appears not in determining whether visualization precedes spatial cognitive performance but in assessing the significance and limits of visual and tactile/haptic (or even auditory) performance in their spatial ability. Certainly, both visual and nonvisual actions are useful. For example, the tactile/haptic dimension allows blind learners to develop their understanding of depth relationships and surface texture,

[8] Morin became fully blind at age 6.

while the visual dimension helps them obtain a quick access to, and a better understanding of, form, shape detection, and coordinate systems (Heller & Kennedy, 1990).

Heller and Kennedy (1990) investigated the impact and significance of visual experiences on two spatial cognitive tasks in 27 congenitally blind, late blind, and sighted subjects. The shape-matching task tested their understanding of space, while the perspective-taking task assessed their understanding of vantage points and various viewpoints (top, side) relative to a given geometric object. In the shape-matching task, the authors found that visual imagery and visual experience did not play a significant role when the subjects were asked to match one figure to another from among a given set of eight simple embossed two-dimensional patterns. In the perspective-taking task, the subjects were first asked to draw views of a cube, a cone, and a sphere from different vantage points with each figure represented haptically in a three-dimensional array. They were then asked to identify the vantage point of pictures drawn from different viewpoints. The authors found that, despite the absence of a visual experience, the congenitally blind subjects were as successful as the sighted and the late blind in perspective taking and interpreting. Further, most subjects in all three groups found the top vantage view much easier to draw than the side views.

While results of the above two tasks seem to suggest that visual imagery is not necessary and crucial in spatial cognition and representation and tactual performance, Heller and Kennedy (1990) articulate benefits in having a visual experience, in particular, in matters involving processing time, memory load, and time-constrained performance. They observed that the congenitally blind subjects in their study needed more time to construct spatial representations and relationships, unlike the sighted and late blind subjects who took less time to accomplish the perspective-taking task (perhaps due to visual familiarity with the drawings). Also, it is likely that blind learners will experience difficulty when their memory is loaded or when performance on a task requires speed, unlike sighted learners who possess visual imagery that, say, allows them to easily interpret tactual impressions.

Veraart and Wanet-Defalque (1987) investigated whether the representations of locomotor space by 16 blindfolded-sighted, early-, and late-blinded subjects could be explained in terms of the functions of their visual experiences. The basic task had them locating objects relative to particular positions or landmarks in a given room. While results of the actual study are not crucial here, the authors' reflection concerning the use of external cues as reference points in visual experiences has useful implications in mathematics learning for the sighted. In today's mathematics classrooms, concrete and virtual manipulatives and graphical technologies are everyday visual tools that teachers use to help students learn mathematical objects, concepts, and processes. For Veraart and Wanet-Defalque (1987), visual experiences enable those who have them to use external cues as reference points instead of having to construct an image of a spatial configuration. That is, "early visual experience gives the ability to create a fictive frame of reference in relation to which kinesthetic and vestibular information is coded" (Veraart & Wanet-Defalque, 1987, p. 137), which is what visualizing activity in school mathematics is intended to be in part when students begin to learn mathematics with manipulatives and other learning tools.

Drawing on her extensive analysis of the research literature on blind subjects and their developing spatial knowledge and her longitudinal studies with two groups of totally and partially blind subjects, Millar (1994) obtained similar results and echoed the same recommendations. But her insight concerning the convergence and overlap of various sensory modalities in (spatial) representation further deepens our understanding of the significance of "redundancy" in mathematical knowledge acquisition processes. Redundancy does not mean repetition of the same experience over and over again until some knowledge is obtained. What it means is that redundant cues taken and recruited from other sensory modalities in the case of a loss in one major modality (say, seeing) tend to facilitate depth, greater connection, and more precise information and representation relative to the information being targeted for acquisition. For sighted learners, this means to say that, at the very least, "having more than one source of concurrent information can actually facilitate processing" (Millar, 1994, p. 46) with the important additional condition that the sources be allowed to converge and overlap, "deliberately producing the coincidence (in space and time) of different forms of the 'same' information (touch, movement, sound, and explanation)" (Millar, 1994, p. 237). Further, Millar notes that while many interventions conducted with blind subjects involve sensory substitution, however,

> substitute information alone is not enough, even if it is made sufficiently redundant. It is essential also to ensure that what is substituted does, in fact, converge with existing information, that it not only complements it, but also overlaps sufficiently to restore redundancy.
>
> (Millar, 1994, p. 236)

For example, Veraart and Wanet-Defalque's (1987) finding about the essential role of external cues also needs to be correlated with purposeful informational redundancy drawn from other modalities, which then need to converge and overlap. As a practical matter and keeping in mind the role of visuals as "devices" in mathematics among sighted learners, Millar points out that

> extremely able and knowledgeable blind users can make the links [among the different information drawn from several modality sources] themselves. But for most people, the link between devices which substitute, rather than complement existing information and means of coping, needs to be made obvious.
>
> (Millar, 1994, p. 242)

Some of these strategies include implementing assisted learning and exploration, using hand movements, employing particular kinds of drawings (and tactual-based materials for the blind), and providing helpful haptic feedback. Also, Millar (1994) recommends the use of "verbal description and spatial demonstrations" that "should not be treated as incompatible alternatives" (p. 244). But then again she stresses that they "need to be linked. ... [and] the link should be made explicit and obvious" (p. 244) depending, of course, on the nature, context, and requirements of the task being resolved.

8 Models and Implications of Multisensory Learning

We close this chapter by revisiting Sawada's (1978, 1982) multisensory modality research investigations in mathematical learning, which should provide a useful synthesizing model in thinking about recent samples of interventions done with blind students who learn mathematics from at least two modalities (i.e., auditory and visual). Certainly, the insights of researchers whose findings I have discussed in greater detail above, especially Millar's (1994) ideas involving convergence and redundancy, should also be incorporated in this framework. Toward the end of this section, we provide samples of tools that are currently being used with blind learners and exemplify the use of multisensory modeling in mathematical contexts.

Sawada and Jarman note that while there appears to be a correlation between perceptual abilities and mathematics achievement, students' modal-matching ability, in particular, the auditory–visual connection, could also influence mathematics achievement. For example, learning the concept of "four" among young children involves acquiring the fact that hearing it (auditory), circling a diagram with four objects (visual), and demonstrating it with manipulatives and other concrete aids (haptic or tactile) are all equivalent actions despite the fact that they have been routed in different modalities.

In their experimental study with 180 fourth-grade male students[9] of varying IQ abilities, Sawada and Jarman (1978) inferred that any relationship between modality-matching ability and mathematics achievement seems to depend on the type of matching and the IQ level of the student. Their finding suggests that awareness of the appropriate type of matching could provide important information relevant to effective ways of individualizing mathematics instruction and learning. For example, students with low IQ may likely benefit from an integrated auditory-to-auditory matching compared with students that have medium intelligence. Sawada and Jarman note:

> This finding is somewhat surprising, since much of elementary school mathematics would seem to depend as much on AV or VA or even VV matching abilities. In fact, in the high IQ group, mathematics achievement seems to be uniformly dependent on all four modality matching abilities, but this uniform dependence seems to all but vanish with pupils of medium intelligence.
>
> (Sawada & Jarman, 1978, p. 133)

In a second study, Sawada (1982) added the haptic modality dimension in assessing for predictors of mathematics achievement among 169 third-grade males and females. He found that the haptic dimension plays a significant and useful role in the mathematical learning of the different IQ-level groups. He writes:

> [C]hildren who have difficulty integrating haptic information with other sensory information do less well in learning mathematics. To the extent that a particular classroom was strongly into the use of manipulatives, the force of these conclusions would be even

[9]Females were excluded as the authors' way of coping with less complicated data at the time of the study (i.e., holding the sex variable constant).

> more significant. In summary, if we as mathematics educators continue to recommend the widespread use of manipulative aids (hands-on experience) in providing mathematical learning for children, then it seems important that we realize that children vary in the extent to which they are able to integrate such haptically acquired information with other information in the experience.
>
> (Sawada, 1982, p. 393)

Recent studies and interventions used with blind students in both mathematical and nonmathematical contexts also highlight the significant role of multiple-modality learning.[10] The images that blind students produce, Tinti and Galanti point out,

> have the essential characteristics of visual objects ... [and while] the images do not contain typical attributes unique to visual objects ... they seem to contain characteristics of objects that could be processed to enhance memory.
>
> (Tinti & Galanti, 1999, p. 579)

Gardner, Gardner, Jones, and Jones (2008), for example, developed a graphical sight/audio/touch accessibility technology with a flexible user interface that helps disabled children learn arithmetic by either receiving or entering data in different ways. Figures 7.5 and 7.6 show two students doing arithmetic with the aid of the arithmetic learning system tool. Depending on the type of disability, students can enlarge an image or distinguish small objects using a tactile embosser. Tactual images are mapped onto a touchpad, thus enabling learners to feel either text or object. Also, labels of text or objects speak when pressed. Low-tech devices are just as useful as the high-tech ones. Gibson and Darron (1999) used cardboard and clay models in helping their blind subject haptically explore tactual displays as a means of visualizing statistical concepts. Ducker (1993) used a combination of Thermoforms and German films with two blind subjects in making sense of bar charts and scatter diagrams.

Brewster (2002) was concerned about the manner in which blind people are presented with data obtained from graphs and tables. Instead of being presented with the actual graphs, they either hear lines of digits that are read to them or feel serial rows of digits in Braille that consequently make it impossible for them to establish patterns and inferences. Karshmer and Farsi (2008), in particular, point out two difficulties in representing mathematical formulas in Braille. First, while ordinary text is linear, mathematical equations are two dimensional. For example, the formula

[10] Ginns's (2005) meta-analysis of 43 experimental studies prior to 2004 shows an overall strong modality effect (with moderation effects in some aspects), that is, "across a broad range of instructional materials, age groups, and outcomes, students who learned from instructional materials using graphics with spoken text outperformed those who learned from a graphics with printed text" (p. 326). A nonmathematical context that involves the use of nonvisual modality modes of learning among blind subjects involves studies in mobility (e.g., use of echoes, guide dogs, and long canes; sound cues; felt cues; electronic travel aids and ultrasonic echolocating prosthesis; cf. Strelow, 1985; Veraart & Wanet-Defalque, 1987).

Fig. 7.5 Doing arithmetic with an arithmetic learning system tool 1 (Gardner et al., 2008, p. 4)

Fig. 7.6 Doing arithmetic with an arithmetic learning system tool 2 (Gardner, et al., 2008, p. 6)

$$a = \sqrt{\frac{x^2 - y}{z}}$$

contains a fraction and a superscript that are both two dimensional. When translated in linear form, it appears to blind subjects in the following form: $a =$ sqrt $(((x$ super $2) - y) / z)$. Equations and formulas that have more parts make them almost impossible to learn due to the complexity of the corresponding linear representation. Second, while ordinary text has a manageable number of characters (uppercase and lowercase letters, 10 digits, punctuation marks, and some special characters), equations and formulas contain all the normal text characters and a larger number of special characters and "escape sequences" (in cases when new characters are introduced, thus necessitating new assignations). The blind mathematician Giroux[11] notes the difficulty of performing long calculations in Braille, which to him explains why most blind mathematicians study geometry since "the information is very concentrated, it's something you can keep in mind (which is not necessarily pictorial)" (Johnson, 2002, p. 1249).

Brewster (2002) then tested and found that nonspeech audio and haptic graphs are effective alternative tools for presenting visual material to blind students. Sonification (sound visualization) is a nonspeech sound tool that is used to present graphical data. For example, sound graphs present line graphs in sound with time mapped onto the x-axis and pitch to the y-axis. Learners hear the shape of a graph in terms of rising and falling notes that are played over time, which then allow them to distinguish between types of graphs and identify extremum points. Haptic tools provide students with an opportunity to feel virtual objects through mechanical actuators. They are more dynamic and appear three dimensional than are graphs drawn on an embossed paper.[12] For example, Brewster (2002) developed a device called PHANToM that enables students to "feel textures and shapes of virtual objects, modulate and deform objects with a very high degree of realism ... [and] feel graphs and tables as if they were really present in front of them" (Brewster, 2002, p. 614). In Brewster's (2002) two-part experimental study, the blind subjects performed significantly well in all aspects that assessed workload, completion time, and number of correct answers. He found that combining speech with sound graphs and using haptic line graphs were effective alternative methods for presenting graphs and tables to blind learners.

[11]Giroux was completely blind at age 11.

[12]While recent haptic interface technologies in virtual reality provide externally induced compensatory strategies for blind learners due to loss in visual ability, they are also meant for them to develop useful and detailed spatial cognitive maps in long-term memory that will help improve their mobility and orientation skills (cf. Lahav, 2006).

9 Overview of Chapter 8

In Chapter 8, we discuss various components of the progressive model shown in Fig. 2.18. We begin with some provisional thoughts on visualization for mathematics education. We also assert a stronger symbolic view of visual actions in mathematical cognitive activity. Then four sections are used to discuss the following aspects of the progressive model: the intersecting circles; context of visualizing in the referential phase; the ascension process within the concentric circles, and; the nature of mathematical content and visualization in the operative phase. The closing section clarifies the educational significance of "pleasure" in mathematical knowing through visualizing.

Chapter 8
Instructional Implications: Toward Visual Thinking in Mathematics

> *Simply put, the visual teaches us to think with the body.*
> (Sherwin, Feigenson, & Spiesel, 2007, p. 147)
>
> *The advantage of verbal formulations is that they can conform to semantic or logical rules, but a preoccupation with syntactical features of representation means that we still lack an understanding of how visual thinking works in conjunction with language-based reasoning.*
> (Gooding, 2006, p. 41)
>
> *I see the growth of mathematical knowledge as a process in which an unrigorous reasoning-practice, a scattered set of beliefs about manipulations of physical objects, gives rise to a succession of multi-faceted practices through rational transitions, leading ultimately to the mathematics of today.*
> (Kitcher, 1983, p. 226)

In this closing chapter, I should point out that we certainly have come a long way since the time Klotz (1991) asserted that "visualization has a more important role to play in mathematics education" and the need to be "willing to ask hard questions about this approach" (p. 103). Research data from a variety of sources at least in the last 20 years or so should help us deal with a nagging issue on visualization that Taussig (2009) has succinctly captured for us in the following way: "But to get to basics, why draw?[1]" (p. 265).

---

[1] Taussig's (2009) reflection about the picture/text hierarchy below is worth noting in light of our goal in this book of suturing the split:

> [T]he hiatus or no-man's land between picture and text in the anthropological tradition raises a further question as to the general devaluation of drawing in relation to reading and writing in modern Western cultures and maybe in many other cultures as well. We do everything to get children to read, write, and speak well. *But why not draw too?* Shortly after I wrote this I drove to the supermarket close to where I live in upstate New York past a sign on the road. It read: "Summer Reading Camp."
> (Taussig, 2009, p. 268; italics added for emphasis)

F.D. Rivera, *Toward a Visually-Oriented School Mathematics Curriculum*, Mathematics Education Library 49, DOI 10.1007/978-94-007-0014-7_8,
© Springer Science+Business Media B.V. 2011

The perspective, *toward visual thinking in mathematics,* moves beyond the perennial view that associates visuals and visual action with merely using and conjuring "pretty pictures" that suffer a mild form of "illustration fallacy" (Tucker, 2006, p. 117).[2] It embraces Arnheim's (1971) "seeing as seeing in relation" and puts to work the "inseparable suturing" (Mitchell, 1994, p. 95) of the visual and the alphanumeric in the construction of mathematical knowledge. Our response in this book to the relevant hard questions, especially, Presmeg's (2006) 13th Big Question – "*What is the structure and what are the components of an overarching theory of visualization for mathematics education?*" – involves the notion of progressive modeling. In the Introduction, I referenced my own work (Rivera, 2007a) in a precalculus class to demonstrate my initial orientation toward, and grounding in, this perspective. Also, in the Introduction, I introduced the term *progression* in the context of more recent work by Gravemeijer, Cobb, and colleagues, which basically sees symbolizing, notating, and tool use as processes that grow in abstraction and rigor over classroom time. In Chapter 2, I introduced the term *modeling* in Magnani's sense, which involves developing and manipulating different types of representations as a result of thinking through doing. Consistent with Kitcher's (1983) perspective in the opening epigraph, progressive modeling involves rational transitions along several dimensions. In Chapter 3, we spoke about visual kinds from imaginal to formational and then to transformational. In Chapter 4, we discussed transitions in modes of signification from iconic to indexical to symbolic. In Chapter 5, we compared early with mature abductions in the context of progressive abduction. In Chapter 6, we discussed progressive diagrammatization.

Certainly, the term *rational* in Kitcher's rational transitions needs to be clarified as well in the progressive modeling view. In Chapter 2, I emphasized the distributed nature of cognition, a conceptual molting out of mind–brain perspectives in favor of systemic embodiment, which involves orchestral actions involving the whole body, the mind, and the relevant external representations. In Chapter 3, we sourced the nature of individual mathematical cognition in an "always-already" lifeworld-dependent context, where the developing rational actions of each learner are seen as confronting and at the same time guided by the rational actions of the mathematics community in which he or she participates. This distributed phenomenon would then enable transitions in sign use, which has been significantly pursued in some detail in Chapter 4. Chapter 5 provided an exemplar of this distributed transition in the context of pattern generalization, from the visual to the numeric and then to visuoalphanumeric as a consequence of individual and social mathematical cognitive activity. In Chapter 6, we pointed out the necessity of making transitions in diagram use from the representational to the diagrammatic. In Chapter 7, we

[2] I borrowed these terms from Tucker (2006) who used them in the context of the history of visual representations in science, in particular, nineteenth century scientific photographs. The use of *pretty pictures* conveys the need to popularize scientific discourses and "the packaging of scientific concepts for mass audiences" (p. 117). By *illustration fallacy*, it refers to "the mistake of assuming that illustrations produced 'outside' of professional science lack scientific significance or value" (p. 117).

dealt with this issue of distributed contexts more forcefully by articulating macro conditions such as sociocultural influences in the practice of mathematics and the central neurophysiological functioning of multimodal processes in the extreme case of learning (mathematics) among blind subjects.

In our progressive modeling structure, we have also pointed out the changing intra-semiotic roles and content of visualization, from the visuals of imaginals to the visual representations of abstract forms, and then again. Such changes are a necessary component of the progressive model since "mathematical visual images" are "concerned with particular forms of experiential meaning" (O'Halloran, 2005, p. 142). The book is replete with examples of such changes that we have drawn from several sources, in particular, in the number sense and algebra strands of the school mathematics curriculum that for so long has been associated with an almost exclusive preoccupation with alphanumeric symbolizing.[3].

Visuals and visual actions in this book are not (merely) perceived in terms of providing fun-filled activity in the mathematics classrooms. Also, they are not simply situated as being prior to abstraction and abstract actions as the visual/verbal binary would perhaps have us think. They are, in fact, central to school mathematical practices and should, thus, be seen as a conceptual molting from the hegemony of the visual/verbal oppositional view. Of course, we have been explicit in saying that visualization alone does not offer a thaumaturgical route to better mathematics learning. Nevertheless, in this book, we attempted to provide a sustained traveling model in terms of ways in which visualization in mathematical cognitive activity could be made to mean beyond typical, short-term touristic interventions and practices. Just as diagrams in ancient Greece were the metonym of their mathematics, visuoalphanumeric representations could be ours in the history of the present.

In fact, the common thread that binds all visual studies in school mathematics education, from the progressive modeling projects of Gravemeijer and colleagues to instrumental genesis in technologies in mathematics learning and to the diagrammatic model of Dörfler, is the experience of reasoning and understanding that comes with visual image manipulation, discernment, and construction. Such visuals are treated as instantiations of the concepts and processes that should provide learners with an opportunity to develop mathematical habits relevant to developing

---

[3]I have intentionally left out research results from studies involving statistics, geometry, and technology in mathematics learning. But there are interesting overlaps in terms of findings and implications. Statistics and geometry in the school mathematics curriculum fundamentally rely on visual representations, so the visual status in these content strands is not as problematic as the case with the algebra and number sense strands. Certainly, there is much instructional and psychological knowledge that could be gained from research done in these two areas, so I refer readers to exemplary syntheses on geometry understanding (which includes spatial understanding) and statistics learning (e.g., Battista, 2007; Clements & Battista, 1992; Gattis, 2001; Owens & Outhred, 2006; Shaughnessy, 1992, 2007). Concerning technology in mathematics learning, I refer readers to several syntheses of research that also discuss the central role of visual representations in mathematical knowledge acquisition (Ferrara, Pratt, & Robutti, 2006; Laborde, Kynigos, Hollebrands, & Strässer, 2006; Kaput, 1992; Mariotti, 2002; Yerushalmy & Chazan, 2002; Zbiek, Heid, Bume, & Dick, 2007).

logical inferences. Thus, visuals play a central role in mathematical understanding and practices. Perhaps Raju (2001) was correct after all. In Chapter 7, we referred to his historical interpretation of an emerging industrial civilization, characterized by growth and development in machine technology and application of mechanical rules, which may have inspired mathematicians like Hilbert to disavow empirical content in favor of more synthetic and abstract or verbal approaches. But Netz's (1998) cognitive history of ancient Greek mathematics would have us see visuals in a different light, a point that is central to the progressive modeling perspective we have explored in this book. According to Netz,

> In many of the manuscript traditions of Euclid, the definitions at the start of the book are accompanied by diagrams. The text is "a point is that which has no part," and next to this Greek sentence there is a point drawn in the manuscript, and so on with all the rest of the definitions. You will never even get a hint of this in modern editions [of the same text]. This is the axiomatic shrine, the holiest of the holiest. Here, we the modern expect the Word to be enshrined alone; no other presence should defile the purity of the abstract verbal formulation. But when we enter the temple we see there enshrined the picture, the diagram. This, and not the Word, is the central object of Greek mathematics.
>
> (Netz, 1998, p. 38)

Each section below addresses various aspects of the progressive model shown in Fig. 2.18. Section 1 summarizes my thoughts on the role of visualization in mathematics education, which I organize around three systemic standpoints that establish it as a conceptual field. In Section 2, we discuss the importance of the intersecting circles shown in the progressive model. Here we underscore the influence of the social context in individual visual actions. In Section 3, we assert a stronger perspective about visualizing in mathematical cognitive activity, which involves situating it in the third mode of signification, that is, in the symbolic order in which acts of seeing and interpreting necessarily reproduce institutional rules and codes. Section 4 clarifies the context of visualizing in the referential phase, which involves understanding the significance of visual actions such as "visual presence" and "seeing-in." In Section 5, we explain how the ascension process within the concentric circles conveys different aspects of progression, from increasing conceptualizing actions to gradual abstracting and generalizing to conceptual restructuring. Section 6 completes the discussion surrounding the progressive model by clarifying the nature of mathematical content and the role of visualization in the operative phase. Section 7 closes the chapter by clarifying the educational significance of "pleasure" in mathematical knowing through visualizing.

1 Some Final Thoughts on Visualization for Mathematics Education

Based on findings drawn from my own work, I have found it useful to think of the triad of experience, learning, and culture as forming, to borrow a term from Vergnaud (2009), a conceptual field that engenders the development of, and growth in, visual thinking in mathematics among students. The meanings they associate with visual cognitive action in mathematical activity ought to be viewed not from "one situation only but from a variety of situations and that, reciprocally, situation[s] cannot be analyzed with one concept alone, but rather with several concepts, forming systems" (Vergnaud, 2009, p. 86). Such a view is compatible with the distributed context of visualization that I have talked about in several places in this book. The triad provides "situations, schemes, and symbolic tools of representations" (Vergnaud, 2009, p. 87) in varying degrees (contrasting ones, especially) depending on interactions that take place in a lifeworld-dependent context. In the following paragraphs, we address issues in each element of the triad.

From a practical standpoint, visualizing in mathematics education provides students with "an immediacy of access" (Sherwin et al., 2007, p. 147) that is not clearly visible with alphanumeric symbols alone. For example, the paper-and-scissors activity in Fig. 2.1, which involves factoring a difference of two squared polynomial terms, enabled my Algebra 1 students to construct a vivid pictorial understanding of why the corresponding algebraic factored form took shape that way and not some other way. Consequently, the visual representation influenced the manner in which they developed their mathematical understanding of the represented knowledge. This particular illustration, in fact, echoes a convergent perspective about the influential role of external representations in individuals' thoughts relative to some represented knowledge (see, e.g., Karmiloff-Smith, 1992; Lave, 1988; Ng & Lee, 2009; Luria, 1976; Vygotsky, 1962, 1978).

Another practical engagement with the visual is the convergent view drawn from various empirical studies dealing with everyday, scientific, and mathematical objects, concepts, and processes that shows individuals' capability in keeping their visual experiences longer in memory and recalling them easily when prompted. For example, my Algebra 1 students found the visual process in solving a quadratic equation (Fig. 3.12b) much easier to recall than the quadratic formula. Another example is drawn from my Grade 2 class whereby the students easily tapped onto their visual knowledge when they needed to add and subtract whole numbers of up to three digits with and without regrouping in alphanumeric form (Fig. 1.1 and 3.1).

A third practical consideration deals with solid empirical evidence that shows significant relationships between visual perceptual skills and mathematical ability, between visual perception and mathematical achievement, and between visual representations (i.e., the use of either pictures or diagrams) and success in mathematical problem solving. For example, the empirical study conducted with 171 second, fourth, and sixth grade students by Kulp, Earley, Mitchell, Timmerman, Frasco, and Geiger (2004) shows in clear terms the significance of visual skills in mathematics learning, as follows:

> [P]oor visual perceptual ability is significantly related to poor achievement in mathematics, even when controlling for verbal cognitive ability. Therefore, visual perceptual ability, and particularly visual memory, should be considered to be amongst the skills that are significantly related to mathematics achievement.
>
> (Kulp et al., 2004, p. 50)

From both instrumental and psychological standpoints, visualizing in mathematics education occurs in the context of the epistemological model shown in Fig. 2.18, which conveys the dynamic and embedded nature of growth in mathematical knowledge resulting from purposeful progression. Under normal circumstances, various signs that are used to express mathematical objects and their relationships in whatever mode of signification (icon, index, symbol) progress from the referential to the operative, from situated to institutional, from concrete instances to a collection to hypostatic abstractions, from the imaginal to the structured form of visuoalphanumeric, and so on. With each transition, we associate inter- and intra-semiotic changes in inferential habits, mathematical explanations, and symbol notation and use that occur as a result of a developing proficiency and competence with the relevant visual tools that are used to reify mathematical objects, including varying levels of mediation from emerging, shared, and institutional mathematical practices that endow the tools with meaning and purpose.

It is also worth noting that mathematical objects, in particular, alphanumeric expressions and diagrams, emerge but only in the context of an interpretive structure. Hence, one source of student difficulty is when symbols are pursued as singular entities apart from some structure that give them specificity and relevance in the first place. This observation was noted in Chapter 5 in connection with progressive abductive reasoning in pattern generalization activity. We underscore the significance of an algebraically useful generalization in Fig. 5.7 resulting from meaningful actions of abduction and induction of patterns. Another source of student difficulty is when a target structure is initially approached either as an instance of the operative domain or as situated in the abstract phase of the referential domain (Fig. 2.8). A more fundamental source of student difficulty is when structures and their symbols are not situated within a reasonable sense-making context that allows them to be interpretively reified in a meaningful manner.

From both anthropological and cultural semiotic standpoints, visualizing in mathematics education is not a simple matter of categorizing it as simply one among many processes of knowledge acquisition. It should not be perceived merely as a means for concretizing objects, concepts, and processes. Prevailing views among teachers indicate this to be the case and nothing else (e.g., Cai & Moyer, 2008; Moyer, 2001). The analysis offered by Tiragallo (2007) in relation to the weaving practices of the women in Sardinia captures the purpose in which I implemented and supported various visual thinking approaches in my own studies with my middle school and Grade 2 students. Weaving processes seem to share many similarities with mathematical practices. The mathematical knowledge found in textbooks is like carpets, blankets, linen chest covers, and bedsheets that are considered finished products. In creating the latter, weavers, in fact, require having the expertise to coordinate various elements in an assemblage that consists of yarns, pegs, designs, and braiding strategies. Tiragallo points out that the "visual experience" of weaving "intertwines with that of the other senses in a multisensorial environment" (Grasseni quoted in Tiragallo, 2007, p. 207). She writes:

> The technical and perceptive skills ... come together with the corporeal capabilities to bring together the rhythms and the recollections ... with the resources of [the weaver's] particular environment and relations with the members of her community and other expert communities.
>
> (Tiragallo, 2007, p. 207)

Based on my classroom experiences with different age- and grade-level groups, a similar phenomenon occurs whenever we ask students to visualize mathematical objects, concepts, and processes. Especially in formational and transformational contexts, their structured images and models are not mere isolated entities but exist within a larger process in which seeing and acting on them contribute to gaining access to the intended knowledge in a lifeworld-dependent activity. Thus, effective and meaningful visual thinking in mathematics could be aptly described in the same manner that Tiragallo theorized expert weavers' visual skills. Tiragallo's further thoughts below about weaving action capture an important aspect of visualizing about mathematical objects, concepts, and processes. They should remind us of the many arguments in various places in the book concerning the distributed, multimodal, and material dimensions of visual action in mathematical activity.

> A weaver's visual skill does not appear as separated from the other sensorial skills that go into weaving. ... [The weaver] ... operat[es] in a visual world full of objects and surfaces, which are for her the subject of her technical actions. In such a context, the perceptions of her eyes are added to a complex of other sensorial stimuli ..., which in turn combine with the sensorial memory and the whole of spatial references belonging to the visual world. It is a world in which the things that make it up have a meaning for the person acting in it. [The weaver's] visual skill does not live in the weaving as a skill *per se*: all the spaces and objects involved in her actions ... are elements in a corpus of experiences and they live in her, taken from the immediate visual perception, which, if anything, confirms this wealth of past experience.
>
> (Tiragallo, 2007, pp. 206–207)

Consequently, the mathematical knowledge that students obtain, like expert weavers' knowledge, is not found in "the project or the relationship between the structure of the mind and that of the world," but "is immanent in the life of those who know and develops in the context of practices that are established" (Tiragallo, 2007, p. 209).

2 Signifying Intersecting Circles

Considering the openly sociocultural orientation of the teaching experiments on visualization that I developed and implemented in my own studies, I underscore the culturally constituted nature – in O'Halloran's (2005) words, the "interpersonally orientating function" – of visual construction in mathematics that engenders particular ways of representing, interpreting, and apprehending images. The intersecting circles in Fig. 2.18 symbolize the "always-already" complex relationship between individuals and their communities in various stages of progression. "Perceiving and imaging," Sherwin et al. (2007) point out, and rightly so, "are not merely processes of identification brought about by looking and listening but active performances in which specific intentions, purposes, and actions need to be fulfilled" (p. 156). The intersection conveys the ethnographic fact that while some students find it easy to perform visual thinking in mathematics, for many others it is an acquired skill resulting from the appropriation and internalization of various attentional and intentional resources. In this book, culturally constituted visual representations in school mathematics are refinements that are drawn from personally constructed imaginals, various visual-images-in-the-wild, and previously acquired formal visual representations.

For example, when my middle school and Grade 2 students used learning tools such as graphing technologies and manipulatives in mathematics learning, they interpretively reproduced valued structures and "cultural protocols" (Deger, 2007) of mathematical knowledge. In fact, one indication of competence in the use of such tools is when students are able to grasp the intended mathematical knowledge (short-term effect) and institutional ways of seeing, knowing, and reasoning (long-term effect). Another interesting example that highlights the importance of mathematical ways of visualizing is drawn from a recent study by Mulligan, Prescott, and Mitchelmore (2003) with 109 Year 1 Australian students. The children were briefly

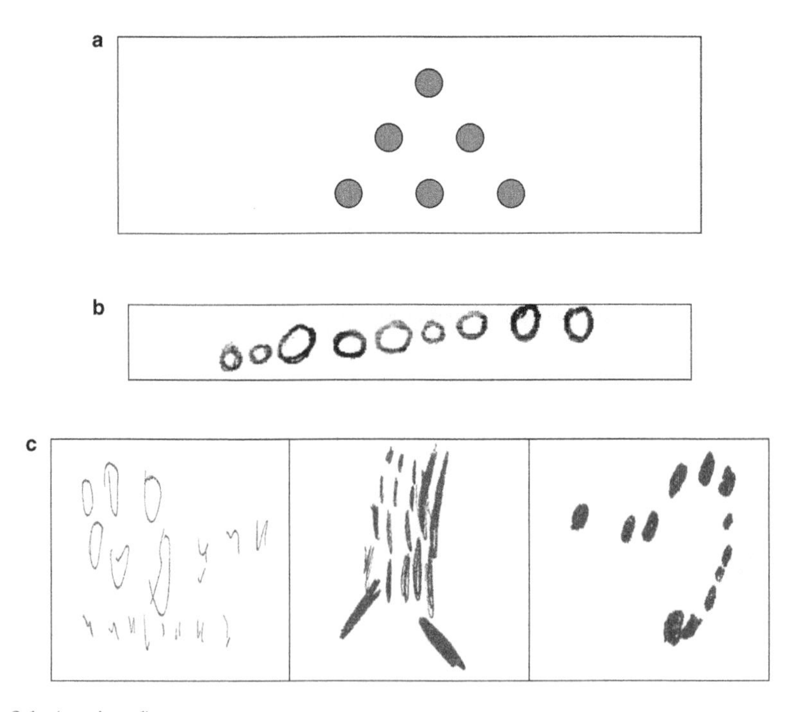

Fig. 8.1 (continued)

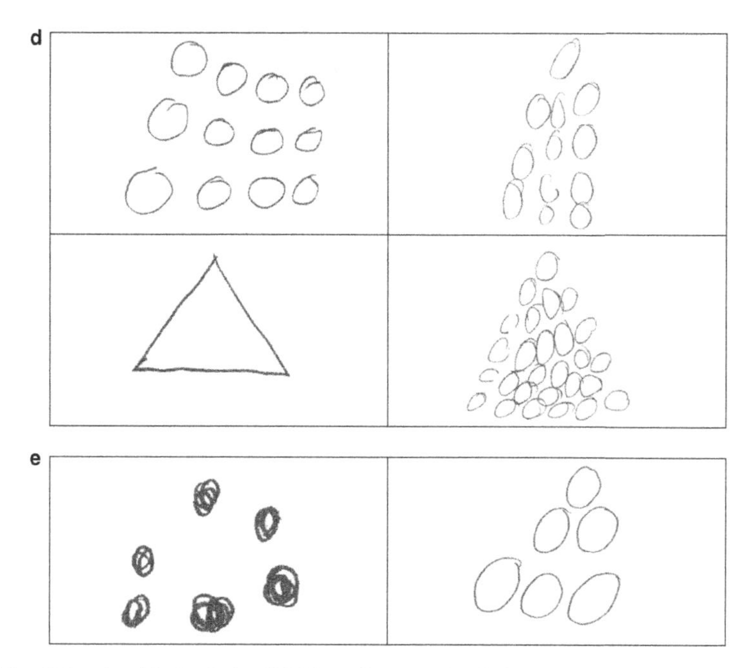

Fig. 8.1 **a** Triangle with six circles (Mulligan, Prescott, & Mitchelmore, 2003, p. 24). **b** An example of a no structure response from a Year 1 student (Mulligan et al., 2003, p. 25). **c** Three examples of an emergent structure response from Year 1 students (Mulligan et al., 2003, p. 25). **d** Four examples of a partial structure response from Year 1 students (Mulligan et al., 2003, pp. 25–26). **e** Two examples of a (full) structure response from Year 1 students (Mulligan et al., 2003, pp. 25–26)

shown a triangular pattern consisting of six circles (Fig. 8.1a). When then asked to draw what they saw on a separate sheet of paper, only 20% produced the same triangular pattern. Figure. 8.1b–e illustrates four categories of responses on the basis of their level of visual experiences, namely: no structure; emergent; partial; and full structure. The classification was justified on the basis of what the teachers saw as important aspects of the shown figure (shape, arrangement, quantity) that the students should have demonstrated in their pictorial responses. In this book, what Mulligan et al. (2003) interpreted as "no structure" could be imaginals for the students since they saw circles to begin with and nothing else. They exhibited mere object or sense (versus cognitive) perception (cf. Chapter 5). Further, the categories of emergent and partial structures fall within visual-images-in-the-wild, while full structure conveys an interpreted visual representation (in Goldin's sense). Hence, from a sociocultural standpoint, the skill of visualizing mathematical objects, concepts, and processes is not a simple matter of everyday visualizing as it is also about constructing and justifying structural features, properties, attributes, and relationships that are mathematically viable and valid from an institutional standpoint.

3 Past the Iconic and the Indexical: The Symbolic Mediated Nature of Visual Thinking in Mathematics (a Stronger Version)

Everything commences in a lifeworld-dependent context, which means that mental and physical images are constructed as signs in cognitive activity and, thus, are interpreted phenomena. Further, when individual learners employ visual thinking in mathematics in the lifeworld, the images they construct and manipulate work within rules in cognitive activity that are needed for those images to make sense. In other words, mathematical ways of seeing involve skilled and learned ways of seeing in a particular way.

The foregoing view shares the perspectives of Miller (1990) and Goodman (1976) about the cognitive subtleties of picture depiction.[4] *Contra* the view that pictures resemble the objects they depict, Miller and Goodman foreground mediation by rules, conventions, symbol systems, functions, and intentions in picture depiction processes. In mathematics, this means that students can only begin to recognize and manipulate images and objects once they understand the intent of/and the underlying symbolic codes that shape those images and objects. We should add to the symbol-mediated view the significance of tactile and gestural action, including individual and social input, in gaining a better understanding of the nature and content of such images and objects. For example, in my Algebra 1 class, the students' fascination with, and aesthetic appreciation of, the quadratic process in Fig. 3.12a, b emerged when they visually grasped the necessary symbolic relationships that came with manipulating the relevant objects.

4 Visual Presence and Seeing in the Referential Phase

Except in Chapter 7 where we explicitly dealt with a few issues concerning multimodality mathematics learning among blind individuals, one aspect that I did not fully pursue in this book involves the "absent presence" (following Gergen,[5] 2002) of gesturing, touching, and hearing that accompany visual action. That is, together

---

[4]Gregory (2005) also surfaces the continuous relationship between image and knowledge and construction (i.e., what "it" is and what "it" is for). He writes:

> We see a table as something to support things, and as made of hard scratchable inflammable wood we have learned about from years of interactive experiments. A *picture* of a table calls up this object-knowledge so it looks almost real – yet pictures are very odd. ... When we look at a familiar object we interact with it by appreciating its potential uses or functions. Its picture calls up the knowledge of functions; but at the same time we are warned, by noting that it is merely a picture, not to expect anything materially useful from it.
>
> (Gregory, 2005, p. 120)

[5]I am using the term here in a metaphorical manner. But as an aside I refer readers to Gergen's (2002) very interesting article in which he explored the phenomenon of absent presence – "present but simultaneously rendered absent ... erased by an absent presence" (p. 227) – resulting from

with visual thinking, the tactile, gestural, and auditory contribute in ways that provide students with the experience of grasping, controlling, and conceptualizing the relevant structure of a mathematical object, concept, or process (cf. Radford, 2009; Roth & Thom, 2009). "The golden rule of didactics," Komenskii (1955) reminds us, involves

> everything that can be perceived by the senses, namely: the visible – for perception by vision, the audible – for hearing, odors – for smell, that which can be tasted – for taste, and that which can be touched – for touch. If any objects can be perceived by several senses at once, let them be immediately sensed in different ways.
>
> (pp. 302–303; quoted in Leushina, 1991, pp. 163–164)

For example, in Fig. 5.5, Dexter (Grade 2, 7 years old) used his right hand to describe and ostensively indicate the structural parts of his emergent generalization, which illustrates representational (versus everyday) gesture (cf. Goldin-Meadow, Alibali, & Church, 1993). In the case of blind learners, for example, the use of multiple sensory modalities assists them in developing visuospatial ability and representations. In the case of students with learning disabilities, the auditory appears to play an important role in their mathematical learning. In the case of sighted learners, when they use their hands to either draw a figure or a diagram or play with algeblocks and other physical manipulatives, the interanimated experience allows them to develop an appropriate cognitive perception of the relevant conceptual elements at play.

The referential phase in Fig. 2.18 applies to mathematical objects, concepts, and processes whose representations are either concrete (e.g., fractions) or abstract (e.g., irrational numbers). Visual-relevant actions in this phase help individual learners develop a systematic way of conceptualizing or hypostasizing a target object, concept, or process. In most cases such a process begins rather idiosyncratically (i.e., imaginals) that then undergoes classroom reconstruction (in formational or transformational contexts) reflective of shared cultural views and practices. With more experience and learning, the nonnatural process gives the appearance of everydayness in which all visual images transition from personal to classroom to cultural.

In the referential phase, students are initially provided with an empirical experience of either a general nature or some particular occurrences relative to an intended mathematical knowledge. This learning opportunity, interanimated in intent, allows them to observe and establish conceptual interpretations. By either working through the details in the case of a general phenomenon (i.e., "ascending to the concrete" in Davydov's sense; Chapter 3) or working with particular cases or situations in a visual format leading to a target generalization of a mathematical object, concept, or process, they begin to develop a sense of the representational scope of the visual image or model despite limitations in representational power.

In other words, the image or model is the metonym of the target knowledge. That is, while the image or model conveys a particular case or a domain set, it

---

our "diverted or divided consciousness invited by communication technology" (p. 227) and cyber-driven culture.

exemplifies both the case or set itself and the relevant general phenomenon. Having representational scope of the model is a necessary action in transitioning to the operative phase. Further, underlying representational scope is a phenomenon that I refer to as *visual presence*, a term inspired from Noë's (2004) notion of perceptual presence. Noë underscores a most subtle human perceptual phenomenon called presence that predisposes us to sense the totality of features of an object. For example, when we see a face of a rectangular polyhedron, we actually have a sense of the whole polyhedron despite not having access to the partially hidden or occluded parts (which, as I have observed with my Grade 2 students, is a developing visual skill especially when the polyhedron is drawn on paper with the missing parts). In my Algebra 1 class, the students' success, say, in factoring a difference of two squares in Fig. 2.1 or in establishing an algebraic generalization for the pattern in Fig. 2.6a could be explained in part by the visual presence they have about, and impose on, those objects.

Related to the notion of visual presence is Veldeman's (2008) phenomenological concept of *seeing-in.* He notes that when we see objects in pictures, we are "simultaneously aware of the picture surface and the depicted subject" (p. 493). By picture surface, it means either the design features that make an image possible or the design features "in virtue of which it represents" (p. 493). In the case of mathematical objects, concepts, and processes, when students *see in* a particular image or model a depiction of a more general or abstract phenomenon, it means they are consciously attending to particular features of the image or model in terms of how they convey an interpretive general or abstract relationship or structure.

Visual thinking also applies to situations with an emergent structure. For example, my Algebra 1 students made sense of simplifying rational algebraic expressions only after having visualized the process in the context of an empty fraction strip (Figs. 6.11a, b and 6.12), which helped them develop a structure in thinking about the different operations. The empty strip as conveying the unit in any rational expression exercise provided a powerful visual explanation that allowed them to conceptually organize and think about an effective way to model the operations. Hence, visual representations at the referential phase are not necessarily derived from observations drawn from an external context. They could also be emergent depending on the nature of the target regularity or structure, as "cognitive resources for interpretation and modeling" with an epistemic function "that could provide pictorial evidence" (Gooding, 2006, p. 60) leading to a more general and abstract phenomena.

5 A Circling Ascent to the Operative

There are at least three interconnected interpretations of the concentric circles in Fig. 2.18. *First*, the concentric circles convey *increasing conceptualizing actions* that take place in progressive modeling, from personal to classroom to cultural *and* then again. With more experience and learning, an interpreted structure emerges leading to formal mathematical knowledge. The circles also indirectly assert the necessity of meaningful repetition. Here I use the term *repetition* in the following context described by Bier in relation to her account of the material meaning of algorithm in the history of geometry of ornament in Islamic art:

> Geometry [in ornament making] is not entirely present at the outset, but rather it becomes emergent through process. ... (W)hen an artist employs a particular technology to repeat a specific design, no matter how complex the design unit that is to be repeated, it is the process of repetition that carries the artisan from the initial step to the completion of a pattern. Through the process of repetition, geometry emerges with the relationship of one shape to another. The process is not necessarily deliberative (although it may be), but rather it relies upon the interaction of a technology ... and a medium.
>
> (Bier, 2008, p. 492)

For Bier, Islamic artisans' theoretical knowledge of patterns emerged as a consequence of an "empirical understanding of the nature of pattern making" (p. 498), which she considers to be a precursor to our current mathematical understanding of algorithms. Algorithms in Islamic art, Bier (2008) notes, have been drawn from artists' analysis of "designs [that were] repeated to form patterns" (p. 508), a practical motive that differs from the nonreferential context in which algorithms in school mathematics are oftentimes conveyed to students. Her claim reminds us of Sherry's (2009) claim about treating diagrams as instances of mathematical concepts. That is, the more we learn about the diagrams we manipulate, the more we learn processes relevant to making inferences and rules that are central to mathematical thinking and work. Hence, in light of both views of Sherry and Bier, the concentric circles in Fig. 2.18 represent increasing theoretical/mathematical manipulations and processes with the relevant visual tools. Further, the growing circles symbolize a conceptual progression from the referential to formal thought.

Second, the concentric circles can be seen as modeling the process of *gradual abstracting or generalizing*. Cognitive mathematical activity is seen in terms of a sequence of (hypostasized) abstractions or generalizations that occurs over the course of acquiring mathematical knowledge about an object, a concept, or a process. We point out that each circle conveys some conceptual content, which means to say that the relevant concepts are (hypostasized) abstractions as well. For example, when my Algebra 1 students transitioned from the iconic to the symbolic phase in solving a quadratic equation (Fig. 3.12b), the iconic content at the referential phase already conveys an abstract conceptualization of the relevant process. Different types of examples that exemplify complexity in variable assignment (e.g., the leading coefficient is not equal to 1; fractional coefficients) signal the necessity of growth in the circling process up to the point when the quadratic formula emerges at the operative phase that can handle all cases. Thus, among my Algebra 1 students, their mathematical understanding of the process of solving a quadratic equation could be described in terms of the activity of building the operative through a sequence of abstracting processes (cf. Brook, 1997, p. 80).

Third, the concentric circles symbolize *a series of conceptual restructuring* (Stern & Mevarech, 1996) that occurs among students' mathematical thinking as they move from the referential to the operative phase. The restructuring involves either realigning new knowledge to fit into the old or "acquiring a new ability" that is "able to do something that could not be done before" (i.e., an effect of instrumentalization in instrumental genetic terms; Brook, 1997, p. 83).

6 The Operative Content Phase

There are at least three ways to describe the operative phase.

First, while the operative phase indicates cognitive activity that focuses on operative content, it could also signify finality or closure in a gradual abstracting activity.[6]

Second, while the operative phase refers to mathematical content in which the attributes and properties are either generalizations or (hypostasized) abstractions drawn from particular or general instances, it can also be perceived as having, in Tiragallo's (2007) words, the "trace of expert presence" (p. 215). Generalizations involve interpretively establishing commonalities from known cases that are then applied to the projected instances corresponding to the unknown cases. (Hypostasized) Abstractions involve further generalizations that target essential characteristics or qualities independent of any particular case.

Third, the operative phase does not imply the absence of visual thinking. It could mean either seeing the visual in the symbolic or engaging in mathematical ways of seeing that goes beyond everyday ways of seeing.

Also, the operative phase could refer to any of the following situations in operative content whose components may or may not be grounded in concrete contexts:

- Context-derived: A generalization or (hypostasized) abstraction that represents the totality or essence of its original source data through processes of leaving out or setting apart and summarizing.
- Context-independent: A generalization or abstraction that has no association with any concrete or physical source data, but is constructed in some other sign-mediated cognitive activity.
- Meta-context: A generalization or abstraction in context through processes of comparing and/or extending particular context.

In all three situations above, learners work with rules and prototypes and exemplars. They use an operative content either to deepen their understanding of the referential content (in the case of context-derived) or as a means to explore other content.

What role does visualizing accomplish in the operative phase? It can support the underlying general or abstract rule or process that is central to the relevant operative content. For example, the account of visual templates that my Algebra 1 class manifested in patterning activity in Chapter 5 summarizes the content of my

[6]The terms *generalization, abstraction,* and *gradual abstracting* are conceptually related to the three types of abstracting activity offered by Brook (1997), that is, *abstracting out, abstracting away,* and *building abstraction,* respectively.

students' visuoalphanumeric experiences with patterns. In the context of free pattern construction activity, in particular, they articulated various visual-driven processes relevant to pattern generalization (multiplicative thinking, grouping via the concept of a unit, etc.) that mattered significantly beyond their empirical experiences with particular examples.

Visualizing can also support understanding when students make vertical transitions from one operative content level to the next. For example, my Algebra 1 students' visual experiences with everyday fractions (as parts of wholes) and rules using fraction strips gave them an operative context that allowed them to make sense of rational algebraic expressions that are at a higher level of abstraction. Hence, when they simplified rational expressions, the manner in which they made sense of the relevant rules was based on their operative understanding of rules governing rational numbers (Figs. 6.11a–6.13). However, vertical transitions are sometimes difficult in cases when there is a cut between two content levels. A classic example is drawn from Filloy and Rojano's (1989) study in which the students' experiences were fraught with difficulties as they were transitioning from the simple case of $Ax + B = C$ to the more complex case $Ax + B = Cx + D$. As I have discussed in Chapter 3, my Algebra 1 students used their combined knowledge of algeblocks and prior knowledge in developing a visuoalphanumeric method that enabled them to transform the latter equation into some equivalent form corresponding to the simpler equation.

7 The Pleasure of Visual Forms in Mathematical Knowing

It is an established empirical fact (drawn from qualitative, quantitative, and, more recently, neuroscientific data) that scientists, mathematicians, some professionals in various disciplines, and different age-level groups of individuals in imaginal, formational, and transformational contexts employ visual thinking in various aspects of their cognitive activity. Apparently, there is *pleasure* that comes with visualizing, a *sensuous semiotic mode of knowing* that today still remains underutilized and underemphasized in most institutional practices that value formal knowledge in alphanumeric form, which is meant to signify resistance to ambiguous meaning. By sensuous semiotic mode of knowing, we mean it in the same sense that Peirce (1949) referred to abduction (or hypothesis or hypothetical inferences) as that which "produces the *sensuous* element of thought," unlike induction, which provides the "*habitual* element," and deduction, "the *volitional* element of thought" (p. 152). Like any construction that generates hypotheses, visualizing images, in particular, forming visual representations, relative to mathematical knowledge construction does carry with it "a peculiar sensation," "emotion," or "complicated feeling . . . of greater intensity" (Peirce, 1949, pp. 151–152).[7]

[7]Radford's (2009) notion of *sensuous cognition* encompasses the emotional dimension that Peirce has noted about the term sensuous relative to abductive action. As an "alternative approach to classical mental [and rational] views of cognition," Radford characterizes sensuous modes of thinking in mathematics in "multimodal material" terms, which involve "a sophisticated coordination"

Peirce's point above concerning the sensuous element of thought reminds me of Wise's (2006) account of Maxwell, famous Scottish mathematician and theoretical physicist, who, in his 1870 address to the mathematicians and physicists of the British Association for the Advancement of Science, shared how his colleagues "felt" about "the results" of their "scientific inquiry" in the following way:

> They learn at what a rate the planets rush through space, and they experience a delightful feeling of exhilaration. They calculate the forces with which the heavenly bodies pull at one another, and they feel their own muscles straining with the effort. To such men momentum, energy, mass are not mere abstract expressions of the results of scientific inquiry. They are words of power, which stir their souls like the memories of childhood.
>
> (Maxwell, 1965, pp. 218, 220; quoted in Wise, 2006, p. 82)

Wise's (2006) view of the "power of embodied mathematics," which fuses intellectual knowledge and its "material and sensual" (p. 81) dimensions together, is one characteristic of this "momentous historical shift toward visualization" (Stafford quoted in Wise, 2006, p. 82).

We provisionally close this book with the significance of "pleasure" relative to visual thinking in mathematics. We articulate three senses of pleasure based on how we perceive it in this book. *First*, pleasure can be seen as a marked indication that one has achieved expert knowledge. Tiragallo describes the women in Sardinia who associate their expert weaving practices with pleasure in the following manner:

> (T)he doing . . . is connected to this pleasure in proceeding. A pleasure that is recognized is a distinctive feature of the art of weaving, one of the signs of the successful and full entry of the subject into that expert community.
>
> (Tiragallo, 2007, p. 210)

In both my Algebra 1 and Grade 2 classes, the students' expert knowledge of mathematical content has come to mean the ability to construct and justify mathematical meaning by comfortably oscillating between individual and social ways of seeing. Further, it is true that the act of interpreting with a visual form "lends it sensuously to concrete instances"; however, it is not, to borrow Davey's (2005) words, "conceived as a solitary monologue on private pleasure but, rather, as an integral part of a shared historical discourse concerning the realization of meaning" (p. 136). Like Tiragallo's (2007) weavers, many of my students who consistently employed visuoalphanumeric representations expertly negotiated their personal meanings with institutional

---

of speech and symbols, gestures, body, and actions performed on cultural tools (p. 111) in order to deepen our understanding of how such "ephemeral symptoms [do not merely] announc(e) the imminent arrival of abstract thinking, but *genuine* constituents of it" (p. 123). A good example of this sensuous cognitive action to pattern generalization is shown in Fig. 5.5. Second grader Dexter explained the structure of his generalization by pointing out the necessary parts using his right hand. In this book, I addressed and emphasized various aspects of what I might classify as visual rational cognitive action in the context of sociocultural learning despite the appearance of similarities in the manner Radford (2009) developed his sensuous cognitive perspective. Both perspectives are complementary, of course. Finally, I note Thagard's (2010) initial exploration of what he calls *embodied abduction* that appears to share many of the same features of sensuous cognition. Embodied abduction involves the generation of explanatory hypotheses by a convolutary conceptual process that combines information drawn from two or more modality sources.

practices. Suffice it to say, those student-generated imaginals and visual-images-in-the-wild that are not effectively brought into a relationship with institutional knowledge in a lifeworld suffer limited generalities and tend to be unproductive. What is desirable, however, involves reconciling subjective and institutional forms of expression through situations that enable all learners

> to bring into language that which is held within an image, not to the end of surpassing the visual but with the aim of enabling to see more of what has yet to be seen.
>
> (Davey, 2005, p. 136)

For example, I am reminded of my Algebra 1 students' experiences in relation to the institutional practice of factoring simple second- and third-degree polynomial expressions, which they consistently and interpretively associated with finding dimensions of rectangles and solids due to their individual visual experiences with algeblocks. Another example is drawn from my Grade 2 class. Figure 3.11a–c illustrates powerful transitions in the students' developing understanding of subtracting with regrouping from the visual to the numeric. Suffice it to say, institutional mathematical knowledge represents valued knowledge that oftentimes appear in "linguistic" or symbolic modes of signification. The pleasure dimension occurs when students are able to link and see the visual in the symbolic.

Second, there is pleasure when progressive schematization is achieved in a conceptually smooth manner. Gooding's (2006) interpretive account of the many important scientific discoveries of Faraday, English chemist and physicist who discovered benzene and contributed significantly to the study of electromagnetism, illustrates the central and productive role of visual reasoning and thinking in Faraday that led him to his theories in linguistic form (see also Miller, 1997 and Nersessian, 1994). Gooding's point below nicely summarizes the encompassing nature of visual representations in relation to discovery and rational thought:

> (W)hile successful science does require a stable linguistic formulation, creative research cannot be conducted solely with well-formed linguistic representations. There are non-visual ways of forging an isomorphism of words, images or symbols to what they denote, but images are particularly conducive to the essential, dialectical movement between the creative stages of discovery and the deliberative, rational stages in which rules and evaluative criteria are introduced to fix meanings and turn images from interpretations into evidence.
>
> (Gooding, 2006, p. 60)

Gooding captures an essential characteristic of progressive schematization in mathematical thinking whereby visual forms (mental images, concrete models) assist in the transition from the creative stage of discovery, which involves interpretations, to the rational stage of rule formation, which involves evidence and explanation. Certainly, there are didactical challenges in cases when mathematical objects, concepts, or processes appear to have no corresponding visual representations, including those mathematical situations that Poincaré, French mathematician, categorized under *sensible intuitions* (i.e., "something else than pure logic" and "not necessarily founded on the evidence of the senses" (Miller, 1997, p. 56)).

Third, pleasure can signify "control, agency, and meaningfulness . . . [and] create deep learning" (Gee, 2005, p. 4). Gee's (2005) reflections on pleasure and learning in video games provide an appropriate closure in this book on visualization in/and mathematics learning. Video games consist of story elements with underlying abstract systems of rules about how objects are to be manipulated, what moves should and should not be performed, and how objects and movements should be interpreted to produce winning strategies. The initial stage in a video game activity involves understanding the story element and appropriately assigning meanings to the objects, movements, and other relevant features. While any video game is certainly constrained by the intent of its designer, the foregoing description shares fundamental similarities with mathematical activity. Gee further points out that the pleasure one gains in playing video games is rooted in the human predisposition for pattern recognition, including the desire

> to solve problems by finding patterns inside a safe world in which there is a clear and comforting underlying order. We see the order (simplicity, pattern) clearly and we safely play among the surprising complexity the game generates always knowing that simplicity and order is there.
>
> (Gee, 2005, p. 15)

Also, the story elements in video games exist not for the mere purpose of providing a context as it is more about making the process of playing more "profoundly meaningful" (Gee, 2005, p. 20) to the gamer. Gee writes:

> Humans find story elements profoundly meaningful and are at loss when they cannot see the world in terms of story elements. We try to interpret everything that happens as if it were part of some story, even if we don't know the whole story. We are not just avid pattern recognizers, we human prefer story-like patterns . . .. When something happens and we can find no cause or explanation, we are at a loss and deeply unsatisfied.
>
> (Gee, 2005, pp. 20–21)

Thus, one derives pleasure in a video game activity when he or she is able to grasp and coordinate both the story elements and the underlying pattern structure. Once the story is understood, the relevant constructed patterns of objects and movements are easily and more meaningfully acquired through an evolving process of interacting with the game. The same type of pleasure is what is desired in the case of visualizing activity in school mathematics.

References

Abe, A. (2003). Abduction and analogy in chance discovery. In Y. Ohsawa & P. McBurney (Eds.), *Chance discovery* (pp. 231–248). New York: Springer.

Ainsworth, S., & Loizou, A. T. (2003). The effects of self-explaining when learning with text or diagrams. *Cognitive Science, 27*, 669–681.

Alcock, L., & Simpson, A. (2004). Convergence of sequences and series: Interactions between visual reasoning and the learners' beliefs about their own role. *Educational Studies in Mathematics, 57*, 1–32.

Alexander, A. (2006). Tragic mathematics: Romantic narratives and the refounding of mathematics in the early nineteenth century. *Isis, 97*, 714–726.

Antonietti, A. (1999). Can students predict when imagery will allow them to discover the problem solution? *European Journal of Cognitive Psychology, 11*(3), 407–428.

Antonietti, A., Cerana, P., & Scafidi, L. (1994). Mental visualization before and after problem presentation: A comparison. *Perceptual and Motor Skills, 78*, 179–189.

Arcavi, A. (1994). Symbol sense: Informal sense-making in formal mathematics. *For the Learning of Mathematics, 14*(3), 24–35.

Arcavi, A. (2003). The role of visual representations in the learning of mathematics. *Educational Studies in Mathematics, 52*, 215–241.

Arnheim, R. (1971). *Visual thinking*. Berkeley, CA: University of California Press.

Arp, R. (2008). *Scenario visualization: An evolutionary account of creative problem solving.* Cambridge, MA: MIT Press.

Artigue, M. (2002). Learning mathematics in a CAS environment: The genesis of a reflection about instrumentation and the dialectics between technical and conceptual work. *International Journal of Computers for Mathematical Learning*, *7*(3), 245–274.

Arzarello, F., Micheletti, C., Olivero, F., & Robutti, O. (1998). A model for analyzing the transition to formal proofs in geometry. In A. Olivier & K. Newstead (Eds.), *Proceedings of the 22nd annual conference of the international group for the psychology of mathematics education* (Vol. 2, pp. 24–31). Stellenbosch, South Africa: University of Stellenbosch.

Atkin, A. (2005). Peirce on the index and indexical reference. *Transactions of the Charles S. Peirce Society, XLI*(I), 161–188.

Bakker, A. (2007). Diagrammatic reasoning and hypostasized abstraction in statistics education. *Semiotica, 164*(1/4), 9–29.

Barmby, P., Harries, T., Higgins, S., & Suggate. J. (2009). The array representation and primary children's understanding and reasoning in multiplication. *Educational Studies in Mathematics, 70*, 217–241.

Bartolini-Bussi, M., & Boni, M. (2003). Instruments for semiotic mediation in primary school classrooms. *For the Learning of Mathematics, 23*(2), 15–22.

F.D. Rivera, *Toward a Visually Oriented School Mathematics Curriculum*, Mathematics Education Library 49, DOI 10.1007/978-94-007-0014-7,

© Springer Science+Business Media B.V. 2011

Barwise, J., & Etchemendy, J. (1991). Visual information and valid reasoning. In W. Zimmerman & S. Cunningham (Eds.), *Visualizing in teaching and learning mathematics* (pp. 9–24). Washington, DC: Mathematical Association of America.

Bastable, V., & Schifter, D. (2008). Classroom stories: Examples of elementary students engaged in early algebra. In J. Kaput, D. Carraher, & M. Blanton (Eds.), *Algebra in the early grades* (pp. 165–184). New York: Erlbaum.

Batt, N. (2007). Diagrammatic thinking in literature and mathematics. *European Journal of English Studies, 11*(3), 241–249.

Battista, M. (2007). The development of geometric and spatial thinking. In F. Lester, Jr. (Ed.), *Second handbook of research on mathematics teaching and learning* (pp. 843–908). Charlotte, NC: National Council of Teachers of Mathematics and Information Age Publishing.

Baudrillard, J. (1981/1972). *For a critique of the political economy of the sign.* St. Louis, MO: Telos Press.

Bautista, E., & Garces, I. J. (2010). *Mathematical excusion: A problem solving primer for trainers and olympiad enthusiasts.* Quezon City, Philippines: C&E Publishing, inc.

Bavelier, D., Dye, M., & Hauser, P. (2006). Do deaf individuals see better?. *Trends in Cognitive Sciences, 10*(11), 512–518.

Beckmann, S. (2008). *Activities manual: Mathematics for elementary teachers* (3rd ed.). Boston, MA: Pearson and Addison Wesley.

Begle, E. (1979). *Critical variables in mathematics education: Findings from a survey of the empirical literature.* Washington, DC: Mathematical Association of America and the National Council of Teachers of Mathematics.

Belting, H. (2008). Perspective: Arab mathematics and renaissance Western art. *European Review, 16*(2), 183–190.

Berkeley, I. (2008). What the <0.70, 1.17, 0.99, 1.07> is a symbol? *Minds and Machines, 18*, 93–105.

Bialystok, E., & Codd, J. (1996). Developing representations of quantity. *Canadian Journal of Behavioural Science, 28*(4), 281–291.

Bier, C. (2000). Choices and constraints: Pattern formation in oriental carpets. *Forma, 15*, 127–132.

Bier, C. (2006). Number, shape, and the nature of space: An inquiry into the meaning of geometry in Islamic art. In J. B. White (Ed.), *How should we talk about religion? Perspectives, contexts, particularities* (pp. 246–278). Notre Dame, IN: University of Notre Dame Press.

Bier, C. (2008). Art and Mithãl: Reading geometry as visual commentary. *Iranian Studies, 41*(4), 491–509.

Bishop, A. (1979). Visualizing and mathematics in a pre-technological culture. *Educational Studies in Mathematics, 10*, 135–146.

Boero, P., Garuti, R., & Mariotti, M. (1996). Some dynamical mental processes underlying producing and proving conjectures. In L. Puig & A. Gutierrez (Eds.), *Proceedings of the 20th annual conference of the international group for the psychology of mathematics education* (Vol. 2, pp. 121–128). Valencia, Spain: University of Valencia.

Booth, R., & Thomas, M. (2000). Visualization in mathematics learning: Arithmetic problem-solving and student difficulties. *Journal of Mathematical Behavior, 18*(2), 169–190.

Brewster, S. (2002). Visualization tools for blind people using multiple modalities. *Disability and Rehabilitation, 24*(11–12), 613–621.

Brook, A. (1997). Approaches to abstraction: A commentary. *International Journal of Educational Research, 27*, 77–88.

Brown, J. R. (1997). Proofs and pictures. *British Journal for the Philosophy of Science, 48*, 161–180.

Brown, J. R. (1999). *Philosophy of mathematics: An introduction to the world of proofs and pictures.* New York: Routledge.

Brown, R. (2002). Forum discussion: How art can help the teaching of mathematics. In C. Bruter (Ed.), *Mathematics and art: Mathematical visualization in art and education* (pp. 155–159). Berlin: Springer.

Brown, J. R. (2005). Naturalism, pictures, and platonic intuitions. In P. Mancosu, K. Jorgensen, & S. Pedersen (Eds.), *Visualization, explanation, and reasoning styles in mathematics* (pp. 57–73). Dordrecht, Netherlands: Springer.

Brown, D., & Presmeg, N. (1993). Types of imagery used by elementary and secondary school students in mathematical reasoning. In I. Hirabayashi, N. Nohda, K. Shigematsu, & F. Lai-Lin (Eds.), *Proceedings of the 17th international conference for the psychology of mathematics education* (Vol. 2, pp. 137–145). Tsukuba, Japan: University of Tsukuba.

Brown, E., & Jones, E. (2006). Understanding conic sections using alternate graph paper. *Mathematics Teacher, 99*(5), 322–327.

Bruner, J. (1968). On cognitive growth. In J. Bruner, R. Olver, & P. Greenfield (Eds.), *Studies in cognitive growth* (pp. 1–29). New York: Wiley.

Bruner, J. (1978). Prologue to the English edition. In R. Rieber & A. Carton (Eds.), *The collected works of L. S. Vygotsky, volume 1: Problems of general psychology* (pp. 1–16). New York: Plenum Press.

Bruner, J. (1990). *Acts of meaning.* Cambridge, MA: Harvard University Press.

Buchwald, J. (1989). *The rise of the wave theory of light: Optical theory and experiment in the early nineteenth century.* Chicago, IL: University of Chicago Press.

Butcher, K. (2006). Learning from text with diagrams: Promoting mental model development and inference generation. *Journal of Educational Psychology, 98*(1), 182–197.

Byrne, R., Barnard, P., Davidson, I., Janik, V., McGrew, W., Miklósi, A., et al. (2004). Understanding culture across species. *Trends in Cognitive Sciences, 8*(8), 341–346.

Cai, J. (2000). Mathematical thinking involved in U.S. and Chinese students' solving of process-constrained and process-open problems. *Mathematical Thinking and Learning, 2*(4), 309–340.

Cai, J., & Moyer, J. (2008). Developing algebraic thinking in earlier grades: Some insights from international comparative studies. In C. Greenes & R. Rubenstein (Eds.), *Algebra and algebraic thinking in school mathematics* (pp. 169–182). Reston, VA: National Council of Teachers of Mathematics.

Campbell, K. J., Collis, K., & Watson, J. (1995). Visual processing during mathematical problem solving. *Educational Studies in Mathematics, 28*, 177–194.

Campos, D. (2007). Peirce on the role of poietic creation in mathematical reasoning. *Transactions of the Charles S. Peirce Society, 43*(3), 470–489.

Carlson, W. B., & Gorman, M. (1990). Understanding invention as a cognitive process: The case of Thomas Edison and early motion pictures, 1888–91. *Social Studies of Science, 20*, 387–430.

Carpenter, P., & Eisenberg, P. (1978). Mental rotation and the frame of reference in blind and sighted individuals. *Perception and Psychophysics, 23*, 117–124.

Carter, J. (2010). Diagrams and proofs in analysis. *International Studies in the Philosophy of Science, 24*(1), 1–14.

Cattaneo, Z., Vecchi, T., Cornoldi, C., Mammarella, I., Bonino, D., Ricciardi, E., et al. (2008). Imagery and spatial processes in blindness and visual impairment. *Neuroscience and Biobehavioral Reviews, 32*, 1346–1360.

Cellucci, C. (2008). The nature of mathematical explanation. *Studies in the History and Philosophy of Science, 39*, 202–210.

Chambers, D. (1993). Images are both depictive and descriptive. In B. Roskos-Ewoldson, M. Intons-Peterson, & R. Anderson (Eds.), *Imagery, creativity, and discovery: A cognitive perspective* (pp. 77–97). Netherlands: Elsevier.

Cheng, P. (2002). Electrifying diagrams for learning: Principles for effective representational systems. *Cognitive Science, 26*(6), 685–736.

Chiu, M. M. (2001). Using metaphors to understand and solve arithmetic problems: Novices and experts working with negative numbers. *Mathematical Thinking and Learning, 3*(2/3), 93–124.

Cifarelli, V. (1999). Abductive inference: Connections between problem posing and solving. In O. Zaslavsky (Ed.), *Proceedings of the 23rd annual conference of the international group for the psychology of mathematics education* (Vol. 2, pp. 217–224). Haifa, Israel: University of Technion.

Cifarelli, V., & Saenz-Ludlow, A. (1996). Abductive processes and mathematical learning. In E. Jakubowski, D. Watkins, & H. Biske (Eds.), *Proceedings of the 18th annual meeting of the North American chapter of the international group for the psychology of mathematics education* (Vol. 1, pp. 161–166). Columbus, Ohio: Ohio State University.

Clark, F., & Kamii, C. (1996). Identification of multiplicative thinking in children in grades 1–5. *Journal for Research in Mathematics Education, 27*(1), 41–51.

Clark, J., & Campbell, J. (1991). Integrated versus modular theories of number skills and acalculia. *Brain and Cognition, 17*, 204–239.

Clark, J., & Paivio, A. (1991). Dual coding theory and education. *Educational Psychology Review, 3*(3), 149–170.

Clements, D., & Battista, M. (1992). Geometry and spatial reasoning. In D. Grouws (Ed.), *Handbook of research on mathematics teaching and learning* (pp. 420–464). New York: Macmillan.

Cobb, P. (2007). Putting philosophy to work: Coping with multiple theoretical perspectives. In F. Lester, Jr. (Ed.), *Second handbook of research on mathematics teaching and learning* (pp. 3–38). Charlotte, NC: National Council of Teachers of Mathematics and Information Age Publishing.

Coffey, M. (2001). Irrational numbers on the number line: Perfectly placed. *Mathematics Teacher, 94*(6), 453–455.

Cordes, S., Williams, C., & Meck, W. (2007). Common representations of abstract quantities. *Current Directions in Psychological Science, 16*(3), 156–161.

Corsaro, W. (1992). Interpretive reproduction in children's peer cultures. *Social Psychology Quarterly, 55*(2), 160–177.

Corsaro, W., Molinari, L., & Brown Rosier, K. (2002). Zena and Carlotta: Transition narratives and early education in the United States and Italy. *Human Development, 45*(5), 323–348.

Cox, R., & Brna, P. (1995). Supporting the use of external representations in problem solving: The need for flexible learning environments. *Journal of Artificial Intelligence in Education, 6*(2/3), 239–302.

Crafter, S., & de Abreu, G. (2010). Constructing identities in multicultural learning contexts. *Mind, Culture, and Activity, 17*, 102–118.

Curriculum Planning and Development Division. (2008). *Primary mathematics textbooks.* New Industrial Road, Singapore: Marshall Cavendish Education.

Damarin, S. (1993). Schooling and situated knowledge: Travel or tourism? In H. McLellan (Ed.), *Situated learning perspectives* (pp. 77–88). Englewood Cliffs, NJ: Educational Technology Publications.

Davey, N. (2005). Aesthetic f(r)iction: The conflicts of visual experience. *Journal of Visual Art Practice, 4*(2–3), 135–149.

Davis, P. (1974). Visual geometry, computer graphics, and theorems of perceived type. *Proceedings of the symposia in applied mathematics: Volume 20* (pp. 113–127). Washington, DC: American Mathematical Society.

Davis, P. (1993). Visual theorems. *Educational Studies in Mathematics, 24*, 333–344.

Davydov, V. (1990). *Types of generalization in instruction: Logical and psychological problems in the structuring of school curricula.* Reston, VA: National Council of Teachers of Mathematics.

De Abreu, G. (2000). Relationships between macro and micro sociocultural contexts: Implications for the study of interactions in the mathematics classroom. *Educational Studies in Mathematics, 41*, 1–29.

De Beni, R., & Cornoldi, C. (1988). Imagery limitations in totally congenitally blind subjects. *Journal of Experimental Psychology: Learning, Memory, and Cognition, 14*(4), 650–655.

De Hevia, M. D., Vallar, G., & Girelli, L. (2008). Visualizing numbers in the mind's eye: The role of visuo-spatial processes in numerical abilities. *Neuroscience and Biobehavioral Reviews, 32*, 1361–1372.

Dea, S. (2006). "Merely a veil over the living thought:" Mathematics and logic in Peirce's forgotten spinoza review. *Transactions of the Charles S. Peirce Society, 42*(4), 501–517.

Deger, J. (2007). Viewing the invisible: Yolngu video as revelatory ritual. *Visual Anthropology, 20*, 103–121.

Dehaene, S. (1997). *The number sense*. Oxford, UK: Oxford University Press.

Dehaene, S. (2001). Précis of the number sense. *Mind & Language, 16*(1), 16–36.

DeLoache, J. (2005). The Pygmalion problem in early problem use. In L. Namy (Ed.), *Symbol use and symbolic representation* (pp. 47–67). Mahwah, NJ: Erlbaum.

Detienne, M. (1996). *The masters of truth in archaic Greece*. New York: Zone Books.

Dickson, S. (2002). Tactile mathematics. In C. Bruter (Ed.), *Mathematics and art: Mathematical visualization in art and education* (pp. 213–222). Berlin: Springer.

Donaghue, E. (2003). Algebra and geometry textbooks in twentieth-century America. In G. Stanic & J. Kilpatrick (Eds.), *A history of school mathematics* (Vol. 1, pp. 329–398). Reston, VA: National Council of Teachers of Mathematics.

Dörfler, W. (1991). Meaning: Image schemata and protocols. In F. Furinghetti (Ed.), *Proceedings of the 15th annual conference of the psychology of mathematics education* (Vol. 1, pp. 17–32). Assisi, Italy: PME.

Dörfler, W. (2001a). Instances of diagrammatic reasoning. Paper presented to the discussion group on semiotic and mathematics education at the 25th PME conference, University of Utrecht, Netherlands. Retrieved from http://www.math.uncc.edu/~sae/dg3/dorfler1.pdf

Dörfler, W. (2001b). How diagrammatic is mathematical reasoning? Paper presented to the discussion group on semiotic and mathematics education at the 25th PME conference, University of Utrecht, Netherlands. Retrieved from http://www.math.uncc.edu/~sae/dg3/dorfler2.pdf

Dörfler, W. (2007). Matrices as Peircean diagrams: A hypothetical learning trajectory. *CERME, 5*, 852–861.

Dörfler, W. (2008). En route from patterns to algebra: Comments and reflections. *ZDM, 40*, 143–160.

Dretske, F. (1969). *Seeing and knowing*. Chicago, IL: Chicago University Press.

Dretske, F. (1990). Seeing, believing, and knowing. In D. Osherson, S. Kosslyn, & J. Hollerback (Eds.), *Visual cognition and action: An invitation to cognitive science* (pp. 129–148). Cambridge, MA: MIT Press.

Ducker, L. (1993). Visually impaired students drawing graphs. *Mathematics Teaching, 144*, 23–26.

Duval, R. (1998). Geometry from a cognitive point of view. In C. Mammana, & V. Villani (Eds.), *Perspectives on the teaching of geometry for the 21st century* (pp. 29–83). Netherlands: Kluwer.

Duval, R. (2000). Basic issues for research in mathematics education. In T. Nakahara & M. Koyama (Eds.), *Proceedings of the 24th annual conference of the international group for the psychology of mathematics education* (Vol. 1, 55–69). Hiroshima, Japan.

Duval, R. (2006). A cognitive analysis of problems of comprehension in a learning of mathematics. *Educational Studies in Mathematics, 61*(1–2), 103–131.

Ebersbach, M., Van Dooren, W., Van den Noortgate, W., & Resing, W. (2008). Understanding linear and exponential growth: Searching for the roots in 6- to 9-year-olds. *Cognitive Development, 23*, 237–257.

Eco, U. (1983). Horns, hooves, insteps: Some hypotheses on three types of abduction. In U. Eco & T. Sebeok (Eds.), *The sign of three: Dupin, Holmes, Peirce* (pp. 198–220). Bloomington, IN: Indiana University Press.

Eisenberg, T., & Dreyfus, T. (1991). On the reluctance to visualize in mathematics. In W. Zimmermann & S. Cunningham (Eds.), *Visualization in teaching and learning mathematics* (pp. 25–38). Washington, DC: Mathematical Association of America.

ETA/Cuisenaire® (2005). *Algeblocks® for high school*. Vernon Hills, IL: ETA/Cuisenaire®.

Farah, M. (1988). Is visual imagery really visual? Overlooked evidence from neuropsychology. *Psychological Review, 95*(3), 307–317.

Feigenson, L., Dehaene, S., & Spelke, E. (2004). Core systems of number. *Trends in Cognitive Science, 8*(7), 307–314.

Ferrara, F., Pratt, D., & Robutti, O. (2006). The role and uses of technologies for the teaching of algebra and calculus. In A. Gutierrez & P. Boero (Eds.), *Handbook of research on the psychology of mathematics education: Past, present, and future* (pp. 237–274). Rotterdam, Netherlands: Sense Publishers.

Ferreira, M. (1997). When 1 + 1 ≠ 2: Making mathematics in central Brazil. *American Ethnologist, 24*(1), 132–147.

Fey, J. (1990). Quantity. In L. Steen (Ed.), *On the shoulders of giants: New approaches to numeracy* (pp. 61–94). Washington, DC: National Academy Press.

Filloy, E., & Rojano, T. (1989). Solving equations: The transition from arithmetic to algebra. *For the Learning of Mathematics, 9*(2), 19–25.

Finke, R. (1980). Levels of equivalence in imagery and perception. *Psychological Review, 87*(2), 113–132.

Fischbein, E. (1977). Image and concept in learning mathematics. *Educational Studies in Mathematics, 8*, 153–165.

Fischbein, E. (1993). The theory of figural concepts. *Educational Studies in Mathematics, 24*, 139–162.

Fitzsimons, G. (Ed.). (2002). Special issue: Cultural aspects of mathematics education. *Journal of Intercultural Studies, 23*(2), 109–228.

Flach, P. (1996). Abduction and induction: Syllogistic and inferential perspectives. In P. Flach, & A. Kakas (Eds.), *Contributing papers to the ECAI 96 workshop on abductive and inductive reasoning* (pp. 31–35). Budapest, Hungary: ECAI.

Foucault, M. (1982). *The archaelogy of knowledge and the discourse on language* (A. M. Sheridan-Smith, Trans.). New York: Pantheon.

Freudenthal, H. (1981). Major problems of mathematics education. *Educational Studies in Mathematics, 12*, 133–150.

Frisch, K. von (1967). *The dance language and orientation of bees.* Cambridge, MA: Harvard University Press.

Gal, H., & Linchevski, L. (2010). To see or not to see: Analyzing difficulties in geometry from the perspective of visual perception. *Educational Studies in Mathematics, 74*(2), 163–183.

Galison, P. (2002). Images scatter into data, data gathers into images. In B. Latour, & P. Weibel (Eds.), *Iconoclash: Beyond the image wars in science, religion, and art* (pp. 300–323). Cambridge, MA: MIT Press.

Gardner, J., Gardner, C., Jones, B., & Jones, E. (2008). Making arithmetic accessible to blind mainstream children. In K. Miesenberger, J. Klaus, W. Zagler, & A. Karshmer (Eds.), *Lecture notes in computer science (Vol. 5105): Proceedings of the 11th international conference on computers helping people with special needs* (pp. 900–906). Netherlands: Springer.

Gattis, M. (2002). Structure mapping in spatial reasoning. *Cognitive Development, 17*, 1157–1183.

Gattis, M. (2004). Mapping relational structure in spatial reasoning. *Cognitive Science, 28*, 589–610.

Gattis, M. (Ed.) (2001). *Spatial schemas and abstract thought.* Cambridge, MA: MIT Press.

Gattis, M., & Holyoak, K. (1996). Mapping conceptual relations in visual reasoning. *Journal of Experimental Psychology: Learning, Memory, and Cognition, 22*(1), 231–239.

Gee, J. P. (2005). *Why video games are good for your soul: Pleasure and learning.* Australia: Common Ground Publishing Pty Ltd.

Gelman, R., & Gallistel, C. (1978). *The child's understanding of number.* Cambridge, MA: Harvard University Press.

Gentner, D. (1983). Structure mapping: A theoretical framework for analogy. *Cognitive Science, 7*, 155–170.

Gerdes, P. (2004). Mathematical thinking and geometric exploration in Africa and elsewhere. *Diogenes, 202*, 107–122.

Gergen, K. (2002). The challenge of absent presence. In J. Katz & M. Aakhus (Eds.), *Perceptual contact: Mobile communication, private talk, public performance* (pp. 227–241). New York: Cambridge University Press.

Giaquinto, M. (1994). Epistemology of visual thinking in elementary real analysis. *The British Journal for the Philosophy of Science, 45*, 789–813.

Giaquinto, M. (2005). Mathematical activity. In P. Mancosu, K. Jorgensen, & S. Pedersen (Eds.), *Visualization, explanation, and reasoning styles in mathematics* (pp. 75–90). Dordrecht, Netherlands: Springer.
Giaquinto, M. (2007). *Visual thinking in mathematics: An epistemological study.* Oxford, UK: Oxford University Press.
Gibson, J. (1986/1979). *The ecological approach to visual perception.* Hillsdale, NJ: Lawrence Erlbaum Associates.
Gibson, W., & Darron, C. (1999). Teaching statistics to a student who is blind. *Teaching of Psychology, 26*(2), 130–131.
Giere, R., & Moffatt, B. (2003). Distributed cognition: Where the cognitive and the social merge. *Social Studies of Science, 33*(2), 301–310.
Ginns, P. (2005). Meta-analysis of the modality effect. *Learning and Instruction, 15*, 313–331.
Ginsburg, H., Lin, C.-L., Ness, D., & Seo, K.-H. (2003). Young American and Chinese children's everyday mathematical activity. *Mathematical Thinking and Learning, 5*(4), 235–258.
Glasgow, J., Narayanan, N., & Chandrasekaran, B. (1995). *Diagrammatic reasoning: Cognitive and computational perspectives.* Cambridge, MA: MIT Press.
Goldin, G. (2002). Representation in mathematical learning and problem solving. In L. English (Ed.), *Handbook of International Research in Mathematics Education* (pp. 197–218). Mahwah, NJ: Erlbaum.
Goldin-Meadow, S., Alibali, M., & Church, R. (1993). Transitions in concept acquisition: Using the hand to read the mind. *Psychological Review, 100*, 279–297.
Gooding, D. (2004). Cognition, construction, and culture: Visual theories in the sciences. *Journal of Cognition and Culture, 4*(3), 551–593.
Gooding, D. (2006). From phenomenology to field theory: Faraday's visual reasoning. *Perspectives on Science, 14*(1), 40–65.
Goodman, N. (1976). *Languages of art: An approach to a theory of symbols.* Cambridge, MA: Hackett.
Gravemeijer, K., Lehrer, R., van Oers, B., & Verschaffel, L. (2002). *Symbolizing, modeling, and tool use in mathematics education.* Dordrecht, Netherlands: Kluwer.
Greenes, C., Cavanagh, M., Dacey, L., Findell, C., & Small, M. (2001). *Navigating through algebra in prekindergarten – Grade 2.* Reston, VA: National Council of Teachers of Mathematics.
Gregory, R. (2005). Images of mind in brain. *Word & Image, 21*(2), 120–123.
Grelland, H. H. (2007). Physical concepts and mathematical symbols. In G. Adenier, A.Y. Khrennikov, P. Lahti, V. I. Mak'ko, & T. M. Niewenhuizen (Eds.), *CP962: quantum theory, reconsideration of foundations* (Vol. 4, pp. 258–260). Washington, DC: American Institute of Physics.
Haciomeroglu, E., Aspinwall, L., & Presmeg, N. (2010). Contrasting cases of calculus students' understanding of derivative graphs. *Mathematical Thinking and Learning, 12*, 152–176.
Harel, G. (2007). The DNR system as a conceptual framework for curriculum development and instruction. In R. Lesh, J. Kaput, & E. Hamilton (Eds.), *Foundations for the future in mathematics education* (pp. 263–280). Mahwah, NJ: Erlbaum.
Hegarty, M., Carpenter, P., & Just, M. (1991). Diagrams in the comprehension of scientific text. In R. Barr, M. Kamil, P. Mosenthal, & P. Pearson (Eds.), *Handbook of reading research* (Vol. 2, pp. 641–688). New York: Longman.
Hegarty, M., & Kozhevnikov, M. (1999). Types of visual-spatial representations and mathematical problem solving. *Journal of Educational Psychology, 91*, 684–689.
Heller, M., & Kennedy, J. (1990). Perspective taking, pictures, and the blind. *Perception and Psychophysics, 48*(5), 459–466.
Hershkowitz, R., Friedlander, A., & Dreyfus, T. (1991). LOCI and visual thinking. In F. Furinghetti (Ed.), *Proceedings of the 15th international conference for the psychology of mathematics education* (Vol. 2, pp. 181–188). Assisi, Italy: PME Program Committee.
Hoffmann, M. (1999). Problems with Peirce's concept of abduction. *Foundations of Science, 4*, 271–305.

Hoffmann, M. (2007). Learning from people, things, and signs. *Studies in Philosophy Education, 26*, 185–204.

Houser, N. (1987). The importance of pictures. *Sciences, 27*(3), 17–18.

Humphries, L. (2010). *Six jumping frogs and the aesthetics of problem solving*. Retrieved April 23, 2010, from http://www.thinkingapplied.com/frog-folder/frog.htm

Hutchins, E. (1995). *Cognition in the wild*. Cambridge, MA: MIT Press.

Jacob, P., & Jeannerod, M. (2003). *Ways of seeing: The scope and limits of visual cognition*. Great Clarendon Street, Oxford, UK: Oxford University Press.

Jaffe, A., & Quinn, F. (1993). Theoretical mathematics: Toward a cultural synthesis of mathematics and theoretical physics. *Bulletin of the American Mathematical Society, 29*(1), 1–13.

Jeannerod, M., & Jacob, P. (2004). Visual cognition: A new look at the two-visual systems model. *Neuropsychologia, 43*, 301–312.

Johnson, A. (2002). The world of blind mathematicians. *Notices of the AMS, 49*(10), 1246–1251.

Josephson, J. (1996). Inductive generalizations are abductions. In Flach, P. & Kakas, A. (Eds.), *Contributing Paper to the ECAI '96 Workshop on Abductive and Inductive Reasoning*. Budapest, Hungary: ECAI.

Josephson, J. (2000). Smart inductive generalizations are abductions. In P. Flach & A. Kakas (Eds.), *Abduction and induction: Essays on their relation and integration* (pp. 31–44). Netherlands: Kluwer.

Josephson, J., & Josephson, S. (1994). *Abductive inference: Computation, philosophy, technology*. New York: Cambridge University Press.

Kaput, J. (1992). Technology and mathematics education. In D. Grouws (Ed.), *Handbook of research on mathematics teaching and learning* (pp. 515–556). New York: Macmillan.

Kaput, J., Blanton, M., & Moreno, L. (2008). Algebra from a symbolization point of view. In J. Kaput, D. Carraher, & M. Blanton (Eds.), *Algebra in the early grades* (pp. 19–56). New York: Erlbaum.

Karmiloff-Smith, A. (1992). *Beyond modularity: A developmental perspective on cognitive science*. Cambridge, MA: MIT Press.

Karshmer, A., & Farsi, D. (2008). Access to mathematics by blind students: A global problem. *Journal of Systemics, Cybernetics, and Informatics, 5*(6), 77–81.

Katz, V. (1999). Algebra and its teaching: An historical survey. *Journal of Mathematical Behavior, 16*(1), 25–38.

Katz, V. (2007). Stages in the history of algebra with implications for teaching. *Educational Studies in Mathematics, 66*(2), 185–201.

Kaufmann, G. (1985). A theory of symbolic representation in problem solving. *Journal of Mental Imagery, 9*(2), 51–70.

Kaufmann, G., & Helstrup, T. (1988). Mental imagery and problem solving: Implications for the educational process. In M. Denis, J. Engelkamp, & J. Richardson (Eds.), *Cognitive and neuropsychological approaches to mental imagery* (pp. 113–144). Dordrecht, Netherlands: Martinus Nijhoff.

Kennedy, P. (2000). Concrete representations and number line models: Connecting and extending. *Journal of Developmental Education, 24*(2), 2–6.

Kerr, N. (1983). The role of vision in "visual imagery" experiments: Evidence from the congenitally blind. *Journal of Experimental Psychology: General, 112*(2), 265–277.

Kerr, N., Foulkes, D., & Schmidt, M. (1982). The structure of laboratory dream reports in blind and sighted subjects. *Journal of Nervous and Mental Disease, 170*, 286–294.

Kirshner, D., & Awtry, T. (2004). Visual salience of algebraic transformations. *Journal for Research in Mathematics Education, 35*(4), 224–257.

Kitcher, P. (1983). *The nature of mathematical knowledge*. Oxford, MA: Oxford University Press.

Klotz, E. (1991). Visualization in geometry: A case study of a multimedia mathematics education project. In W. Zimmermann & S. Cunningham (Eds.), *Visualization in teaching and learning mathematics* (pp. 95–104). Washington, DC: Mathematical Association of America.

Kolers, P., & Smythe, W. (1979). Images, symbols, and skills. *Canadian Journal of Psychology, 33*(3), 158–184.

Kosslyn, S., Ganis, G., & Thompson, W. (2001). Neural foundations of imagery. *Nature Reviews, 2*, 635–642.

Kotsopoulos, D., & Cordy, M. (2009). Investigating imagination as a cognitive space for learning mathematics. *Educational Studies in Mathematics, 70*, 259–274.

Kuhn, T. (1970). *The structure of scientific revolutions* (2nd ed.). Chicago, IL: University of Chicago Press.

Kulp, M., Earley, M., Mitchell, G., Timmerman, L., Frasco, C., & Geiger, M. (2004). Are visual perceptual skills related to mathematics ability in second through sixth grade children? *Focus on Learning Problems in Mathematics, 26*(4), 44–51.

Kulpa, Z. (2009). Main problems of diagrammatic reasoning. Part I: The generalization problem. *Foundations of Science, 14*(1–2), 75–96.

Laborde, C., Kynigos, C., Hollebrands, K., & Strässer, R. (2006). Teaching and learning geometry with technology. In A. Gutierrez & P. Boero (Eds.), *Handbook of research on the psychology of mathematics education: Past, present, and future* (pp. 275–304). Rotterdam, Netherlands: Sense Publishers.

Lahav, O. (2006). Using virtual environment to improve spatial perception by people who are blind. *Cyberpsychology and Behavior, 9*(2), 174–177.

Landau, B., Gleitman, H., & Spelke, E. (1981). Spatial knowledge and geometric representation in a child blind from birth. *Science, 213*, 1275–1278.

Larkin, J., & Simon, H. (1987). Why a diagram is (sometimes) worth ten thousand words. *Cognitive Science, 11*, 65–99.

Lave, J. (1988). *Cognition in practice: Mind, mathematics, and culture in everyday life.* Cambridge, UK: Cambridge University Press.

Lee, L. (1996). An initiation into algebraic culture through generalization activities. In N. Bednarz, C. Kieran, & Lee, L. (Eds.), *Approaches to algebra: Perspectives for research and teaching* (pp. 87–106). Dordrecht, Netherlands: Kluwer.

Legg, C. (2008). The problem of the essential icon. *American Philosophical Quarterly, 45*(3), 208–232.

Leinhardt, G., Zaslavsky, O., & Stein, M. (1990). Functions, graphs, and graphing: Tasks, learning, and teaching. *Review of Educational Research, 60*, 1–64.

Lerman, S. (2001). Cultural, discursive psychology: A sociocultural approach to studying the teaching and learning of mathematics. *Educational Studies in Mathematics, 46*, 87–113.

Leslie, A., Xu, F., Tremoulet, P., & Scholl, B. (1998). Indexing and the object concept: Developing "what" and "where" systems. *Trends in Cognitive Science, 2*(1), 10–18.

Leushina, A. (1991). *Soviet studies in mathematics education (volume 4): The development of elementary mathematical concepts in preschool children.* (J. Teller, Trans.). Reston, VA: National Council of Teachers of Mathematics.

Levy, S. (1997). Peirce's theoremic/corollarial distinction and the interconnections between mathematics and logic. In N. Houser, D. Roberts, & J. Van Evra (Eds.), *Studies in the logic of Charles Sanders Peirce* (pp. 85–110). Bloomington, IN: Indiana University Press.

Lewis, K. (2009). Patterns with checkers. *Mathematics Teaching in the Middle School, 14*(7), 418–422.

Leyton, M. (2002). *Symmetry, causality, mind.* Cambridge, MA: MIT Press.

Lithner, J. (2008). A research framework for creative and imitative reasoning. *Educational Studies in Mathematics, 67*, 255–276.

Livingston, E. (1999). Cultures of proving. *Social Studies of Science, 29*(6), 867–888.

Lohse, G., Biolsi, K., Walker, N., & Rueler, H. (1994). A classification of visual representations. *Communications of the ACM, 37*(12), 36–49.

Lomas, D. (2002). What perception is doing, and what it is not doing, in mathematical reasoning. *British Journal for Philosophy of Science, 53*, 205–223.

Löwy, I. (2008). Ways of seeing: Ludwik Fleck and Polish debates on the perception of reality, 1890–1947. *Studies in History and Philosophy of Science, 39*, 375–383.

Luria, A. (1976). *Cognitive development: Its cultural and social foundations.* Cambridge, MA: Harvard University Press.

Magnani, L. (2001). *Abduction, reason, and science: Processes of discovery and explanation.* Dordrecht, Netherlands: Kluwer and Plenum.
Magnani, L. (2004). Conjectures and manipulations: Computational modeling and the extra-theoretical dimension of scientific discovery. *Minds and Machines, 14*, 507–537.
Maienschein, J. (1991). From presentation to representation in E. B. Wilson's "The Cell." *Biology and Philosophy, 6*, 227–254.
Malaty, G. (2008). The role of visualization in mathematics education: Can visualization promote causal thinking? Paper presented at the 11th international congress in mathematical education, Merida, Mexico.
Mandler, J. (2007). On the origins of the conceptual system. *American Psychologist, 62*(8), 741–751.
Mandler, J. (2008). On the birth and growth of concepts. *Philosophical Psychology, 21*(2), 207–230.
Mariotti, M. A. (2000). Introduction to proof: The mediation of a dynamic software environment. *Educational Studies in Mathematics, 44*, 25–53.
Mariotti, M. A. (2002). The influence of technological advances on students' mathematics learning. In L. English (Ed.), *Handbook of international research in mathematics education* (pp. 695–724). Mahwah, NJ: Erlbaum.
Markman, A. B., & Gentner, D. (2000). Structure-mapping in the comparison process. *American Journal of Psychology, 113*(4), 501–538.
Marmor, G., & Zabeck, L. (1976). Mental rotation by the blind: Does mental rotation depend on visual imagery? *Journal of Experimental Psychology: Human Perception and Performance, 2*(4), 515–521.
Mason, J. (1980). When is a symbol symbolic? *For the Learning of Mathematics, 1*(2), 8–12.
Mason, J. (1987). What do symbols represent? In C. Janvier (Ed.), *Problems of representation on the teaching and learning of mathematics* (pp. 73–81). Hillsdale, NJ: Erlbaum.
Mason, J. (2008). Making use of children's powers to produce algebraic thinking. In J. Kaput, D. Carraher, & M. Blanton (Eds.), *Algebra in the early grades* (pp. 57–94). New York: Erlbaum.
Mason, J., Stephens, M., & Watson, A. (2009). Appreciating mathematical structure for all. *Mathematics Education Research Journal, 21*(2), 33–49.
Mathematics in Context. (2006). *Expressions and formulas*. Chicago, IL: Britannica.
Mayer, R., & Massa, L. (2003). Three facets of visual and verbal learners: Cognitive ability, cognitive style, and learning preference. *Journal of Educational Psychology, 95*(4), 833–846.
McCormick, B., DeFanti, T., & Brown, M. (1987). Definition of visualization. *ACM SIGGRAP Computer Graphics, 21*(6), 1–3.
Metzger, W. (2006). *Laws of seeing*. Cambridge, MA: The MIT Press.
Michalowicz, K., & Howard, A. (2003). Pedagogy in text: An analysis of mathematics texts from the nineteenth century. In G. Stanic & J. Kilpatrick (Eds.), *A history of school mathematics* (Vol. 1, pp. 77–112). Reston, VA: National Council of Teachers of Mathematics.
Millar, S. (1994). *Understanding and representing space: Theory and evidence from studies with blind and sighted children.* Oxford, UK: Oxford University Press.
Miller, A. (1997). Cultures of creativity: Mathematics and physics. *Diogenes, 45*(1), 53–72.
Miller, J. (1990). The essence of images. In H. Barlow, C. Blakemore, & M. Weston-Smith (Eds.), *Images and understanding* (pp. 1–4). New York: Cambridge University Press.
Mishra, R., Singh, S., & Dasen, P. (2009). Geocentric dead reckoning in Sanskrit- and Hindi-medium school children. *Culture & Psychology, 15*(3), 386–408.
Mitchell, W. (1984). What is an image? *New Literary History, XV*(3), 503–537.
Mitchell, W. (1986). *Iconology: Image, text, ideology.* Chicago, IL: University of Chicago Press.
Mitchell, W. (1994). *Picture theory: Essays on verbal and visual representation.* Chicago, IL: University of Chicago Press.
Mitchell, W. (2005). *What do pictures want? The lives and loves of images*. Chicago, IL: University of Chicago Press.

Miura, I., Okamoto, Y., Vlahovic-Stetic, V., Kim, C., & Han, J. H. (1999). Language supports for childrens' understanding of numerical fractions: Cross-national comparisons. *Journal of Experimental Child Psychology, 74*, 356–365.

Moyer, P. (2001). Are we having fun yet? How teachers use manipulatives to teach mathematics. *Educational Studies in Mathematics, 47*, 175–197.

Mulligan, J., Prescott, A., & Mitchelmore, M. (2003). Taking a closer look at young students' visual imagery. *Australian Primary Mathematics, 8*(4), 23–27.

Murata, A. (2008). Mathematics teaching and learning as a mediating process: The case of tape diagrams. *Mathematical Thinking and Learning, 10*, 374–406.

Neisser, U. (1976). *Cognitive psychology*. New York: Meredith Publishing Company.

Nemirovsky, R., & Ferrara, F. (2009). Mathematical imagination and embodied cognition. *Educational Studies in Mathematics, 70*, 159–174.

Nemirovsky, R., & Noble, T. (1997). On mathematical visualization and the place where we live. *Educational Studies in Mathematics, 33*, 99–131.

Nersessian, N. (1994). *Faraday to Einstein: Constructing meaning in scientific theories*. Dordrecht, Netherlands: Kluwer.

Netz, R. (1998). Greek mathematical diagrams: Their use and their meaning. *For the Learning of Mathematics, 18*(3), 33–39.

Netz, R. (1999). *Shaping of deduction in Greek mathematics: A study in cognitive history*. Cambridge, MA: Cambridge University Press.

Ng, S. F., & Lee, K. (2009). The model method: Singapore children's tool for representing and solving algebraic word problems. *Journal for Research in Mathematics Education, 40*(3), 282–313.

Nickson, M. (1992). The culture of the mathematics classroom: An unknown quantity? In D. Grouws (Ed.), *Handbook of research on mathematics teaching and learning* (pp. 101–114). New York: Simon & Schuster Macmillan and National Council of Teachers of Mathematics.

Nisbett, R., Peng, K., Choi, I., & Norenzayan, A. (2001). Culture and systems of thought: Holistic versus analytic cognition. *Psychological Review, 108*, 291–310.

Noë, A. (2004). *Action in perception*. Cambridge, MA: MIT Press.

Norman, J. (2006). *After Euclid: Visual reasoning and the epistemology of diagrams*. Stanford, CA: CSLI Publications.

Novick, L. (2006). Understanding spatial diagram structure: An analysis of hierarchies, matrices, and networks. *Quarterly Journal of Experimental Psychology, 59*, 1826–1856.

Nunes, T. (1992). Ethnomathematics and everyday cognition. In D. Grouws (Ed.), *Handbook of research on mathematics teaching and learning* (pp. 557–574). New York: Simon & Schuster Macmillan and National Council of Teachers of Mathematics.

O'Brien, E. (n.d.). *A discovery activity for the algebra classroom using the transformation application*. Retrieved October 19, 2010 from http://education.ti.com/calculators/downloads/US/Activities/Details?ID=6637.

O'Daffer, P., & Clemens, S. (1992). *Geometry: An investigative approach*. New York: Addison Wesley.

O'Halloran, K. (2005). *Mathematical discourse: Language, symbolism, and visual images*. New York: Continuum.

O'Regan, J. K. (2001). The "feel" of seeing: An interview with J. Kevin O'Regan. *Trends in Cognitive Sciences, 5*(6). 278–279.

Otte, M. (2003). Complementarity, sets, and numbers. *Educational Studies in Mathematics, 53*, 203–228.

Otte, M. (2007). Mathematics history, philosophy, and education. *Educational Studies in Mathematics, 66*, 243–255.

Owens, K. (1999). The role of visualization in young students' learning. In O. Zaslavsky (Ed.), *Proceedings of the 23rd international conference of the psychology for mathematics education* (Vol. 1., pp. 220–234). Haifa, Israel: University of Technion.

Owens, K., & Outhred, L. (2006). The complexity of learning geometry and measurement. In A. Gutierrez & P. Boero (Eds.), *Handbook of research on the psychology of mathematics education: Past, present, and future* (pp. 83–115). Rotterdam, Netherlands: Sense Publishers.

Özdural, A. (2000). Mathematics and arts: Connections between theory and practice in the medieval Islamic world. *Historia Mathematica, 27*, 171–201.

Paivio, A. (1971). *Imagery and verbal processes*. New York: Holt.

Paivio, A. (1986). *Mental representations*. New York: Oxford University Press.

Paivio, A. (2006). Dual coding theory and education. Paper presented at the pathways to literacy achievement for high poverty children, University of Michigan School of Education. Retrieved October 19, 2010 from http://www.umich.edu/~rdytolrn/pathwaysconference/presentations/paivio.pdf

Paivio, A., & Desrochers, A. (1980). A dual coding approach to bilingual memory. *Canadian Journal of Psychology, 34*, 390–401.

Palais, R. (1999). The visualization of mathematics: Towards a mathematical exploratorium. *Notices of the AMS, 46*(6), 647–658.

Panofsky, E. (1991). *Perspective as symbolic form* (C. S. Wood, Trans.). New York: Zone Books.

Pantziara, M., Gagatsis, A., & Elia, I. (2009). Using diagrams as tools for the solution of nonroutine mathematical problems. *Educational Studies in Mathematics, 72*(1), 39–60.

Papert, S. (2002). Afterword: After how comes what. In R. Sawyer (Ed.), *Cambridge handbook of the learning sciences* (pp. 581–586). New York: Cambridge University Press.

Parker, M. (2004). Reasoning and working proportionally with percent. *Mathematics Teaching in the Middle School, 9*(6), 326–330.

Parker, T., & Baldridge, S. (2004). *Elementary mathematics for teachers*. Okemos, MI: Sefton-Ash Publishing.

Parker, T., & Baldridge, S. (2008). *Elementary geometry for teachers*. Okemos, MI: Sefton-Ash Publishing.

Pedemonte, B. (2007). How can the relationship between argumentation and proof be analyzed? *Educational Studies in Mathematics, 66*(1), 25–41.

Peirce, C. (1934). *Collected papers of Charles Sanders Peirce: Volume V*. In C. Hartshorne & P. Weiss (Eds.), Cambridge, MA: Harvard University Press.

Peirce, C. (1949). *Chance, love, and logic: Philosophical essays*. In M. Cohen (Ed.), New York: Harcourt, Brace and Company.

Peirce, C. (1956). The essence of mathematics. In J. Newman (Ed.), *The world of mathematics: Volume 3* (pp. 1773–1786). New York: Simon & Schuster.

Peirce, C. (1957). *Essays in the philosophy of science*. In V. Thomas (Ed.), New York: The Bobbs-Merrill Company

Peirce, C. (1958a). *Collected papers of Charles Sanders Peirce: Volume VIII*. In A. Burks (Ed.), Cambridge, MA: Harvard University Press.

Peirce, C. (1958b). Lessons of the history of science. In P. Wiener (Ed.), *Charles S. Peirce: Selected writings (Values in a universe of chance)* (pp. 227–232). New York: Dover.

Peirce, C. (1960). *Collected papers of Charles Sanders Peirce: Volume I and volume II*. In C. Hartshorne & P. Weiss (Eds.), Cambridge, MA: The Belknap Press of Harvard University Press.

Peirce, C. (1976). *The new elements of mathematics: Volume IV mathematical philosophy*. In C. Eisele (Ed.), Atlantic Highlands, NJ: Humanities Press.

Perkins, D. (1997). Epistemic games. *International Journal of Educational Research, 27*, 49–61.

Piaget, J. (1951). *Play, dreams, and imitations in childhood*. New York: Norton.

Piaget, J. (1987). *Possibility and necessity*. Minneapolis, MN: University of Minnesota Press.

Pica, P., Lemer, C., Izard, V., & Dehaene, S. (2004). Exact and approximate arithmetic in an Amazonian indigene group. *Science, 306*, 499–503.

Pinker, S. (1990). A theory of graph comprehension. In R. Freedle (Ed.), *Artificial intelligence and the future of testing* (pp. 73–126). Mahwah, NJ: Erlbaum.

Pitta-Pantazi, D., Christou, C., & Zachariades, T. (2007). Secondary school students' levels of understanding in computing exponents. *Journal of Mathematical Behavior, 26*(4), 301–311.

Ploetzner, R., Lippitsch, S., Galmbacher, M., Heuer, D., & Scherrer, S. (2009). Students' difficulties in learning from dynamic visualizations and how they may be overcome. *Computers in Human Behavior, 25*, 56–65.

Polya, G. (1988). *How to solve it: A new aspect of mathematical method.* Princeton, NJ: Princeton University Press.

Ponce, G. (2008). Using, seeing, feeling, and doing absolute value for deeper understanding. *Mathematics Teaching in the Middle School, 14*(4), 234–240.

Poplu, G., Ripoll, H., Mavromatis, S., & Baratgin, J. (2008). How do expert soccer players encode visual information to make decisions in simulated game situations? *Research Quarterly for Exercise and Sport, 79*(3), 392–398.

Presmeg, N. (1986). Visualization in high school mathematics. *For the Learning of Mathematics, 6*(3), 42–46.

Presmeg, N. (1991). Classroom aspects which influence use of visual imagery in high school mathematics. In F. Furinghetti (Ed.), *Proceedings of the 15th International group for the psychology of mathematics education* (Vol. 3, pp. 191–198). Assisi, Italy: PME Committee.

Presmeg, N. (1992). Prototypes, metaphors, metonymies, and imaginative rationality in high school mathematics. *Educational Studies in Mathematics, 23*, 595–610.

Presmeg, N. (2006). Research on visualization in learning and teaching mathematics. In A. Gutierrez & P. Boero (Eds.), *Handbook of research on the psychology of mathematics education: Past, present, and future* (pp. 205–236). Rotterdam: Sense Publishers.

Presmeg, N. (2008). *An overarching theory for research in visualization in mathematics education.* Paper presented at the 11th international congress in mathematical education, Merida, Mexico.

Pylyshyn, Z. (1973). What the mind's eye tells the mind's brain: A critique of mental imagery. *Psychological Bulletin, 80*, 1–24.

Pylyshyn, Z. (2006). *Seeing and visualizing: It's not what you think.* Cambridge, MA: MIT Press.

Pylyshyn, Z. (2007). *Things and places: How the mind connects with the world.* Cambridge, MA: MIT Press.

Radford, L. (2002). Signs and meanings in students' emergent algebraic thinking: A semiotic analysis. *Educational Studies in Mathematics, 42*, 237–268.

Radford, L. (2008). Iconicity and contraction: A semiotic investigation of forms of algebraic generalizations of patterns in different context. *ZDM: International Journal in Mathematics Education, 40*(1), 83–96.

Radford, L. (2009). Why do gestures matter? Sensuous cognition and the palpability of mathematical meanings. *Educational Studies in Mathematics, 70*, 111–126.

Raju, C. K. (2001). Computers, mathematics education, and the alternative epistemology of the calculus in the Yuktibhasa. *Philosophy East & West, 51*(3), 325–362.

Rakoczy, H., Tomasello, M., & Striano, T. (2005). How children turn objects into symbols: A cultural learning account. In L. Namy (Ed.), *Symbol use and symbolic representation: Developmental and comparative perspectives* (pp. 69–97). Mahwah, NJ: Erlbaum.

Reber, A. (1989). Implicit learning and tacit knowledge. *Journal of Experimental Psychology: General, 118*(3), 219–235.

Reed, S. (2010). *Thinking visually.* London: Psychology Press.

Resnick, L. (1989). Developing mathematical knowledge. *American Psychologist, 44*(2), 162–169.

Resnik, M. (1997). *Mathematics as a science of patterns.* Oxford, UK: Oxford University Press.

Rieser, J., Guth, M., & Hill, E. (1986). Sensitivity to perspective structure while walking without vision. *Perception, 15*, 173–188.

Rival, I. (1987). Picture puzzling: Mathematicians are rediscovering the power of pictorial reasoning. *Sciences, 27*(1), 41–46.

Rivera, F. (2005). An anthropological account of the emergence of mathematical proof and related processes in technology-based environments. In W. Masalski (Ed.), *Technology-supported math learning environments: NCTM 67th yearbook* (pp. 125–136). Reston, VA: National Council of Teachers of Mathematics.

Rivera, F. (2006). Changing the face of arithmetic: Teaching children algebra. *Teaching Children Mathematics, 12*(6), 306–311.

Rivera, F. (2007a). Accounting for students' schemes in the development of a graphical process for solving polynomial inequalities in instrumented activity. *Educational Studies in Mathematics, 65*(3), 281–307.

Rivera, F. (2007b). Visualizing as a mathematical way of knowing: Understanding figural generalization. *Mathematics Teacher, 101*(1), 69–75.

Rivera, F. (2008). On the pitfalls of abduction: Complexities and complicities in patterning activity. *For the Learning of Mathematics, 28*(1), 17–25.

Rivera, F. (2009). Visuoalphanumeric mechanisms that support pattern generalization. In I. Vale & A. Barbosa (Eds.), *Patterns: Multiple perspectives and contexts in mathematics education* (pp. 123–136). Portugal: Escola Superior de Educação do Instituto Politécnico de Viana de Coastelo.

Rivera, F. (2010a). Visual templates in pattern generalization activity. *Educational Studies in Mathematics, 73*, 297–328.

Rivera, F. (2010b). There is more to mathematics than symbols. *Mathematics Teaching, 218*, 42–47.

Rivera, F. (2010c). Second grade students' preinstructional competence in patterning activity. In M. Pinto & T. Kawasaki (eds.), *Proceedings of the 34th conference of the international group for the psychology of mathematics education* (Vol. 4, pp. 81–88). Belo Horizante, Brazil: PME.

Rivera, F. (2010d). Spicing up counting through geometry. *Mathematics Teacher, 104*(4), 319–324.

Rivera, F., & Becker, J. (2007a) Abductive–inductive (generalization) strategies of preservice elementary majors on patterns in algebra. *Journal of Mathematical Behavior, 26*(2), 140–155.

Rivera, F., & Becker, J. (2007b). Ethnomathematics in the global episteme: Quo vadis? In B. Atweh, A. C. Barton, M. Borba, N. Gough, C. Keitel, & R. Vithal (Eds.), *Internationalization and globalization in mathematics and science education* (pp. 209–226). Dordrecht, Netherlands: Springer.

Rivera, F., & Becker, J. (2008). Middle school children's cognitive perceptions of constructive and deconstructive generalizations involving linear figural patterns, *ZDM, 40*(1), 65–82.

Rivera, F., & Becker, J. (2010). Formation of linear pattern generalization in middle school children: Results of a three-year longitudinal study. In J. Cai & E. Knuth (Eds.), *Early algebraization.* Dordrecht, Netherlands: Springer.

Rodd, M. (2000). On mathematical warrants: Proof does not always warrant, and a warrant may be other than a proof. *Mathematical Thinking and Learning, 2*(3), 221–244.

Roth, W.-M., & Thom, J. (2009). Bodily experience and mathematical conceptions: From classical views to a phenomenological reconceptualization. *Educational Studies in Mathematics, 70*, 175–189.

Rotman, B. (1995). Thinking diagrams: Mathematics, writing, and virtual reality. *South Atlantic Quarterly, 94*(2), 389–415.

Sawada, D. (1982). Multisensory information matching ability and mathematics learning. *Journal for Research in Mathematics Education, 13*(5), 390–394.

Sawada, D., & Juryman, R. F. (1978). Information matching within and between auditory and visual sense modalities and mathematics achievement. *Journal for Research in Mathematics Education, 9*(2), 126–136.

Saxe, G. (1991). *Culture and cognitive development: Studies in mathematical understanding.* Hillsdale, NJ: Erlbaum.

Schnepp, M., & Chazan, D. (2004). Incorporating experiences of motion into a calculus classroom. *Educational Studies in Mathematics, 57*, 309–313.

Sebeok, T. (1983). One, two, three spells U B E R T Y. In U. Eco & T. Sebeok (Eds.), *The sign of three: Dupin, Holmes, Peirce* (pp. 1–10). Bloomington, IN: Indiana University Press.

Shaughnessy, J. M. (1992). Research in probability and statistics: Reflections and directions. In D. Grouws (Ed.), *Handbook of research on mathematics teaching and learning* (pp. 465–494). New York: Macmillan.

Shaughnessy, J. M. (2007). Research on statistics learning and reasoning. In F. Lester, Jr. (Ed.), *Second handbook of research on mathematics teaching and learning* (pp. 957–1010). Charlotte, NC: National Council of Teachers of Mathematics and Information Age Publishing.

Sherry, D. (2009). The role of diagrams in mathematical arguments. *Foundations of Science, 14*, 59–74.

Sherwin, R., Feigenson, N., & Spiesel, C. (2007). What is visual knowledge, and what is it good for? Potential ethnographic lessons from the field of legal practice. *Visual Anthropology, 20*, 143–178.

Shin, S. (1994). *The logical status of diagrams*. New York: Cambridge University Press.

Simon, T. (1997). Reconceptualizing the origins of number knowledge: A "non-numerical" account. *Cognitive Development, 12*, 349–372.

Skemp, R. (1987/1971). *The psychology of learning mathematics: Expanded American edition*. Hillsdale, NJ: Erlbaum.

Solomon, Y. (1989). *The practice of mathematics*. London: Routledge.

Sophian, C. (2007). *The origins of mathematical knowledge in childhood*. Mahwah, NJ: Erlbaum.

Soto-Andrade, J. (2008). *Mathematics as the art of seeing the invisible*. Paper presented at the 11th international congress in mathematical education, Merida, Mexico.

Spelke, E., Breinlinger, K., Macomber, J., & Jacobson, K. (1992). Origins of knowledge. *Psychological Review, 99*, 605–632.

Stephan, M., Bowers, J., Cobb, P., & Gravemeijer, K. (2003). Supporting students' development of measuring conceptions: Analyzing students' learning in social context. *Journal for Research in Mathematics Education Monograph Number 12*. Reston, VA: National Council of Teachers of Mathematics.

Stern, E., & Mevarech, Z. (1996). Children's understanding of successive divisions in different contexts. *Journal of Experimental Child Psychology, 61*(2), 153–172.

Stevenson, H., & Stigler, J. (1992). *The learning gap: Why our schools are failing and what can we learn from Japanese and Chinese education*. New York: Summit Books.

Stjernfelt, F. (2000). Diagrams as centerpiece of a Peircean epistemology. *Transactions of the Charles S. Peirce Society, XXXVI*(3), 357–384.

Stjernfelt, F. (2007). *Diagrammatology: An investigation on the borderlines of phenomenology, ontology, and semiotics*. Dordrecht, Netherlands: Springer.

Strelow, E. (1985). What is needed for a theory of mobility: Direct perception and cognitive maps – Lessons from the blind. *Psychological Review, 92*(2), 226–248.

Suh, J., Johnson, C., Jamieson, S., & Mills, M. (2008). Promoting decimal number sense and representational fluency. *Mathematics Teaching in the Middle School, 14*(1), 44–50.

Swafford, J., & Langrall, B. (2000). Grade 6 students' preinstructional use of equations to describe and represent problem situations. *Journal for Research in Mathematics Education, 31*(1), 89–112.

Tall, D., & Dinner, S. (1981). Concept image and concept definition in mathematics with particular reference to limits and continuity. *Educational Studies in Mathematics, 12*, 151–169.

Tang, Y., Zhang, W., Chen, K., Feng, S., Ji, Y., Shen, J., et al. (2006). Arithmetic processing in the brain shaped by cultures. *Proceedings of the National academy of sciences of the United States of America, 103*(28), 10755–10780.

Tarlow, L. (2008). Sense-able combinatorics: Students' use of personal representations. *Mathematics Teaching in the Middle School, 13*(8), 484–489.

Taussig, M. (2009). What do drawings want? *Culture, Theory, & Critique, 50*(2–3), 263–274.

Taylor, M., Pountney, D., & Malabar, I. (2007). Animation as an aid for the teaching of mathematical concepts. *Journal of Further and Higher Education, 31*(3), 246–261.

Teaching Committee of the Mathematical Association (2001). *Are you sure? Learning about proof*. In D. French & C. Stripp (Eds.), Leicester, UK: Teaching Committee of the Mathematical Association.

Thagard, P. (1978). Semiosis and hypothetic inference in C. S. Peirce. *Versus Quaderni Di Studi Semiotici, 19/20*, 163–172.

Thagard, P. (2010). How brains make mental models. In L. Magnani, W. Carnielli, & C. PIzzi (Eds.), *Model-based reasoning in science and technology: Abduction, logic, and computational discovery* (pp. 447–461). Berlin: Springer.
Thurston, W. (1994). On proof and progress in mathematics. *Bulletin of the American Mathematical Society, 30*(2), 161–177.
Tinti, C., & Galanti, D. (1999). Interactive auditory and visual images in persons who are totally blind. *Journal of Visual Impairment & Blindness, 93*(9), 579–583.
Tiragallo, F. (2007). Embodiment of the gaze: Vision, planning, and weaving between filmic ethnography and cultural technology. *Visual Anthropology, 20*, 201–219.
Tokyo Shoseki (2006). *Mathematics for elementary school 5B.* Tokyo, Japan: Tokyo Shoseki.
Torren, C. (1990). *Making sense of hierarchy: Cognition as social process in Fiji (LSE monograph on social anthropology No. 61).* London and Atlantic Highlands, NJ: Athlone Press.
Trouche, L. (2003). From artifact to instrument: Mathematics teaching mediated by symbolic calculators. *Interacting with Computers, 15*(6), 783–800.
Trouche, L. (2005). Instrumental genesis, individual and social aspects. In D. Guin, K. Ruthven, & L. Trouche (Eds.), *The didactical challenge of symbolic calculators: Turning a computational device into a mathematical instrument* (pp. 197–230). New York: Springer.
Tucker, J. M. (2006). The historian, the picture, and the archive. *Isis, 97*, 111–120.
Tucker, J. M. (2010). A lesson on the slopes of perpendicular lines. *Mathematics Teacher, 103*(8), 603–608.
Tversky, B. (2001). Spatial schemas in depictions. In M. Gattis (Ed.), *Spatial schemas and abstract thought* (pp. 79–112). Cambridge, MA: MIT Press.
Uttal, D., Scudder, K., & DeLoache, J. (1997). Manipulatives as symbols: A new perspective on the use of concrete objects to teach mathematics. *Journal of Applied Developmental Psychology, 18*, 37–54.
Van Garderen, D. (2003). Visual–spatial representation, mathematical problem solving, and students of varying abilities. *Learning Disabilities Research & Practice, 18*, 246–254.
Van Garderen, D. (2006). Spatial visualization, visual imagery, and mathematical problem solving of students with varying abilities. *Journal of Learning Disabilities, 39*(6), 496–506.
Van Garderen, D. (2007). Teaching students with LD to use diagrams to solve mathematical word problems. *Journal of Learning Disabilities, 40*(6), 540–553.
Veldeman, J. (2008). Reconsidering pictorial representation by reconsidering visual experience. *Leonardo, 41*(5), 493–497.
Veraart, C., & Wanet-Defalque, M. C. (1987). Representation of locomotor space by the blind. *Perception & Psychophysics, 42*(2), 132–139.
Vergnaud, G. (1996). The theory of conceptual fields. In P. Cobb, G. Goldin, & B. Greer (Eds.), *Theories of Mathematical Learning* (pp. 219–239). Mahwah, NJ: Erlbaum.
Vergnaud, G. (2009). The theory of conceptual fields. *Human Development, 52*, 83–94.
Vlassis, J. (2008). The role of mathematical symbols in the development of number conceptualization: The case of the minus sign. *Philosophical Psychology, 21*(4), 555–570.
Vygotksy, L. S. (1962). *Thought and language.* Cambridge, MA: MIT Press.
Vygotsky, L. S. (1978). *Mind in society.* Cambridge, MA: MIT Press.
Ward, J., & Meijer, P. (2010). Visual experiences in the blind induced by an auditory sensory substitution device. *Consciousness and Cognition, 19*, 492–500.
Warren, B., Ogonowski, M., & Pothier, S. (2005). "Everyday" and "scientific:" Rethinking dichotomies in modes of thinking in science learning. In R. Nemirovski, A. Rosebery, J. Solomon, & B. Warren (Eds.), *Everyday matters in science and mathematics: Studies of complex classroom events* (pp. 119–152). Mahwah, NJ: Erlbaum.
Wartofsky, M. (1978). Rules and representation: The virtues of constancy and fidelity put in perspective. *Erkenntnis, 12*, 17–36.
Wassmann, J. (1994). The Yupno as post-Newtonian scientists: The question of what is "natural" in spatial description. *Man, 29*(3), 645–666.

Watson, H. (1990). Investigating the social foundations of mathematics: Natural number in culturally diverse forms of life. *Social Studies of Science, 20*, 283–312.

Weber, K. (2002). Developing students' understanding of exponents and logarithms. In D. Mewborn, P. Sztajn, D. White, H. Wiegel, R. Bryant, & K. Nooney (Eds.), *Proceedings of the 24th annual meeting of the North American chapter of the international group for the psychology of mathematics education* (Vol. 4, pp. 1019–1027). Athens, GA: University of Georgia.

Were, G. (2003). Objects of learning: An anthropological approach to mathematics education. *Journal of Material Culture, 8*(1), 25–44.

Wise, M. N. (2006). Making visible. *Isis, 97*, 75–82.

Wittgenstein, L. (1961). *Tractacus logico-philosophicus* (D. Pears & B. McGuiness, Trans.). London: Routledge.

Wittgenstein, L. (1973). *Letters to C. K. Ogden*. In G. H. von Wright (Ed.), Oxford, UK: Blackwell.

Wood, T., Williams, G., & McNeal, B. (2006). Children's mathematical thinking in different classroom cultures. *Journal for Research in Mathematics Education, 37*(3), 222–255.

Yerushalmy, M., & Chazan, D. (2002). Flux in school algebra: Curricular change, graphing technology, and research on student learning and teacher knowledge. In L. English (Ed.), *Handbook of international research on mathematics education* (pp. 725–755). Hillsdale, NJ: Erlbaum.

Yip, K. (1991). Understanding complex dynamics by visual and symbolic reasoning. *Artificial Intelligence, 51*, 179–221.

Zahner, D., & Corter, J. (2010). The process of probability problem solving: Use of external visual representations. *Mathematical Thinking and Learning, 12*, 177–204.

Zazkis, R., Dubinksky, E., & Dautermann, J. (1996). Coordinating visual and analytic strategies: A study of students' understanding of the group D4. *Journal for Research in Mathematics Education, 27*(4), 435–457.

Zbiek, R. M., & Heid, M. K. (2009). Using computer algebra systems to develop big ideas in mathematics. *Mathematics Teacher, 102*(7), 540–544.

Zbiek, R., Heid, M., Bume, G., & Dick, T. (2007). Research on technology in mathematics education: The perspective of constructs. In F. Lester, Jr. (Ed.), *Second handbook of research on mathematics teaching and learning* (pp. 1169–1208). Charlotte, NC: National Council of Teachers of Mathematics and Information Age Publishing.

Zebian, S. (2008). Number conceptualization among Lebanese micro-business owners who engage in orally-based versus paper-based numeracy practices: An experimental cognitive ethnography. *Journal of Cognition and Culture, 8*, 359–385.

Zhang, J. (1997). The nature of external representations in problem solving. *Cognitive Science, 21*(2), 179–217.

Zhang, J., & Norman, D. (1994). Representations in distributed cognitive tasks. *Cognitive Science, 8*, 87–122.

Zimmermann, W., & Cunningham, S. (1991). *Visualization in teaching and learning mathematics. MAA Notes Series 19.* Washington, DC: Mathematical Association of America.

Author Index

Subject Index

D

E

F

G

W

Z

Lightning Source UK Ltd.
Milton Keynes UK
UKOW06n1957110615

253345UK00001B/27/P